**Single Molecule Dynamics
in Life Science**

*Edited by
Toshio Yanagida and Yoshiharu Ishii*

Further Reading

Feringa, B. L. (Ed.)

Molecular Switches

Second Edition
2 volumes
2009
ISBN: 978-3-527-31365-5

Anselmetti, D. (Ed.)

Single Cell Analysis

Technologies and Applications

2008
ISBN: 978-3-527-31864-3

Miller, L. W. (Ed.)

Probes and Tags to Study Biomolecular Function

for Proteins, RNA, and Membranes

2008
ISBN: 978-3-527-31566-6

Mirkin, C. A., Niemeyer, C. M. (Eds.)

Nanobiotechnology II

More Concepts and Applications

2007
ISBN: 978-3-527-31673-1

Daunert, S., Deo, S. K. (Eds.)

Photoproteins in Bioanalysis

2006
ISBN: 978-3-527-31016-6

Goeldner, M., Givens, R. (Eds.)

Dynamic Studies in Biology

Phototriggers, Photoswitches and Caged Biomolecules

2005
ISBN: 978-3-527-30783-8

Single Molecule Dynamics in Life Science

Edited by
Toshio Yanagida and Yoshiharu Ishii

WILEY-VCH

WILEY-VCH Verlag GmbH & Co. KGaA

The Editors

Prof. Dr. Toshio Yanagida
Osaka University
Laboratory for Nanobiology
1-3 Yamada-oka, Suita
Osaka 562-0871
Japan

Dr. Yoshiharu Ishii
CREST
Soft nano-machine Project JST
1-3 Yamada-oka, Suita
Osaka 562-0871
Japan

Library of Congress Card No.: applied for

British Library Cataloguing-in-Publication Data
A catalogue record for this book is available from the British Library.

Bibliographic information published by the Deutsche Nationalbibliothek
Die Deutsche Nationalbibliothek lists this publication in the Deutsche Nationalbibliografie; detailed bibliographic data are available on the Internet at http://dnb.d-nb.de.

© 2009 WILEY-VCH Verlag GmbH & Co. KGaA, Weinheim

Typesetting Thomson Digital, Noida, India
Printing Strauss Gmbh, Mörlenbach
Bookbinding Litges & Dopf GmbH, Heppenheim-

Printed in the Federal Republic of Germany
Printed on acid-free paper

ISBN: 978-3-527-31288-7

Contents

Single Molecule Dynamics in Life Science. Edited by T. Yanagida and Y. Ishii
Copyright © 2009 WILEY-VCH Verlag GmbH & Co. KGaA, Weinheim
ISBN: 978-3-527-31288-7

Preface

Biomolecules are responsible for various dynamic biological functions ranging from the intracellular to the whole body and brain. Recently developed molecular and cellular biology techniques have allowed biomolecules involved and their roles to be identified. Snapshots of atomic structures and molecular organization of these identified biomoledules, which have been determined using structural biology techniques, have been foundation for understanding the mechanism of their functions. The dynamic behavior of biomolecules, which is essential for the functios, is often inferred from static and average data.

Alternatively, single molecule measurements have been developed to directly measure the dynamic behavior of single biomolecules, which is hidden in average signals that originate from a large number of molecules, which is the case for previous ensemble measurements. Single molecules can be traced while they are working in biological environments. Furthermore, high resolution, reaching the scale of nm, msec and pN, has enabled the unitary events within a molecular function to be measured. For example, the step movement of molecular motors was observed while the motors hydrolyzed single ATP molecules. Single molecule imaging has been used to monitor the dynamic changes in location, structure and interactions of biomolecules in several complicated systems such as molecular machines, organelles and cells leading to significant progress in clarifying the fundamental principles driving cellular functions. With continued progress, we believe that single molecule measurements offer the promise of disclosing the fundamental principles behind biological dynamics.

It has been more than a decade since both single fluorophores attached to biomolecules were detected in aqueous solution and single biomolecules were manipulated. Since then, single molecule measurements have become common tools in a large number of laboratories. However, there is still much that needs to be accomplished. The technology is still immature as new techniques, such as new probes, are needed. Forging single molecule measurements with other fields like nanotechnology will be instrumental to further develop single molecule detection and potentially enable single molecule detection to be applied throughout the whole body. We hope that this book serves as a steppingstone for the next generation of

Single Molecule Dynamics in Life Science. Edited by T. Yanagida and Y. Ishii
Copyright © 2009 WILEY-VCH Verlag GmbH & Co. KGaA, Weinheim
ISBN: 978-3-527-31288-7

single molecules measurements, as readers will find information that will guide them to developing future techniques, subjects, strategies and goals.

Finally, we would like to thank all the authors for their valuable contributions to this book. Each is a pioneer for single molecule measurements in their respective biological field. We would also like to thank the staff of Wiley-VCH, in particular Dr. Rainer Muenz, for their help and encouragement.

August 2008

Toshio Yanagida
Yoshiharu Ishii

List of Contributors

Toshio Ando
Kanazawa University
Department of Physics
Kanazawa 920-1192
Japan

and

CREST
JST
Sanban-cho
Chiyoda-ku
Tokyo 102-0075
Japan

Donna J. Arndt-Jovin
Max Planck Institute for Biophysical
Chemistry
Laboratory of Cellular Dynamics
Am Faßberg 11
37077 Göttingen
Germany

Jasha Brujic
New York University
Department of Physics
New York, NY 10003
USA

Wouter Caarls
Max Planck Institute for
Biophysical Chemistry
Laboratory of Cellular Dynamics
Am Faßberg 11
37077 Göttingen
Germany

Christophe Danelon
Ecole Polytechnique Fédérale de
Lausanne (EPFL)
Laboratoire de Chimie Physique des
Polymères et Membranes
1015 Lausanne
Switzerland

Lorna Dougan
Columbia University
Department of Biological Science
New York, NY 100027
USA

Julio M. Fernandez
Columbia University
Department of Biological Science
New York, NY 100027
USA

Single Molecule Dynamics in Life Science. Edited by T. Yanagida and Y. Ishii
Copyright © 2009 WILEY-VCH Verlag GmbH & Co. KGaA, Weinheim
ISBN: 978-3-527-31288-7

Yale E. Goldman
University of Pennsylvania
Pennsylvania Muscle Institute
School of Medicine
Philadelphia, PA 19104-6083
USA

Taekjip Ha
University of Illinois
Urbana-Champaign and Howard
Hughes Medical Institute
Department of Physics
Urbana, Il 61801
USA

Guy M. Hagen
Department of Cell Biology
Charles University Medical School
Albertor 4
12800 Prague
Czech Republic

Minako Hirano
Osaka University, 1-3
Graduate School of Frontier Biosciences
Laboratories for Nanobiology
Soft Biosystem Group
Yamadaoka
Suita
Osaka 565-0871
Japan

Toru Ide
Osaka University, 1-3
Graduate School of Frontier Biosciences
Laboratories for Nanobiology
Soft Biosystem Group
Yamadaoka
Suita
Osaka 565-0871
Japan

Ryota Iino
Osaka University
The Institute of Scientific and Industrial
Research
Mihogaoka 8-1
Ibaraki 567-0047
Osaka
Japan

Noriyuki Kodera
Kanazawa University
Department of Physics
Kanazawa 920-1192
Japan

and

CREST
JST
Sanban-cho
Chiyoda-ku
Tokyo 102-0075
Japan

Yoshiharu Ishii
CREST
Soft nano-machine Project JST
1-3 Yamada-oka, Suita
Osaka 562-0871
Japan

Thomas M. Jovin
Max Planck Institute for Biophysical
Chemistry
Laboratory of Cellular Dynamics
Am Faßberg 11
37077 Göttingen
Germany

Ted A. Laurence
University of California
Department of Chemistry and
Biochemistry
Los Angeles, CA 90095
USA

Keith A. Lidke
University of New Mexico
Department of Physics
Albuquerque, NM 87131
USA

Diane S. Lidke
University of New Mexico
Department of Pathology
Albuquerque, NM 87131
USA

Atsushi Miyagi
Kanazawa University
Department of Physics
Kanazawa 920-1192
Japan

Hiroyuki Noji
Osaka University
The Institute of Scientific and Industrial
Research
Mihogaoka 8-1
Ibaraki 567-0047
Osaka
Japan

Bernd Rieger
Delft University of Technology
Department of Imaging Science &
Technology
Quantitative Imaging Group
Lorentzweg 1
2628 CJ Delft
The Netherlands

Yasushi Sako
Cellular Informatics Laboratory
RIKEN

Tatsuo Shibata
Hiroshima University
Department of Mathematical and Life
Sciences
Higashi-Hiroshima
Hiroshima 739-8526
Japan

Robert H Singer
Albert Einstein College of Medicine
Gruss-Lipper Biophotonics Center
Department of Anatomy and Structural
Biology
1300 Morris Park Avenue
Bronx, NY 10461
USA

Yuko Takeuchi
Osaka University, 1-3
Graduate School of Frontier Biosciences
Laboratories for Nanobiology
Soft Biosystem Group
Yamadaoka
Suita
Osaka 565-0871
Japan

Masaaki Taniguchi
Kanazawa University
Department of Physics
Kanazawa 920-1192
Japan

Valeria de Turris
Albert Einstein College of Medicine
Department of Anatomy and Structural
Biology
1300 Morris Park Avenue
Bronx, NY 10461
USA

Takayuki Uchihashi
Kanazawa University
Department of Physics
Kanazawa 920-1192
Japan

and

CREST
JST
Sanban-cho
Chiyoda-ku
Tokyo 102-0075
Japan

Masahiro Ueda
Osaka University
Graduate School of Frontier Biosciences
Laboratories for Nanobiology
Suita
Osaka 565-0671
Japan

Horst Vogel
Ecole Polytechnique Fédérale de
Lausanne (EPFL)
Laboratoire de Chimie Physique des
Polymères et Membranes
1015 Lausanne
Switzerland

Shimon Weiss
University of California
Department of Chemistry and
Biochemistry
Los Angeles, CA 90095
USA

Daisuke Yamamoto
Kanazawa University
Department of Physics
Kanazawa 920-1192
Japan

and

CREST
JST
Sanban-cho
Chiyoda-ku
Tokyo 102-0075
Japan

Hayato Yamashita
Kanazawa University
Department of Physics
Kanazawa 920-1192
Japan

Toshio Yanagida
Osaka University
Graduate School of Frontier Bioscience
and Graduate School of Medicine
1-3 Yamada-oka
Suita
Osaka 562-0871
Japan

1
A Road Map to Single Molecule Dynamics

Yoshiharu Ishii

1.1
Visualization of Single Molecules

Signals from single fluorescent dye molecules were first measured in non-biological environments. Some of the earliest studies acquired signals from single fluorescence dyes embedded in a solid matrix at the temperature of liquid helium [1] and from excitation fluorescence signals [2]. In early studies carried out at room temperature, single fluorescence dye molecules spread over a film were observed using near-field scanning fluorescence microscopy [3].

In a biological environment where fluorescence is largely quenched by the surrounding water molecules, single molecule imaging was first accomplished in 1995 using total internal reflected fluorescence (TIRF) and epi-fluorescence microscopy [4] (see Chapter 2). Single fluorophores attached to myosin molecules were visualized in aqueous solution in the vicinity of a glass surface while single ATP turnover from the same myosin molecule was measured using fluorescently-labeled ATP. TIRF microscopy has been widely used for single molecule imaging. Other studies have measured single fluorophores bound to biomolecules in solution while passing through a fixed small volume using confocal microscopy [5].

1.2
Single Molecule Position Tracking

The motion of biomolecules has been tracked by following the visualization of immobilized single biomolecules. The motion of processive motors has been visualized by using large particles or by attaching if fluorescent dyes to the motor. In 1996, the smooth and processive motion of a molecular motor, kinesin attached to a single fluorescence dye was detected along microtubules immobilized on a glass surface using TIRF microscopy [6]. Epi- and TIRF microscopy were used to monitor the 2D diffusion of fluorescently-labeled phospholipids on phospholipid membranes

Single Molecule Dynamics in Life Science. Edited by T. Yanagida and Y. Ishii
Copyright © 2009 WILEY-VCH Verlag GmbH & Co. KGaA, Weinheim
ISBN: 978-3-527-31288-7

[7] while 3D diffusion of single fluorescence molecules was observed in gel structures [8]. One-dimensional diffusion, which was discovered by single molecule measurements, is a biomolecular mechanism used in the search for specific reaction sites on DNA. This mechanism was first observed directly during studies to elucidate the promoter binding site of RNA polymerase on DNA molecules which had been extended with laser traps [9].

It has been demonstrated that the high-resolution analysis of individual fluorescence spots enables the position of a single molecule to be tracked with nanometer accuracy, thus overcoming the diffraction limit of light [10]. In 2003, this nanometer accuracy proved crucial in the detection of the step movement of myosin V, a processive motor, along actin filaments at no load [11]. Taken together with the laser trap data, the step movement of molecular motors containing two motor domains (heads) can then be defined by a hand-over-hand mechanism (see Chapter 3). Thus single molecule detection has been instrumental in establishing the basic operations of molecular motors. This technique was also used to detect the step movement of microtubule-based motors in live cells by tracking cargoes fluorescently labeled with GFP [12] or carrying QDs (quantum dots) introduced via endocytosis [13].

In addition to fluorescence imaging, dark-field imaging of gold nanoparticle-labeled molecular motors provides another method for high sensitivity tracking. Tracking labeled myosin V heads during single step movements to the adjacent landing position with sub-milliseconds accuracy, enabled investigators to directly observe the diffusive search pattern in the step movement [14].

1.3
Single Molecules in Live Cells

Single fluorescently labeled molecules have been visualized in artificial lipid bilayers. The model bilayer was formed at the tip of a pipette near the surface of the glass in a TIRF microscope, which enabled simultaneous imaging and electric measurements of single ion channels [15]. Given that the ion current through a single ion channel has already been reported [16], the visualization of single molecules in the lipid bilayer has opened up the possibility of studying the structure and function of ion channels directly (see Chapter 4).

In 2000 single molecule imaging was successfully applied to living cells using TIRF and epi-fluorescence microscopy [17]. The binding of fluorescently-labeled signaling molecule epidermal growth factor (EGF) to its receptor on the surface of live cells was monitored at the single molecule level, which provided further information relating to the kinetics of binding and the behavior of the EGF–receptor complex (see Chapter 5). Single molecule tracking in live cells has also been achieved using fluorescent labels. In particular quantum dots (QDs) have been frequently used for quantitative and prolonged tracking (see Chapter 6). The trajectory of QDs bearing EGF revealed a variety of transport paths used by the ligand–receptor complex [18]. There have been a large number of papers on both single molecule tracking in live cells and the dynamic properties of the corresponding biomolecules, including diffusion. The motion of single molecules on the cell surface and inside the cell has also been measured using fluorescence correlation spectroscopy (FCS) [19, 20].

The interaction between ligands and receptors has been studied using single molecule imaging and mechanical measurements (see Chapter 7). The binding of ligands to receptors triggers signal transduction processes leading to the activation of whole cells (see Chapter 5). The activation of a cell can be directly related to the binding events of the ligands inside the cell. Such an experiment was first carried out by increasing the cytoplasmic Ca^{2+} concentration to determine the number of peptide-MHC ligands bound to a T-cell [21]. In general, only a small number of binding ligands have been reported in various systems. In fact, results have emphasized that small input signals within large noisy signals can generate consistent output. For example, in dictyosterium cells, in which motion is biased in one direction in response to a small cAMP concentration gradient, the kinetics of cAMP binding was observed to be dependent on location, leading to the polarity of the moving cell [22].

The expression of RNA and protein has also been detected in live cells at the single molecule level. RNA plays a pivotal role in gene expression in which RNA molecules work dynamically in the nucleus and cytoplasm (see Chapter 8). The trafficking of messenger RNA was monitored in real time after being expressed inside the nucleus of live cells [23]. It was demonstrated that diffusion is the basis for mRNA displacement. Regarding proteins, individual proteins expressed in single cells have been counted in real time. YFP fluorescence showed [24] that protein expression occurs in bursts. The stochastic nature of the gene expression has also been studied.

1.4
Fluorescence Spectroscopy and Biomolecular Dynamics

Combined with single molecule imaging, fluorescence spectroscopy can be used to measure structural dynamics and biomolecule interactions in real time (see Chapter 9). In particular, fluorescence resonance energy transfer (FRET) between single fluorophores has become a popular tool of choice following its application in dried DNA using scanning near-field microscopy [25]. Conformational dynamics were observed using chymotrypsin inhibitor 2 [26] and *Tetrahymena thermophila* ribozyme [27]. These conformational studies have been followed by a large number of studies on multiple conformations and rugged energy landscapes. Single molecule FRET has also been used for protein folding studies. Subpopulations of folded and denatured conformations of proteins freely diffusing in solution were directly determined by confocal microscopy [28, 29]. These measurements have provided useful tests for a new view on dynamic structures of biomolecules in which they behave according to the energy landscape. In addition to FRET, spectral shift and fluorescence lifetime polarizations have been used to study protein dynamics. Single molecule fluorescence polarization imaging was first carried out to monitor the axial rotation of actin filaments sliding over myosin immobilized on a glass surface [30]. It was recently demonstrated using polarization imaging from fluorescence dyes attached bifunctionally to the neck domain of myosin V that the orientation of the neck domain is associated with the step movement, in agreement with a walking model of myosin V [31] (see Chapter 3).

By monitoring fluorescence changes associated with enzymatic reactions, the operation of enzymes can be studied. Taking advantage of the cycle between the active fluorescent oxidized form (FAD) and non-fluorescent reduced form (FADH) of flavin adenine dinucleotide, the turnover of cholesterol oxidase was monitored [32]. The reaction was stochastic, consistent with Michaelis–Menten mechanics while a memory effect was observed between several turnover intervals as a result of slow conformational isomerizations. Similar memory effects were observed for RNA enzymes (ribozymes) [27].

Proteins, DNA and RNA fulfill their function by assembling and disassembling dynamically. Single molecule spectroscopy is capable of determining the dynamic structure of particular molecules and molecular interactions in such biomolecular complexes. A more physiological environment has been observed during protein folding in more complex systems such as Gro EL and ES complexes [33]. The dynamic structures of certain molecular complexes such as helicase, DNA and holiday junctions, which appear in gene recombination and repair, have been extensively studied (see Chapter 11).

1.5
Single Molecule Manipulation and Molecular Motors

Laser and magnetic trapping and glass microneedles are common tools used to manipulate single biomolecules. After Ashkin applied optical trapping techniques to trap viruses [34], single kinesin was trapped using an attached dielectric bead to monitor its movement along microtubules immobilized on a glass surface [35], followed by the finding of an 8-nm unit step associated with the hydrolysis of single ATP molecules [36]. Microneedles were first used to manipulate actin filaments in 1988 [37]. This led to the measurement of the mechanical properties of actin filaments and the force generated by both myosin filaments and single muscle myosin heads [38]. Actin filaments have also been manipulated using the laser trap. Actin filaments, in which each end was attached to a trapped bead, were manipulated to interact with single non-processive muscle myosin molecules. The step movement of muscle myosin was measured by monitoring the displacement of the actin filaments [39, 40]. Alternatively, instead of manipulating actin filaments, single myosin molecules have been captured on the tip of a microneedle and manipulated to interact with actin filaments [41]. The movement of myosin was monitored directly by measuring the displacement of myosin. Given that these measurements have sensitivity to detect thermal motion, the effect of thermal motion on the mechanism of molecular motor step movement is discussed in Chapter 2.

Unlike muscle myosin, it is relatively easy to measure step movement in processive motors. Manipulation by laser trap has been used for mechanical studies on several processive myosin motors including myosin V [42], myosin VI [43] and dynein [44, 45] demonstrating that ATP hydrolysis reactions are strain-dependent, suggesting that cooperation between the two heads is necessary for processive movement [46]. When excess load (greater than stall force) was applied, motors such as kinesin showed stepwise movement in the backward direction [47].

1.6
Mechano-Chemical Coupling of Molecular Motors

In order to understand the mechanism of molecular motors, it is critical to describe how mechanical steps are related to ATP hydrolysis (see Chapter 2). Simultaneous imaging of ATP turnover and mechanical measurements using a single molecule optical trap was used to determine the mechano-chemical coupling in myosin [48].

In contrast to ATP-driven but irreversible actomyosin motors, the ATP synthase F_0F_1 is a reversible mechano-chemical machine. Using single molecule detection, ATP synthase F_0F_1 was found to be a rotary motor (see Chapter 10). The rotation of F_1 was first observed by monitoring the rotation of fluorescent actin filaments attached to the rotor [49]. The 120° rotation unit steps generated by a single ATP molecule were detected at low ATP concentrations [50]. It was also demonstrated that ATP is synthesized when F_1 is forced to rotate in the backward direction [51]. This mechanism has been extensively studied.

1.7
DNA-Based Motors

DNA-based motors translocate along a DNA molecule powered by nucleotide hydrolysis while they transcribe gene information. As compared with actin- and microtubules-based motors, the properties of the DNA-based motors are not yet known. Individual DNA molecules were first visualized by staining with fluorescent dyes and using video-enhanced optical microscopy [52]. This led to single DNA molecules being manipulated by magnetic traps [53], laser traps [54], and by pipette [55]. These manipulation techniques have been used to measure various mechanical properties of DNA such as the force–extension relationship and its interaction with motor molecules.

Transcription by a single RNA polymerase molecule has also been observed by laser trap. The displacement and force exerted on a trapped bead attached to the end of a DNA molecule being pulled by an RNA polymerase was measured while the polymerase was immobilized on a slide glass during transcription [56, 57]. In 2005, RNA polymerase was shown to have a step-size of 3.7 Å, equivalent to a DNA base pair [58]. This technique also proved that RNA polymerase moved along a DNA helix [59] as confirmed by monitoring the rotation of a magnetic bead attached to the end of a DNA molecule during transcription while the RNA polymerase was immobilized on the slide glass. Unlike actin filament- and microtubule-based molecular motors, some DNA-based motors change the topology of DNA while they translocate. This change has been used to monitor the movement of DNA-based motors. For example, Topoisomerase II removed supercoils that had been created by twisting DNA through magnetic beads [60]. The removal of the supercoil was measured by monitoring the expansion between the two DNA ends. Another example is DNA polymerase, which converts single-stranded DNA into double-stranded DNA. This DNA conformation change was detected by measuring the extension of DNA at constant load [61]. The unwinding of DNA and RNA as catalyzed by helicase has also been measured using mechanical and optical methods (see Chapter 11)

1.8
Imaging with AFM and Force Measurements

AFM [62] is a unique tool for force measurements and imaging, offering single molecule sensitivity. It has been applied to both single protein [63] and DNA molecules [64]. In contrast to fluorescence imaging, which is based on local information around the fluorophores, AFM provides topological imaging and specific electrostatic interaction mapping without modifying the biomolecules. Recently, both temporal and spatial resolutions have been dramatically improved [65] allowing the dynamic operations of biomolecules and the molecular machines responsible for their assembly to be visualized in real time (see Chapter 12).

The interaction force between single molecules has been measured for streptavidin-biotin using AFM [66, 67] and the forces involved in ligand–receptor complexes, antibody–antigen complexes, adhesion of complementary strands of DNA, carbohydrate–carbohydrate complexes, lectin–carbohydrate complexes, and cell adhesion proteins have all been studied (see Chapter 6). AFM and other manipulation techniques have also been used to study the folding and unfolding of proteins. In 1997, the unfolding and folding of titin which is an extremely large protein, was measured when its two ends were pulled and shortened [68–70]. Force clamp methods of AFM allow the detailed processes of individual protein molecule folding to be scrutinized rather than depending on the stepwise transition observed in ensemble measurements [71] (see Chapter 13).

References

1 Moerner, W.E. and Kador, L. (1989) Optical detection and spectroscopy of single molecules in a solid. *Physical Reviews Letters*, **62**, 2535–2538.

2 Orrit, M. and Bernard, J. (1990) Single pentacene molecules detected by fluorescence excitation in a *p*-terphenyl crystal. *Physical Review Letters*, **65**. 2716–2719.

3 Betzig, E. and Chichester, R.J. (1993) Single molecules observed by near-field scanning optical microscopy. *Science*, **262**, 1422–1424.

4 Funatsu, T., Harada, Y., Tokunaga, M., Saito, K. and Yanagida, T. (1995) Imaging of single fluorescent molecules and individual ATP turnovers by single myosin molecules in aqueous solution. *Nature*, **374**, 555–559.

5 Rigler, R. (1995) Fluorescence correlations, single molecule detection and large number screening. Applications

in biotechnology. *Journal of Biotechnology*, **41**, 177–186.

6 Vale, R.D., Funatsu, T., Pierce, D.W., Ronberg, L., Harada, Y. and Yanagida, T. (1996) Direct observation of single kinesin molecules moving along microtubules. *Nature*, **380**, 451–453.

7 Schmidt, T., Schutz, G.J., Baumgartner, W., Gruber, H.J. and Schindler, H. (1995) Characterization of photophysics and motility of single molecules in a fluid liquid membrane. *J. Phys. Chem.*, **99**, 17662–17668.

8 Dickson, R.M., Norris, D.J., Tzeng, Y.L. and Moerner, W.E. (1996). Three-dimensional imaging of single molecules solvated in pores of poly(acrylamide) gels. *Science*, **274**, 966–969.

9 Harada, Y., Funatsu, T., Murakami, K., Nonomura, Y., Ishihama, A. and Yanagida, T. (1999) Single molecule imaging of RNA polymerase-DNA interactions in real time. *Biophysical Journal*, **76**, 709–715.

10 Thompson, R.E., Larson, D.R. and Webb, W.W. (2002) Precise nanometer localization analysis for individual fluorescent probes. *Biophysical Journal*, **82**, 2775–2783.

11 Yildiz, A., Firjey, J.N., McKinney, S.A., Ha, T., Goldman, Y.E. and Selvin, P.R. (2003) Myosin V walks hand-over-hand; single fluorophores imaging with 1.5 nm localization. *Science*, **300**, 2061-2-65.

12 Kural, C., Kim, H., Syed, S., Goshima, G., Gelfand, V.I. and Selvin, P.R. (2005) Kinesin and dynein move a peroxisome *in vivo*: a tug-of-war or coordinated movement? *Science*, **308**, 1469–1472.

13 Nan, X., Sims, P.A., Chen, P. and Xie, X.S. (2005) Observation of fundamental micro-tubule motor steps in living cell with endo-cytosed quantum dots. *J. Phys. Chem. B.*, **109**, 24220–24224.

14 Dunn, A. and Spudich, J.A. (2007) Dynamics of the unbound head during myosin V processive translocation. *Nature. Structural Molecular Biology*, **14**, 246–248.

15 Ide, T. and Yanagida, T. (1999) An artificial lipid bilayer formed on an agarose-coated glass for simultaneous electrical and optical measurement of single ion channels. *Biochemical and Biophysical Research Communications*, **265**, 595–599.

16 Neher, E. and Sakmann, B. (1976) Single-channel currents recorded from membrane of denervated frog muscle fibres. *Nature*, **260**, 799–802.

17 Sako, Y., Minoghchi, S. and Yanagida, T. (2000) Single-molecule imaging of EGFR signaling on the surface of living cells. *Nature. Cell Biology*, **2**, 168–172.

18 Lidke, D.S., Heinzmann, R., Arndt-Jovin, D., Post, J.N., Grecco, H.E., Jares-Erijiman, E.A. and Jovin, T.M. (2004) Quantum dot ligands provide new insights into erbB/HER receptor-mediated signal transduction. *Nature: Biotechnology*, **22**, 198–203.

19 Schwille, P., Korlach, J. and Webb, W.W. (1999) Fluorescence correlation spectroscopy with single-molecule sensitivity on cell and model membranes. *Cytometry*, **36**, 76–82.

20 Brock, R., Vamosi, G., Vereb, G. and Jovin, T.M. (1999) Rapid characterization of green fluorescent protein fusion proteins on the molecular and cellular level by fluorescence correlation microscopy. *Proceedings of the National Academy of Sciences of the United States of America*, **96**, 10123–10128.

21 Irvine, D.J., Purbhoo, M.A., Krogsgaard, M. and Davis, M.M. (2002) Direct observation of ligand recognition by T cells. *Nature*, **419**, 845–849.

22 Ueda, M., Sako, Y., Tanaka, T., Devreotes, P. and Yanagida, T. (2001) Single-molecule analysis of chemotactic signaling in Dictyostelium cells. *Science*, **294**, 864–867.

23 Shav-Tal, Y., Darzacq, X., Shenoy, S.M., Fusco, D., Janckl, S.M., Spector, D.L. and Singer, R.H. (2004) Dynamics of single mRNPs in nuclei of living cells. *Science*, **304**, 1797–1800.

24 Yu, J., Xiao, J., Ren, X., Lao, K. and Xie, X.S. (2006) Probing gene expression in live cells, one protein molecule at a time. *Science*, **311**, 1600–1603.

25 Ha, T.J., Enderle, T., Ogletree, D.F., Chemla, D.S., Selvin, P.R. and Weiss, S. (1996) Probing the interaction between two single molecules: Fluorescence resonance energy transfer between a single donor and single acceptor. *Proc. Natl. Acad. Sci. U.S.A.*, **93**, 6264–6268.

26 Ha, T., Ting, A.Y., Liang, J., Caldwell, W.B., Deniz, W.A., Chemla, D.S., Schutz, P.G. and Weiss, S. (1999) Single-molecule fluorescence spectroscopy of enzyme conformational dynamics and cleavage mechanism. *Proceedings of the National Academy of Sciences of the United States of America*, **96**, 893–898.

27 Zhuang, X.W., Bartley, L.E., Babcock, H.P., Russell, R., Ha, T.J., Herschlag, D. and Chu, S. (2000) A single-molecule study of RNA catalysis and folding. *Science*, **288**, 2048–2051.

28 Deniz, A.A., Laurence, T.A., Beligere, G.S., Dahan, M., Martin, A.B., Chemla, D.S., Dawson, P.E., Schultz, P.G. and Weiss, S. (2000) Single molecule protein folding: Diffusion fluorescence resonance energy transfer studies of the denaturation of chymotrypsin inhibitor 2. *Proceedings of the National Academy of Sciences of the United States of America*, **97**, 5179–5184.

29 Schuler, B., Lipman, E.A. and Eaton, W.A. (2002) Probing the free-energy surface for protein folding with single-molecule fluorescence spectroscopy. *Nature*, **419**, 743–747.

30 Sase, I., Miyata, H., Ishiwata, S. and Kinosita, K., Jr. (1997) Axial rotation of sliding actin filaments revealed by single-fluorophore imaging. *Proceedings of the National Academy of Sciences of the United States of America*, **94**, 5646–5650.

31 Folkey, J.N., Quinlan, M.E., Shaw, M.A., Corrie, E.T. and Goldman, Y.E. (2003) Three-dimensional structural dynamics of myosin V by single-molecule fluorescence polarization. *Nature*, **422**, 399–404.

32 Lu, H.P., Xun, L. and Xie, X.S. (1998) Cholesterol oxidase, a flavoenzyme catalyzes the oxidation of cholesterol by oxygen. Single-molecule enzymatic dynamics. *Science*, **282**, 1877–1882.

33 Taguchi, H., Ueno, T., Tadakuma, H., Yoshida, M. and Funatsu, T. (2001) Single-molecule observation of protein–protein interactions in the chaperonin system. *Nature. Biotechnology*, **19**, 861–865.

34 Ashkin, A. and Dziedzic, J.M. (1987) Optical trapping and manipulation of viruses and bacteria. *Science*, **235**, 1517–1520.

35 Block, S.M., Blair, D.F. and Berg, H.C. (1989) Compliance of bacterial flagella measured with optical tweezers. *Nature*, **338**, 514–51858.

36 Svoboda, K., Schmidt, C.F., Schnapp, B.J. and Block, S.M. (1993) Direct observation of kinesin stepping by optical trapping interferometry. *Nature*, **365**, 721–727.

37 Kishino, A. and Yanagida, T. (1988) Force measurement of micromanipulation of a single actin filament by glass needles. *Nature*, **334**, 74–76.

38 Ishijima, A., Harada, Y., Kojima, H., Funatsu, T., Higuchi, H. and Yanagida, T. (1994) Single-molecule analysis of the actomyosin motor using nano-manipulation. *Biochemical and Biophysical Research Communications*, **199**, 1057–1063.

39 Finer, J.T., Simmons, R.M. and Spudich, J.A. (1994) Single myosin molecule mechanics; piconewton forces and nanometre steps. *Nature*, **368**, 113–119.

40 Molloy, J.E., Burns, J.E., Kendrick-Jones, J., Tregear, R.T. and White, D.C. (1995) Movement and force produced by a single myosin head. *Nature*, **378**, 209–212.

41 Kitamura, K., Tokunaga, M., Iwane, A.H. and Yanagida, T. (1999) A single myosin head moves along an actin filament with regular steps of 5.3 nanometres. *Nature*, **397**, 129–134.

42 Rief, M., Rock, R.S., Mehta, A.D., Moosker, M.S., Cheney, R.E. and Spudich, J.A. (2000) Myosin-V stepping kinetics: A molecular model for processivity.

Proceedings of the National Academy of Sciences of the United States of America, **97**, 9482–9486.

43 Rock, R.S., Rice, S.E., Wells, A.L., Purcell, T.J., Spudich, J.A. and Sweeney, H.L. (2001) Myosin VI is a processive motor with a large step size. *Proceedings of the National Academy of Sciences of the United States of America*, **98**, 13655–13659.

44 Mallik, R., Carter, B.C., Les, S.A., King, S.J. and Gross, S.P. (2004) Cytoplasmic dynein functions as a gear in response to load. *Nature*, **427**, 649–652.

45 Toba, S., Watanabe, T.M., Yamaguchi-Okamoto, I., Toyoshima, Y.Y. and Higuchi, H. (2006) Overlapped hand-over-hand mechanism of single molecular motility of cytoplasmic dynein. *Proceedings of the National Academy of Sciences of the United States of America*, **102**, 5741–5745.

46 Purcell, T.J., Sweeney, H.L. and Spudich, J.A. (2005) A force-dependent state controls the coordination of processive myosin V. *Proceedings of the National Academy of Sciences of the United States of America*, **102**, 13873–13878.

47 Carter, N.J. and Cross, R.A. (2005) Mechanics of the kinesin step. *Nature*, **435**, 308–318.

48 Ishijima, A., Kojima, H., Higuchi, H., Harada, Y., Funatsu, T. and Yanagida, T. (1998) Simultaneous measurement of chemical and mechanical reaction. *Cell*, **70**, 161–171.

49 Noji, H., Yasuda, R., Yoshida, M. and Kinosita, K., Jr. (1997) Direct observation of the rotation of F1-ATPase. *Nature*, **386**, 299–302.

50 Yasuda, R., Noji, H., Kinosita, K. and Yoshida, M. (1998) F_1-ATPase is a highly efficient molecular motor that rotates with discrete 120° steps. *Cell*, **93**, 1117–1124.

51 Itoh, H., Takahashi, A., Adachi, K., Noji, H., Yasuda, R., Yoshida, M. and Kinosita, K., Jr. (2004) Mechanically driven ATP synthesis by F1-ATPase. *Nature*, **427**, 465–468.

52 Morikawa, K. and Yanagida, M. (1981) Visualization of individual DNA molecules in solution by light microscopy: DAPI staining method. *Journal of Biochemistry*, **89**, 693–696.

53 Smith, S.B., Finzi, L. and Bustamante, C. (1992) Direct mechanical measurements of the elasticity of single DNA molecules by using magnetic beads. *Science*, **258**, 1122–1126.

54 Perkins, P.P., Smith, D.E. and Chu, S. (1994) Direct observation of tube-like motion of a single polymer chain. *Science*, **264**, 819–822.

55 Smith, S.B., Cui, Y. and Bustamante, C. (1996) Overstretching B-DNA: The elastic response of individual double-stranded and single-stranded DNA molecules. *Science*, **271**, 795–799.

56 Yin, H., Wang, M.D., Svoboda, K., Landick, R., Block, S.M. and Gells, J. (1995) Transcription against an applied force. *Science*, **270**, 1653–1657.

57 Wang, M.D., Schnitzer, M.J., Yin, H., Landick, R., Gelles, J. and Block, S.M. (1998) Force and velocity measured for single molecules of RNA polymerase. *Science*, **282**, 902–907.

58 Abbondanzieri, E.A., Greenleaf, W.J., Shaevits, J.W., Landick, R. and Block, S.M. (2005) Direct observation of base-pair stepping by RNA polymerase. *Nature*, **438**, 460–465.

59 Harada, Y., Ohara, O., Takatsuki, A., Ito, H., Shimamoto, N. and Kinosita, K., Jr. (2001) Direct observation of DNA rotation during transcription by *Escherichia coli* RNA polymerase. *Nature*, **409**, 113–115.

60 Strick, T.R., Croquette, V. and Bensimmon, D. (2000) Single-molecule analysis of DNA uncoiling by a type II topoisomerase. *Nature*, **404**, 901–904.

61 Wuite, G.J.L., Smith, S.B., Young, M., Keller, D. and Bustamante, C. (2000) Single-molecule studies of the effect of template tension of T7 DNA polymerase activity. *Nature*, **404**, 103–106.

62 Binnig, G., Quate, C.F. and Gerber, C. (1986) Atomic force microscope. *Physical Review Letters*, **56**, 930–933.

63 Hoh, J.H., Lal, R., John, S.A., Revel, J.-P. and Arnsdorf, M.F. (1991) Atomic force microscopy and dissection of gap junctions. *Science*, **253**, 1405–1408.

64 Hansma, H.G., Vesenka, J., Siegerist, C., Kelderman, G., Morrett, H., Sinsheimer, R.L., Elings, V., Bustamante, C. and Hansma, P.K. (1992) Reproducible imaging and dissection of plasmid DNA under liquid with the atomic force microscope. *Science*, **256**, 1180–1184.

65 Ando, T., Kodera, N., Tajau, E., Maruyama, D., Saito, K. and Toda, A. (2001) A high-speed atomic force microscope for studying biological macromolecules. *Proceedings of the National Academy of Sciences of the United States of America*, **98**, 12468–12472.

66 Lee, G.U., Kidwell, D.A. and Colton, R.J. (1994) Sensing discrete streptavidin–biotin interactions with atomic force microscopy. *Langmuir*, **10**, 354–337.

67 Florin, E.-L., Moy, V.T. and Gaub, H.E. (1994) Adhesion forces between individual ligand–receptor pairs. *Science*, **264**, 415–417.

68 Tskhovrebova, L., Trinick, J., Sleep, J.A. and Simmons, F.M. (1997) Elasticity and unfolding of single molecules of the giant muscle protein titin. *Nature*, **387**, 308–312.

69 Rief, M., Gautel, M., Oesterheilt, F., Ferrandez, J.M. and Gaub, H.E. (1997) Reversible unfolding of individual titin immunoglobulin domains by AFM. *Science*, **276**, 1109–1112.

70 Kellermayer, M.S., Smith, S.B., Granzier, H.L. and Bustamante, C. (1997) Folding–unfolding transitions in single titin molecules characterized with laser tweezers. *Science*, **276**, 1112–1116.

71 Fernandez, J.M. and Li, H. (2004) Force-clamp spectroscopy monitors the folding trajectory of a single protein. *Science*, **303**, 1674–1678.

2
Single Molecule Study for Elucidating the Mechanism Used by Biosystems to Utilize Thermal Fluctuations

Toshio Yanagida

2.1
Introduction

2.1.1
Differences between Man-Made and Biological Molecular Machines

To fulfill their functions, biomolecules assemble to form molecular machines which play essential roles in cellular functions such as cell signaling, energy transduction, motion, and DNA duplication. Collaboration between these machines regulates the activity of biological systems (Figure 2.1a). Molecular machines are complex machines whose functions are not analogous to artificial machines. What are the essential differences between biological molecular machines and man-made machines? Biological molecular machines are nanometers in size, are dynamic and have a soft structure. In addition, their input energy is not much greater than the average thermal energy (k_BT) meaning that they function under strong thermal agitation. This is in contrast to man-made machines, which require an energy input that is much higher than thermal energy and thus avoids the effect of thermal noise. Thus, the operations of biological and man-made machines are vastly different. For example, the operation time of electronic processors in a computer is nanoseconds, while that of molecular machines in the brain is milliseconds, i.e. the operation time of molecular machines is millions-fold slower. Furthermore, the accuracy of molecular machines is far poorer than that of electronic processors. But such slow and inaccurate molecular machines can assemble to form wonderful systems such as muscle and brain, functions that cannot be duplicated by man-made machines (Figure 2.1b).

Because of this, molecular machines likely operate on a principle essentially different from man-made machines. The aim of our research is to illuminate the essential engineering principle behind the flexible and adaptive nature of biological systems by understanding the unique operations of molecular machines.

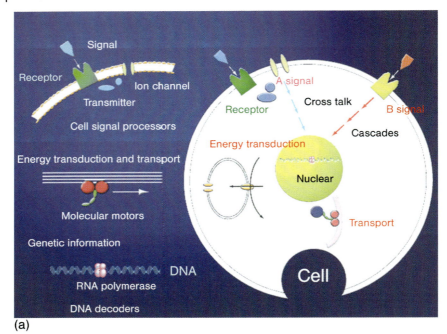

(a)

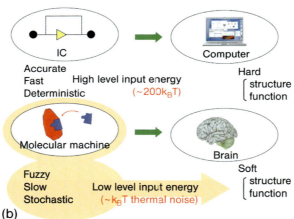

(b)

Figure 2.1 Differences between man-made and biological machines. (a) Biological molecules assemble to form molecular machines that play essential roles such as cell signaling, motion, and DNA duplication. (b) The input energy level of biological machines is not far above average thermal energy and thus their operations are slow and ambiguous. On the other hand, man-made machines are operated at energies much higher than average thermal energy and operate rapidly and accurately.

2.1.2
Single Molecule Imaging and Nano-Detection

To fully comprehend molecular machines, it is necessary to understand the dynamic properties of biomolecules themselves and their interactions with each other. Single molecule detection (SMD) techniques have been developed to directly monitor the dynamics of biomolecules and molecular machines. SMD techniques are based on two key technologies: single-molecule imaging and single-molecule nanomanipulation. The size of biomolecules and even their assemblies are in the order of nanometers, so they are too small to observe by optical microscopy. To overcome this problem, biomolecules can be fluorescently labeled and visualized using fluorescence microscopy. Single fluorophores have been observed in non-aqueous conditions [1] (see Chapter 1). In 1995, we successfully demonstrated that single fluorophores can be seen in aqueous solution by using total internal reflection fluorescence microscopy (TIRFM) and conventional inverted fluorescence microscopy [2]. The major problem to overcome when visualizing single fluorophores in aqueous solution is the huge background noise caused by numerous sources including Raman scattering from water molecules, incident light despite filters, luminescence arising from the objective lens, immersion oil and dust, and the instability of the fluorophores. In our system, the evanescent field was formed when the laser beam was totally reflected by the interface between the solution and the glass. The evanescent field was not restricted to the diffraction limit of light, thus it could be localized close to the glass surface which resulted in a penetration depth ($\sim$150 nm) being several-fold shorter than the wavelength of light. Therefore, the illumination was restricted to fluorophores either bound to the glass surface or located close by, thereby reducing the background light. Furthermore, by careful selection of optical elements, the background noise could be reduced by 2000-fold compared to that of conventional fluorescence microscopy. By adding an oxygen scavenger system, the instability of the fluorophores was significantly reduced. This made it possible to clearly observe single fluorophores in aqueous solution. Fluorescence measurements from single fluorophores attached to biomolecules and ligands have allowed the detection of, for example, the movements [3], enzymatic reactions [2, 4–7], protein-DNA interactions [8] and cell-signal processes [9] at the single molecule level. The single molecule imaging technique has been further extended to detect detailed reactions of biomolecules coupled with (1) detection of position with nanometer accuracy by computer image analysis (FIONA) [10]; (2) distance between two fluorophores with sub-nanometer accuracy by fluorescence resonance transfer (FRET) [11]; and (3) orientation by fluorescence polarization [12] or DOPI [13] (see Chapter 1).

The second key technology is single-molecule nanomanipulation. Biomolecules and even single molecules can be captured on a glass needle [14–17] or on beads trapped by optical tweezers [18, 19]. Optical tweezers are the tools used to trap and manipulate particles of between 25 nm and 25 mm in diameter using the force of the laser radiation pressure [20]. The particle is trapped near the focus of the laser light when focused by a microscope objective with a high numerical aperture. The optical tweezers exert forces in the piconewton range on the particles. Biomolecules are too small to be directly trapped by optical tweezers, so they are generally attached to an optically-trapped bead. Microneedles or a bead trapped by a laser act as a spring that expands in proportion to the applied force. Thus, the force and the displacement caused by the biomolecules can be measured. The displacement of a microneedle and a bead has been determined with sub-nanometer accuracy, which is considerably more sensitive than the diffraction limit of an optical measurement [15–19]. This accuracy of displacement corresponds to the sub-piconewton accuracy in the force measurements. Thus, the mechanical property of biomolecules can be determined directly at the single-molecule level. Furthermore, combined with the single-molecule imaging technique, simultaneous measurements of mechanical and chemical reactions of single biomolecules are possible.

We have used SMD techniques to uncover the unique operation of a typical molecular machine, a molecular motor, and studied how the flexible and adaptive nature specific to biological systems is generated by molecular machines operating under the influence of strong thermal agitation.

2.2
Simultaneous Measurements of Individual ATP Hydrolysis Cycles and Mechanical Events by a Myosin Motor

We have chosen the myosin motor as our model of a molecular machine. There are several types of myosin motors, all of which move along actin filaments by converting the chemical energy produced from ATP hydrolysis into mechanical energy to generate cellular motility such as muscle contraction. Important functions of proteins such as enzymatic action, energy transduction, molecular recognition and self-assembly are integrated into the myosin motor [21]. Therefore, elucidating the molecular mechanism of the myosin motor should provide the general principles used by molecular machines. The most fundamental problem regarding the mechanism of the myosin motor is its conversion of chemical energy into mechanical energy. In order to solve this problem, we simultaneously measured individual ATP hydrolysis cycles and single myosin motor mechanical events.

2.2.1
ATP Hydrolysis Cycles

Individual ATP hydrolysis cycles by a myosin motor were measured by the single-molecule imaging technique TIRFM in combination with the fluorescent ATP analog Cy3-ATP (Figure 2.2) [2, 4, 5]. It was confirmed that Cy3-ATP was hydrolyzed by myosin in the same way as ATP. The biochemical cycle rate of ATP hydrolysis averaged over many events for individual myosin molecules was consistent with that obtained by a conventional biochemical method using a suspension of myosin. Cy3-ATP (or -ADP) free in solution does not produce clear fluorescent spots on a detector, because of its rapid Brownian motion. However, when Cy3-ATP (or -ADP) is

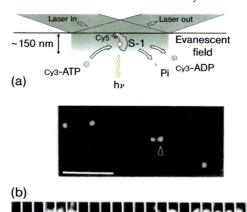

(a)

(b)

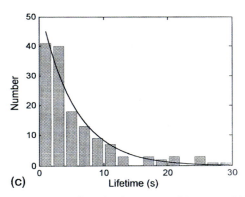

(c)

Figure 2.2 Single molecule imaging of the enzymatic (ATPase) reaction by myosin [2]. (a) TIRFM for the observation of individual ATP turnovers by a single myosin molecule. (b) Fluorescence images of single myosin molecules labeled with Cy-5. Scale bar = 5 mm (top). ATP turnovers by a single myosin molecule; the lower panel shows typical images of fluorescence from a Cy3 nucleotide (ATP or ADP) coming in (down arrows) and out (upper arrows) of focus by associating and dissociating with a myosin molecule. Images were taken every 3 s. (c) A histogram of lifetimes of a Cy3–nucleotide bound to myosin. Solid line shows an exponential fit to the data.

associated with myosin bound to a surface, the Brownian motion ceases and the labeled nucleotide can be observed as a clear fluorescent spot. Thus, association and dissociation of Cy3-nucleotides (ATP or ADP) can be observed by monitoring the flickering of fluorescent spots. Because the affinity of ATP for muscle myosin is $\sim10^5$-fold higher than that of ADP, the nucleotide coming into focus when associating with myosin should be Cy3-ATP and that coming out of focus when dissociating from the myosin should be Cy3-ADP after hydrolysis. Thus, the association and dissociation cycle of Cy3–nucleotides with myosin should be directly coupled to the biochemical cycle of ATP hydrolysis. Myosin has two heads, each of which has ATP and actin binding sites. Therefore, for simplicity, one-headed sub-fragments of myosin were used. It was shown that one-headed sub-fragments of myosin can move actin filaments at the same velocity as two-headed myosin in a myosin-coated surface assay [22].

2.2.2
Mechanical Events

The mechanical events occurring in a myosin motor were measured by optical trapping nanometry. Two ends of an actin filament, which had been visualized by labeling with fluorescent phalloidin [23], were attached to optically-trapped beads. The suspended filament was brought into contact with a single myosin head fixed on the surface of a pedestal made on a glass surface. Individual mechanical events such as displacement and force due to interactions between actin and the myosin head were determined by measuring the displacement of a bead. The bead displacements were determined with nanometer accuracy by projecting the bead onto a quadrant photodiode and measuring its differential outputs [15, 17].

2.2.3
Simultaneous Measurements

Individual ATP hydrolysis cycles and mechanical events associated with a myosin motor were simultaneously measured by combining the single molecule imaging technique with optical trapping nanometry [24] (Figure 2.3a). Figure 2.3b shows time traces of displacements (upper trace) and changes in the fluorescence intensity from Cy3–nucleotides (ATP or ADP) bound to the myosin, which have been measured by a photon counter (lower trace). When an ATP molecule binds to the myosin head, the myosin head dissociates from the actin and the displacement decreases to zero. The myosin head, dissociated from the actin by ATP, rebinds to the actin and generates displacement and force. Thus, each displacement corresponds to a single ATPase turnover.

2.3
Resolving the Process of a Displacement by Scanning Probe Nanometry

Next we attempted to resolve the process of a displacement occurring during a single ATP hydrolysis cycle, in order to make clear how the myosin motor works using the chemical energy provided by ATP [17, 25]. Since the displacements take place rapidly, within 1 ms, it was impossible to resolve the rising phase of a displacement by optical trapping nanometry due to a poor signal to noise ratio. This is because the displacements were measured by observing the movement of actin, not myosin, and thus the compliance (1/stiffness) of the linkage between the optically trapped beads and the actin filament damped the signal to noise ratio. To overcome this

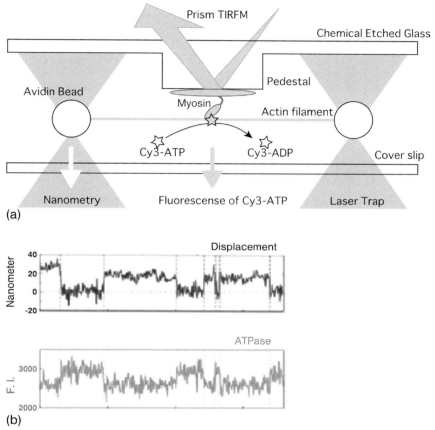

Figure 2.3 Simultaneous observation of individual ATP hydrolysis and mechanical events by single myosin molecules. (a) A schematic drawing of the experimental setup [15]. (b) Time traces of displacements (upper trace) and changes in the fluorescence intensity from Cy3–nucleotides (ATP or ADP) bound to myosin (lower trace).

problem, we developed a more direct method to capture a single myosin molecule and measure its displacements by using a scanning probe. The series stiffness during acto–S1 interaction significantly increased to > 1 pN/nm, compared to that obtained with optical trapping experiments (0.05–0.2 pN/nm). The resulting thermal fluctuations of the probe, namely the noise of the measurements, were reduced from 4–9 nm_{rms} to <2 nm_{rms}. This improvement was critical to the resolution of the process generating the ~10–20-nm displacements. Furthermore, a myosin head rigidly attached to a relatively large scanning probe could steadily interact with actin without diffusing away from the actin filament, as it does in muscle to slow its motion.

2.3.1
Observation and Manipulation of a Single Myosin Motor

Myosin heads were labeled at their tail ends with a fluorescent dye (Cy3) in an almost one to one (0.95) molar ratio. The number of myosin heads captured onto the probe tip was determined from the fluorescence intensity and photobleaching behavior. The fluorescence was observed by an objective type TIRFM [4] (Figure 2.4a). The Cy3-labeled myosin could be clearly observed as fluorescent spots using evanescent field illumination. The Cy3-labeled myosin was captured at its biotinylated tail on the tip of a probe through a biotin–streptavidin bond. Figure 2.4b shows typical fluorescence time trajectories of Cy3-labeled myosin captured on the tip of the probe in which the fluorescence intensities decreased in a single step (upper) and double step (lower). The intensity of the fluorescent spots of single Cy3-labeled myosin captured on the tip of the probe was the same as that of those bound to the glass. These results indicate that it is possible to count the number of S1 molecules on the tip of the probe. For example, the number of Cy3–BDTC–S1 molecules on the tip of the probe in Figure 2.4b (upper and lower panels) was judged to be one and two, respectively. Only the data for single molecules were used for the following analysis.

2.3.2
Displacements

To minimize damage to the myosin caused by interaction with the surface of the probe, the myosin was specifically attached to the tip of a probe at its tail end via a biotin–streptavidin bond. After the number of myosins on the probe tip was confirmed to be one by fluorescence, the captured myosin head was brought into contact with an actin bundle fixed to the glass surface in the presence of ATP (Figure 2.4a). Displacements produced by single myosins were measured with a wide range of needle stiffness from 0.01 to 0.6 pN/nm.

Figure 2.5a and b (upper traces) show typical time courses of displacements at low and high needle stiffness. The myosin–actin interactions could be clearly identified by an increase in stiffness calculated from the reciprocal of the variance of the fluctuations of the probe. Thermal fluctuations occurred when myosin dissociated from the actin bundle. Their amplitude was dependent on the stiffness of the probe. During myosin–actin attachments, the fluctuations decreased to an r.m.s. amplitude of 1.4–2.9 nm which corresponded to a stiffness of 0.5–2 pN/nm. The highest value of stiffness during attachments ($\sim$2 pN/nm) was as large as that of an actomyosin crossbridge in muscle [26].

The mean displacement of myosin was determined by averaging observed events ($n = 274$). The mean displacement was 17 nm at low needle stiffness (0.01–0.1 pN/nm) and 9.2 nm at high needle stiffness (0.1–0.6 pN/nm). The duration of displacements in the presence of 1 mM ATP at 20°C was distributed exponentially. The second-order rate constant for the dissociation of actin–myosin by ATP was deduced from the mean duration and found to be 4–5 $\times 10^6 \, M^{-1} s^{-1}$ at low needle stiffness. This value was consistent with the values obtained for the suspension of actin and

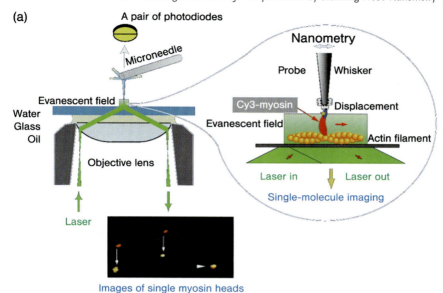

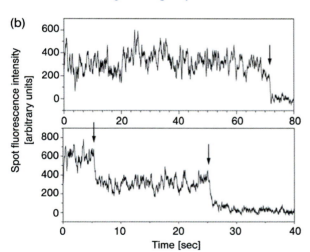

Figure 2.4 Observation and manipulation of a single myosin motor [17]. (a) Schematic drawing of the experimental apparatus. The system was built on an inverted microscope. A ZnO whisker crystal of length 5–10 mm with a radius of curvature at the tip of ~15 nm, was attached to the tip of a very fine glass microneedle, 100 mm long and 0.3 μm in diameter. A myosin head was attached at its tail end to the tip of a ZnO whisker. The glass needle was set perpendicular to the longitudinal axis of the actin filament and the displacements due to actin–myosin interaction along the actin long axis were determined by measuring the bending of the needle with nanometer accuracy. Insert panel, fluorescence images of single myosin heads labeled with Cy3 on a glass surface. A myosin head captured by a probe is marked by an arrow. (b) Time courses of fluorescence intensities of myosin heads captured by a probe.

Low loads (< 1pN)

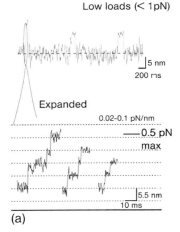

(a)

High loads (> 1 pN)

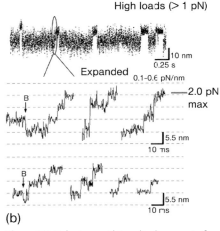

(b)

Figure 2.5 Sub-steps within a displacement of myosin-II [17, 25].
(a) and (b) Time courses of displacements (upper traces) and
rising phases of displacements in expanded time scales
(lower traces) at low loads and high loads, respectively.
(c) Number of sub-steps per displacement at low and high loads.
(d) Ratio of forward and backward sub-steps.

myosin in solution [27] and in optical trapping nanometry [28]. The results indicate that each displacement corresponds to one cycle of ATP turnover, consistent with the observation of Cy3 attached to myosin on the tip of a scanning probe.

2.3.3
Sub-steps within a Displacement

Figure 2.5a and b (lower trace) show the rising phases of the displacements at low (0.01–0.1 pN/nm) and high (0.1–0.6 pN/nm) needle stiffness on an expanded time

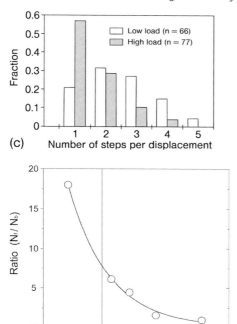

(c)

(d)

Figure 2.5 (*Continued*)

scale, respectively. Displacements were not abrupt but took place in a stepwise fashion. Most steps occurred in the forward direction but a small number were also recorded in the backward direction. The size of the steps was determined by computing the histogram of pairwise distances of all the data points of the stepwise movements in the rising phases [18]. The stepwise data were corrected by using the stiffness after the displacements reached a plateau (maximum) according to the method described earlier. This is because the stiffness of an acto–S1 complex during the rising phases could not be determined quantitatively due to the short dwell phase of the steps, and because the stiffness in the rising phase was assumed to be the same as that at the plateau. The power spectrum of the histogram had an obvious peak at $0.18\,\mathrm{nm}^{-1}$. This corresponded to the periodicity of the peaks in the histogram with a spacing of 5.6 nm. The histogram of pairwise distances had a small peak near 5.5 nm, corresponding to steps in the backward direction. The size of the steps (5.3 nm) at low needle stiffness was almost the same as that (5.6 nm) at high needle stiffness. That is, the size of the steps was independent of the load (0–2 pN), although the overall displacement decreased at high needle stiffness due to a decrease in the total number of steps.

Figure 2.5c shows histograms of the number of steps per displacement at low and high needle stiffness. The number of steps in a displacement varied from one to five with an average of 2.5 steps at low loads (0–0.5 pN) and from one to four with an average of 1.6 steps at high loads (0.5–2 pN).

2.3.4
Nature of Sub-steps

The stepping motion of a myosin head is not always smooth and sometimes moves in the opposite direction along an actin filament. The size of the steps was 5.5 nm for steps in both the forward and backward directions. This step size coincided with the distance between adjacent actin monomers in one strand of an actin filament [21]. Furthermore, the number of steps ranged randomly from one to five during no more than one ATP hydrolysis cycle, i.e. the 5.5-nm steps were not tightly coupled to the ATP hydrolysis cycle. The stochastic features of this stepping motion and the step size strongly suggest that the myosin head walks or slides along the actin monomer repeat using Brownian motion. Because the majority of steps occurred in one direction, they should not result from pure thermal diffusion but rather be biased in one direction (forward).

In order to clarify this point quantitatively, the number of forward (N_f) and backward (N_b) steps were counted at low force levels and the (N_f/N_b) ratio was calculated. The ratio of the number of forward and backward steps near zero force (0–0.5 pN) was $55/9 \approx 6$ (Figure 2.5d). Thus the probability that the 55 forward steps out of total 64 steps were due to pure Brownian motion could be calculated as

$$\frac{64!}{55!9!} \left(\frac{1}{2}\right)^{55} \left(\frac{1}{2}\right)^{9} \approx 10^{-9}.$$

In this calculation, it was assumed that no external force was applied to the myosin. However, because the myosin underwent steps against a force, the actual probability should be even less than the above result. Therefore, the observed steps cannot be due to pure Brownian motion. Rather, myosin Brownian steps are likely biased towards the forward direction.

2.3.5
Comparing the Actions of Individual Myosin Motors with those of Muscle

To quantitatively compare the mechanical properties of individual actomyosin motors with those of muscle, their force–velocity curve was investigated. Displacements started at various force levels due to thermal fluctuations of the needle causing steps to also take place at various force levels. Figure 2.6 shows the force–velocity curve obtained from the step size and the dwell time of single myosin heads. The velocity did not reach zero but a positive value at a large positive force, which is different from that of muscle. This is most likely because the backward steps were not considered in the analysis. Backward steps took place more frequently at larger forces, thus the backward steps need to be considered when calculating the velocity of individual myosin heads. The anisotropy of the stepping direction was defined as $(N_f - N_b)/(N_f + N_b)$, where N_f and N_b are the number of steps in the forward and

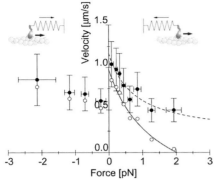

Figure 2.6 Force–velocity curve of individual myosin heads [25]. The velocity was obtained by dividing the step size, 5.5 nm, by the dwell time (filled circles). Bars indicate the standard deviations for 10–30 steps. Open circles indicate the velocity corrected by the anisotropy of the stepping direction. The solid line shows the Hill curve fitted to the corrected velocity.

backward directions at each force level, respectively. The values were corrected by multiplying the velocities by the anisotropy of the stepping direction (Figure 2.6, open circles). The positive force regions of this curve were in close agreement with the data obtained in muscle [29].

The work done per 5.5-nm step (W_{step}) was defined as the energy needed to pull a linear spring (spring constant of a needle, k) over a distance of 5.5 nm from an initial position, x, such that $W_{step} = \frac{1}{2}k((x+5.5)^2 - x^2)$. We measured x during the dwell time of the step from the equilibrium position of the needle. The mean of W_{step} was 7.4 pN/nm ($= 1.8k_BT$) and 4.0 pN/nm ($= 1.0\ k_BT$) at high and low needle stiffness, respectively.

The velocity corrected for the stepping anisotropy had a hyperbolic dependence when a positive force was applied. The data were well described by Hill's equation (solid line in Figure 2.6; $(P + a)(V + b) = (P_0 + a)b$ where P is load, V, velocity, P_0, isometric force, and a and b are constants having the dimensions of force and velocity, respectively). Maximum velocity (V_{max}), isometric force (P_0) and the curvature of the hyperbola (a/P_0) were determined to be 1.0 mm/s, 2.0 pN and 0.7, respectively. V_{max} at zero load was several times smaller than that of rabbit skeletal muscle. This is because the sliding velocity of myosin is slowed by attachment to the large scanning probe. P_0 was consistent with the mean peak force. However, the value of a/P_0, which is a measure of the curvature of the force–velocity curve, was significantly larger than that of shortening muscle (0.2–0.4) [29]. One possible explanation for this difference is that the force-generating step and other steps in the ATPase cycle, such as association and dissociation of actin and myosin, are all involved during measurement of a shortening muscle, giving the steeper force–velocity curve. In individual actomyosin motors, however, only the force-generating step is measured.

The mean work done per 5.5-nm step was determined to be 7.4 pN/nm ($= 1.8k_BT$) at high needle stiffness. The number of steps per displacement was distributed from one to four. Hence the maximum work done during one ATPase cycle was measured as 31 pN/nm ($= 7.2k_BT$). This result indicates that single myosin heads are able to convert chemical energy into mechanical work with a maximum efficiency of $\sim$36% (energy liberated by ATP hydrolysis $\sim$20k_BT). This value is similar to the maximum efficiency of contracting muscle fibers [29, 30]. The similarities between the force–velocity curve and thermodynamic efficiency of single actomyosin molecules to assemble systems such as muscle suggest that the major mechanical and thermodynamic properties of muscle are essentially the effect of the intrinsic characteristics of individual actomyosin motors.

2.3.6
Other Types of Molecular Motors

Class-II myosin (muscle myosin) was found to move along actin filaments by biased Brownian motion. However, there are various classes of molecule motors. Do they work by the same mechanism as myosin-II? Here, we investigated class-V myosin, a processive motor that transports organelles in cells.

Myosin-V has two heads, each with a long neck (6IQ motifs) [31], and moves along an actin filament in large steps of $\sim$36 nm [32]. This step size coincides with an actin half helical pitch. Electron microscopy [33] and single molecule nano-scale imaging (FIONA) [10, 34] have shown that each head of myosin-V moves alternately along an actin filament with a step size ($\sim$72 nm) twice as large as that observed in a two-headed molecule. This result indicates that myosin-V walks along the actin helical repeats using a hand-over-hand mechanism in which the rear head detaches from the actin filament, diffuses to the forward helical pitch and attaches there while the front head is still attached. But how is the diffusion of the rear head biased to the forward target? Also, how does the rear head undergo a Brownian search for the forward target? The diffusion of the rear head occurs very rapidly (<1 ms). Therefore, the process of the diffusion cannot be detected with a sufficient signal-to-noise ratio using optical trapping nanometry [32] and single molecule nano-scale imaging (FIONA) [34]. We again used scanning probe nanometry to resolve the diffusive process of the head which had been slowed by attaching the head to a relatively large probe [35]. To do this, we compared single myosin-V heads with short (2IQ motifs) and long (6IQ motifs) necks to wild-type, two-headed myosin-V. Green fluorescent protein (GFP) and Myc-tag were fused to the N-terminus and the C-terminus, respectively. Single myosin-V molecules were captured on the tip of a scanning probe via c-Myc monoclonal antibody. The number of myosin-V molecules on the tip of a scanning probe was counted by observing the fluorescence of green fluorescent protein (GFP) fused to their N-terminus. We were able to observe the stepwise nature of the rising phases of the displacements of myosin-V 2IQ and 6IQ heads on an expanded time scale. The size of sub-steps was constant, $\sim$5.5 nm, independent of the length of the neck domain (Figure 2.7a,b). Thus, myosin-V heads move along actin monomers in a manner similar to myosin-II heads. A two-headed myosin-V showed processive movement with $\sim$36-nm steps as previously observed by optical trapping nanometry [32]. Furthermore an intermediate step ($\sim$18 nm) within each

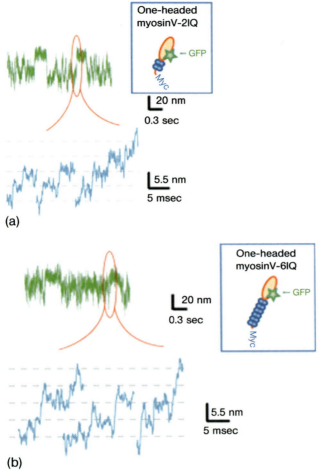

Figure 2.7 Sub-steps within a displacement of myosin-V [35]. (a) One-headed myosin-V with short (2IQ) neck. (b) One-headed myosin-V with long (6IQ) neck. (c) two-headed myosin-V with long (6IQ) necks. (d) The Hopping model for myosin-V stepping. The rear head searches for the actin target in the forward direction by a hopping movement on the actin subunits according to a potential slope along the actin helix. Black closed circle = ATP or ADP-Pi. F = Force. The light gray band of ~19 nm indicates displacements by neck bending and its diffusion, and the dark gray band of ~17 nm indicates directional steps on actin monomers according to the potential slope along the actin helix.

36-nm step was observed. Such intermediate steps have been previously observed using optical trapping nanometry at high loads [32] or in the presence of 2,3-butanedione monoxime [36]. In this experiment, the intermediate step was clearly observed even at low loads. It is likely that the intermediate step was also slowed by attaching the two-headed myosin-V to a large scanning probe. Figure 2.7c shows the rising phases of the intermediate steps on an expanded time scale. The sub-steps of

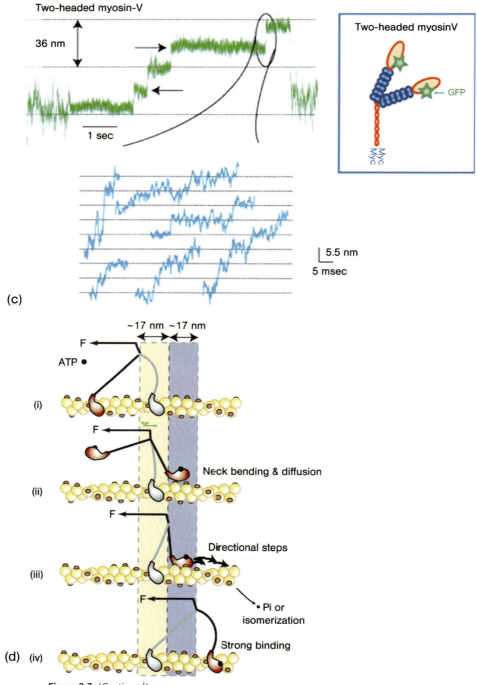

Figure 2.7 (*Continued*)

~5.5 nm could also be clearly observed. Figure 2.7d shows a model for the Brownian search by the myosin-V head for the actin forward target.

2.4
Biased Brownian Step Model

Scanning probe nanometry showed that type-II myosin (muscle myosin) move along actin subunits by Brownian motion. Brownian motion is random, so it should be biased in one direction to generate unidirectional movement of myosin. Here we consider how the Brownian motion of myosin is biased in one direction [25].

2.4.1
Asymmetric Potential

Applying the asymmetric potential model [37, 38] presented in Figure 2.8a, we analyzed the myosin Brownian sub-steps. The activation energy of the forward and backward directions can be described by $(u_+ + Fd_+)$ and $(u_- - Fd_-)$, respectively, where u_+ and u_- are the heights of the potential barrier maximum at zero load, and d_+ and d_- are the characteristic distances. Assuming the Boltzmann energy distribution, the rates in the forward and backward directions will be proportional to $\exp[-(u_+ + Fd_+)/k_BT]$ and $\exp[-(u_- - Fd_-)/k_BT]$, respectively. Differences between the potential barriers for the forward and backward steps at load F is given by $\Delta u - Fd$ $k_BT\ln(N_f/N_b)$, where $\Delta u = u_+ - u_-$ and $d = d_+ + d_- = 5.5$ nm. Figure 2.5b shows the ratios of N_f to N_b at various loads for myosin-II. Using these ratios, Δu is calculated to be 2–3 k_BT (Figure 2.8b). Thus, Brownian steps are biased by a potential energy of 2–3 k_BT at zero load, similar to the experimental results for myosin-V [35]. At $N_f = N_b$, F is calculated to be 2.5 pN, which gives the maximum force at zero velocity, consistent with that measured directly (Figure 2.6). However, this maximum force is smaller than that estimated from the isometric force of muscle [21, 29]. Conformational changes in the myosin head coupled to Pi release may cause additional forces [39, 40] as discussed later.

How do the myosin steps define the potential? So far, several models have been proposed [41–43]. Here, we propose a simple model assuming more realistic situations in which the potential slope is produced by steric compatibility between the orientations of the binding sites of actin and the myosin head. The actin filament has a double helical structure while the protofilament contains seven monomers and rotates 180° per half helical pitch. The tail (neck domain) of the myosin head should not be perfectly rigid so that the myosin head attached to the probe can move along the actin helical pitch. However, the binding sites of actin monomers rotate along the helix relative to the myosin head attached to the probe and hence the steric compatibility between the orientations of the binding sites of the myosin head and the actin should change depending on their relative positions. Thus, this steric compatibility should result in a potential slope along the actin helical pitch. For example, if the binding site of the head faces the right side of the actin filament fixed

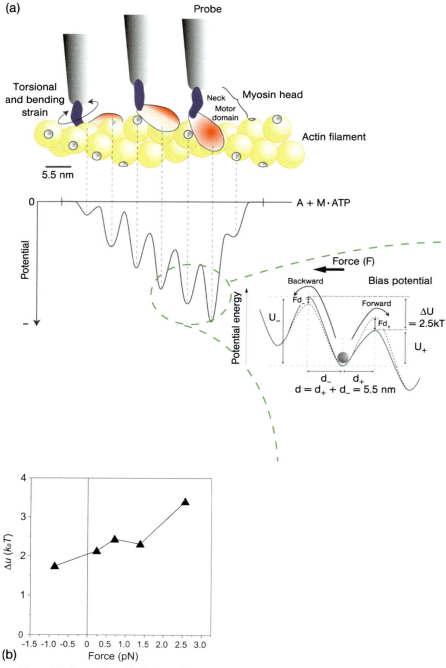

Figure 2.8 Potential profile for biased Brownian steps [25, 35]. (a) Biased Brownian steps of a myosin head along an actin filament (upper) and asymmetric potential of the activation energy (lower). (b) Bias potential energy at various loads.

on a glass surface, the head could favorably bind to the actin on the right side of the filament. But it would be unfavorable for the head to bind to the other sides (up and left) of the actin filament because the head would be required to bend and rotate (Figure 2.8a upper). Thus, the potential slope that declines along the forward direction is produced along the half helical pitch (Figure 2.8a lower). In 1971, A. F. Huxley and R. M. Simmons have proposed a famous model, so called "'71- H-S model" that is uniquo to successfully explain dynamic behaviors of muscle such as mechanical responses of sudden length changes [44]. This model assumes that a myosin head undergoes multiple (2-3). Brownian steps which are biased by the potential energy of 2.5 k_BT per step and cause 4-8 nm displacement per step. This model also assumes the rotation of the head to cause the steps but instead the steps on actin monomers (Fig. 2.8a) could replace it. Thus, our model is eonsistent in a quantitative manner with the '71 H-S model.

Finally, we consider the coupling between the mechanical and ATPase cycles. The myosin head dissociates from actin upon ATP binding. The bound ATP hydrolyzes into ADP and Pi. The myosin head bound to ADP and Pi undergoes rapid attachment and detachment cycles with actin [21, 29]. The myosin head steps toward the forward direction during the attachment and detachment cycles according to the potential slope along the actin helical pitch. When the myosin head reaches the potential base, the head is pulled back. We have proposed a strain sensor model in which a part (likely the neck domain) of the myosin molecule acts as a strain sensor and the backward strain switches the sensor to accelerate Pi release causing a rigor complex to form with actin and thus halting the movement [42, 45, 46]. Coupled to Pi release, conformational changes in the myosin head take place. The changes probably initiate isometric forces [21, 29] which rotate the actin filament [47, 48] (see below).

2.4.2
Comparison with Other Studies

Just half-century ago, H. E. Huxley and Hanson [49] and A. F. Huxley and Niedergerke [50] independently discovered that muscle contraction is caused by the relative sliding motion of actin and myosin filaments. The myosin heads project from the myosin filaments forming links between actin and myosin filaments. These links are referred to as cross-bridges. The cross-bridges attach and detach in a cyclical manner from actin filaments, a process coupled to the biochemical cycles of ATP hydrolysis. H. E. Huxley [51] and A. F. Huxley and Simmons [44] proposed "the cross-bridge rotation model", in which the actin filament is pulled along by the rotating of the cross-bridges. This model had been widely accepted as a working hypothesis for analyzing experimental data. About 10 years later, however, Yanagida [52] and Thomas *et al.* [53] showed by polarized fluorescence microscopy and EPR spectroscopy that the ATP binding site and the SH1 domain, which are located in the motor domain of the myosin head, did not change their orientations during muscle contraction, respectively.

An intramolecular bending model [54–57] was proposed as an alternative to the cross-bridge rotation model, in which the motor domain remains unchanged while the light chain binding domain (neck domain) which is distant from the actin binding

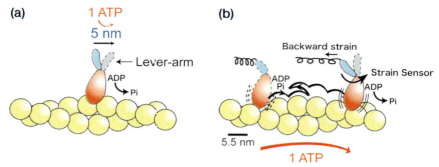

Figure 2.9 Lever arm model and bias Brownian model. (a) Lever arm model. The displacement is developed by a conformational change (lever arm tilting), tightly coupled to the ATP hydrolysis cycle. (b) Bias Brownian model. The myosin head (ADP-Pi) undergoes rapid attachment and detachment cycles with an actin filament in the weakly binding state. The Brownian motion is biased in one direction according to the potential slope along the actin helix (see Figure 2.8a). The neck domain of the myosin head acts as a strain sensor that controls the transition from the weak to strong binding state, coupled with Pi release [79]. When the myosin head approaches the forward actin target by Brownian motion, the neck domain is pulled backward and the strain sensor is then switched on to ensure that the myosin head is strongly bound to the actin. The conformational changes in the neck domain are coupled to the action of a strain sensor which may cause isometric force.

site, bends. In the early 1990s, the crystal structure [58] of the myosin head was elucidated showing that the neck domain which is attached to the motor domain of the head, changed its angle relative to the motor domain depending on the form of the bound nucleotide. Based on these findings, the cross-bridge swinging model has been refined to the "lever-arm swinging model". Here the neck domain acts as a lever arm and a small conformational change in the motor domain causes the lever arm to swing resulting in large displacements of 5 to 10 nm (Figure 2.9a) [59, 60]. Many studies have agreed that the neck domain changes its angle during both muscle contraction [60, 61] and *in vitro* motility assay [62]. The observed conformational changes may contribute to the generation of isometric force at large loads but may not cause the sliding movement at smaller loads. In our bias Brownian step model (Figure 2.9b) the conformational change in the neck domain coupled with Pi release is not the direct cause of the swing movement. The lever arm model hypothesizes that the neck domain swings parallel to the longitudinal axis of the actin filament to directly produce displacements in the forward direction. However, several studies using electron microscopy have suggested that the direction of the neck domain swing is not parallel but diagonal to the longitudinal axis of the actin filament [54, 63]. Therefore, the conformational changes in the neck domain coupled to Pi release may cause rotation of the actin filament, consistent with our model (Figure 2.9b).

Several studies using optical trapping nanometry have reported that the displacements by chemically-modified and genetically-engineered myosin heads with various neck domain lengths are approximately proportional to the neck lengths [64–66], consistent with the lever arm swinging model [59, 60]. As shown in our model (Figure 2.9b), the displacements we observed depended on the interacting length between actin and the myosin head, which may include the power stroke length. The interacting length is determined by the distance between the start and end positions of the actomyosin interaction. Thus, the displacement (interacting length) caused by a single event is not necessarily proportional to the neck length. In the case of other reports, the actin filament is suspended in solution by dual optical traps and the actin filament is able rotate [67]. Therefore, a head with a short neck may bind to actin only when the orientation of the actin binding site rotates into a favorable position for binding and thus the interacting length is small. A head with a long neck may bind to actin even when the orientation of actin is unfavorable because the long neck would have the elasticity to reach the unfavorable binding site, thus leading to a large interacting length.

Tsiavaliaris and his colleagues [68] have demonstrated, using protein engineering, that the orientation of movement is reversed when the orientation of the neck domain is reversed. They have argued that this result strongly supports the lever arm swinging model, but this result is also consistent with our model because the direction of the myosin binding sites relative to the actin helical pitch is reversed and so the potential slope is also reversed.

2.4.3
Computer Simulation: from a Single Molecular Motor to Muscle

How does the stochastic nature of individual myosin motors contribute to the flexible operation of muscle? To answer this question, we performed a computer simulation of the cooperative behavior between multiple myosin motors. We assumed that multiple myosins are bound to ADP and Pi during most of the ATPase cycle; are tethered to a myosin filament via elastic elements (neck domain and S2) (Figure 2.10a,b); and move along the actin helical pitches due to a potential slope generated by the steric compatibility (Figure 2.8a). At some point in time, one of these heads releases Pi to form a rigor complex with actin (Figure 2.10b upper panel). It has been demonstrated that the actin filament is rotated during sliding *in vitro* [69–71] and during force generation in muscle [46, 47]. Since the actin filament is rotated by approximately 90° in muscle [47], we assume that the actin filament is rotated by 90° due to the formation of a rigor complex (Figure 2.10b upper panel). Then, an ATP molecule binds to the rigor head to dissociate it from actin and the actin filament rewinds to its original orientation because one end of it is fixed to the z-line (Figure 2.10b middle panel). ADP-Pi-bound myosin heads near the bottom end of the potential slope dissociate from actin. The unbound myosin will bind to ATP, which is then hydrolyzed into ADP-Pi. The new ADP-Pi myosin heads bind to new actin monomers at the top of the potential slope. Since the potential slope is shifted by about three actin monomers – corresponding to a 90° rotation in the actin filament – the heads previously located at the potential bottom can move the actin filament once more now that they are at a higher position on the potential slope (Figure 2.10b middle and lower panels, vertical arrows). The energy for rotating one end of a 1-μm long actin filament by 90° when the other end is fixed to the z-line is estimated to be 16–32 k_BT based on its torsional rigidity (2.6–6.7 × 10^{-26} Nm^2) [67]. This is similar to the free energy (20 k_BT) produced by the hydrolysis of one ATP molecule.

Based on these features, we first propose a qualitative explanation for the long sliding distance in muscle. It is known that muscle can modulate its motion depending on the conditions. For example, the sliding distance (interacting distance) between actin and myosin filaments at low loads is ~ten-fold as long as that of a single myosin motor [72–74]. How can such a long sliding distance be achieved by the cooperative action of multiple myosin motors? Upon rewinding the actin filament, we consider one head that interacts with the actin filament and moves it until the head moves to the bottom end of the potential slope (Figure 2.10b middle and lower panels).

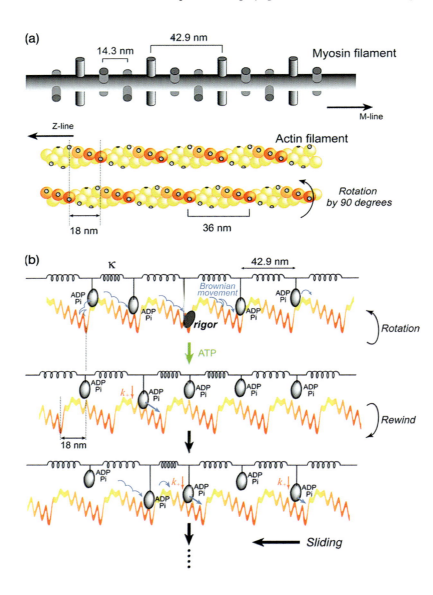

Then two or three heads, dissociated from actin by its rotation, interact with the actin filament and exert force on the actin filament in the forward direction (Figure 2.10b lower panel). Since the potential energy produced by two or three heads is sufficient to move the first head at the potential bottom to the next forward helical pitch, the actin filament is moved further. After that, if more than four heads interact with the actin filament, the filament is moved again. Thus, the actin filament can potentially be moved more than >60 nm per ATP by the cooperative action of multiple heads.

In order to test this model quantitatively, we performed computer simulations describing the cooperative action between multiple myosin heads undergoing biased Brownian motion along the actin helical pitches. Analogously, it has been demonstrated that collective myosin motors connected by rigid bonds to a backbone can conduct dynamic phase transitions, such as hysteretic behavior in sliding velocity against external loads or spontaneous oscillations in a collective manner [75, 76]. In our model, the long (>60 nm) sliding distance per ATP at zero load was successfully simulated by connecting each head to a backbone via a flexible spring (Figure 2.11) [25]. This means that muscle can modulate the sliding distance (interacting distance) per ATP from nearly zero to >60 nm to economize energy depending on loads. Thus, the stochastic nature of individual actomyosin motors is important for the dynamic and adaptive operation of muscle because it can adapt to the ever-changing demands of the muscle.

2.5
Conclusion for the Unique Mechanism of Biological Molecular Machines

The myosin motor does not overcome thermal fluctuation (noise) but rather utilizes it to operate. The origin of motion is Brownian motion and the chemical energy released from ATP hydrolysis is used to bias the Brownian motion or repeat the mechanical cycle. Molecular motors can thus operate at an energy level as small as average thermal energy ($=\sim k_B T$) with high efficiency of energy conversion. This is in sharp contrast to

---

Figure 2.10 Cooperative action of multiple heads undergoing stochastic steps [25]. (a) Schematic diagrams of actin and myosin filaments in skeletal muscle [80]. The actin filament has a helical structure with a half pitch of 36 nm. The myosin filament also has a helical structure with a pitch of 43 nm and a subunit repeat of 14.3 nm. Myosin heads on a myosin filament project toward an actin filament at 43-nm intervals. In skeletal muscle, the actin and myosin filaments are arranged in a hexagonal lattice with one actin surrounded by three myosin filaments. Therefore, the number of myosin molecules that project toward one actin filament 0.7 μm long (length when fully overlapped with myosin filaments) is approximately 50 [80]. When the actin filament is rotated by 90° [47], the relative position between the actin helical pitches and the myosin heads shifts by approximately three actin monomers. The actin slopes along the actin helical pitches are represented by a color gradient. (b) Qualitative explanation of the cooperative action of myosin heads on a thick filament. The myosin filament is equivalently represented by a row of myosin heads connected with springs at intervals of 43 nm. The actin filament is represented by periodic, saw-tooth shape potentials along the half helical pitches. Cooperative action of the myosin heads leads to a long actin filament sliding distance (>60 nm) per ATP (see text for details).

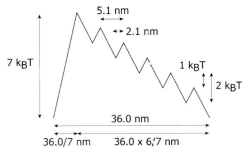

Figure 2.11 Computer simulation of multiple head cooperative activity undergoing stochastic steps [25]. We simulated the cooperative behavior between myosin heads under a periodic and asymmetric potential as shown in Figure 2.10b by numerically solving the Langevin equation,

$$0 = -\rho dx_i/dt - dU(x_i,t)/dx + F(t) - A_i,$$

where x_i is the position of i-th myosin head; $\rho = 8.8 \times 10^3$ pN/ns/nm is a drag coefficient; $F(t)$ is the random force obeying a Gaussian white noise characterized by the ensemble average, $<F(t)> = 0$ and $\langle F(t)F(s) \rangle = 2\,k_B T \rho$ $\delta(t-s)$, where k_B is the Boltzmann constant, T is the absolute temperature, and $k_B T = 4.1$ pN/nm; A_i is the interaction force between the neighboring heads described as $k(x_i - x_{i-1}) - k(x_{i+1} - x_i)$, where k is the spring constant connecting the heads. The potential slope along the actin helical pitch was simplified to be a saw-tooth shaped potential. The drag coefficient was set to be larger than it is in solution so that the velocity of the heads was equal to the maximum velocity in Figure 2.6. Other parameters were chosen such that (1) k was 0.1 pN/nm, which is approximately one tenth as large as that of a rigor

crossbridge; (2) the ratio of the potential rise to decline was 1 to 6 and the depth of the potential at the bottom was 2 $k_B T$; (3) the pitch of the potential and the average intervals of myosin heads were 36 and 43 nm, respectively; (4) the number of heads interacting with the actin filament was 11 (~20% overlap between actin and myosin filaments); (5) the rotation angle of the actin filament was 90°; and (6) the rate constant (k_+) for the rebinding of the heads to actin after the rewinding of the actin filament was 100 s/head. The potential slope was assumed to be smaller than that estimated in the present experiment. The strain exerted on the neck domain would be much smaller during free shortening in muscle because the head is tethered to the myosin filament via a flexible α-helix (S2), while the head is directly attached at its tail end to the probe in the present measurement system. Thus, because the strain is smaller, so too is the potential slope. (a) Schematic graph of the potential function, $U(x)$. (b) A typical time course of the movement of an actin filament. (c) Histogram of the sliding distance of actin filaments per ATP. The average sliding distance of actin filaments was 58.4 nm per ATP.

man-made devices that operate at a much higher input energy level than $k_B T$ to avoid the influence of thermal noise. For example, the input energy level of a transistor is ~200 $k_B T$ and the energy used for processing just one bit data is ~$2 \times 10^7 k_B T$ in a computer [77]. Szilard has shown by his model experiment of Brownian particles that the minimum energy required for processing one bit data is $k_B T \ln 2 = 0.7\ k_B T$ [78]. Thus biological motors likely save energy by utilizing thermal fluctuation. However, the action of molecular motors driven by thermal fluctuation is stochastic and ambiguous. Although this stochastic and ambiguous nature is an impediment for man-made machines, our computer simulations suggested that when stochastic myosin motors assemble to form systems like muscle, they show flexible and adaptive

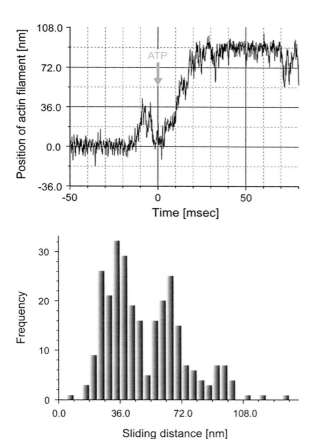

Figure 2.11 (*Continued*)

motion depending on the conditions. In conclusion, molecular machines save energy by utilizing thermal fluctuation and the resulting stochastic nature is skillfully used for the flexible and adaptive operations necessary in biological systems.

Acknowledgment

We are grateful to Drs. A. Iwane, P. Karagiannis and T. Shimokawa for helpful discussions and critical reading of the manuscript.

References

1 Moerner, W.E. and Kador, L. (1989) Optical detection and spectroscopy of single molecules in a solid. *Phys. Rev. Lett.*, **62**, 2535.

2 Funatsu, T., Harada, Y., Tokunaga, M., Saito, K. and Yanagida, T. (1995) Imaging of single fluorescent molecules and individual ATP turnovers by single myosin

molecules in aqueous solution. *Nature*, **374**, 555–559.

3 Vale, R.D., Funatsu, T., Pierce, D.W., Romberg, L., Harada, Y. and Yanagida, T. (1996) Direct observation of single kinesin molecules moving along microtubules. *Nature*, **380**, 451–453.

4 Tokunaga, M., Kitamura, K., Saito, K., Iwane, A.H. and Yanagida, T. (1997) Single molecule imaging of fluorophores and enzymatic reactions achieved by objective-type total internal reflection fluorescence microscopy. *Biochem. Biophys. Res. Commun.*, **235**, 47–53.

5 Lu, H.P., Xun, L. and Xie, X.S. (1998) Single-molecule enzymatic dynamics. *Science*, **282**, 1877–1881.

6 Oiwa, K., Eccleston, J.F., Anson, M., Kikumoto, M., Davis, C.T., Reid, G.P., Ferenczi, M.A., Yamada, A., Nakayama, H. and Trentham, D.R. (2000) Comparative single-molecule and ensemble myosin enzymology: sulfoindocyanine ATP and ADP. *Biophys. J.*, **78**, 3048–3071.

7 Iwane, A.H., Funatsu, T., Harada, Y., Tokunaga, M. and Yanagida, T. (1997) Single molecular assay of the individual ATP turnovers by a myosin–GFP fusion protein expressed *in vitro*. *FEBS Lett.*, **407**, 235–258.

8 Harada, Y., Funatsu, T., Murakami, K., Nonoyama, Y., Ishijima, A. and Yanagida, T. (1999) Single molecule imaging of RNA polymerase–DNA interactions in real time. *Biophys. J.*, **76**, 709–715.

9 Sako, Y., Minoghchi, T. and Yanagida, T. (2000) Single-molecule imaging of EGFR signaling on the surface of living cells. *Nature Cell Biol.*, **2**, 168–172.

10 Yildiz, A. and Selvin, P.R. (2005) Fluorescence imaging with one nanometer accuracy: application to molecular motors. *Acc. Chem. Res.*, **38**, 574–582.

11 Ha, T., Enderle, T., Ogletree, D.F., Chemla, D.S., Selvin, P.R. and Weiss, S. (1996) Probing the interaction between two single molecules: fluorescence resonance energy transfer between a single donor and a single acceptor. *Proc. Natl Acad. Sci. USA*, **93**, 6264–6268.

12 Forkey, J.N., Quinlan, M.E., Shaw, M.A., Corrie, J.E.T. and Goldman, Y.E. (2003) Three-dimensional structural dynamics of myosin V by single-molecule fluorescence polarization. *Nature*, **422**, 399–404.

13 Toprak, E., Enderlein, J., Syed, S., McKinney, S.A., Petschek, R.G., Ha, T., Goldman, Y.E. and Selvin, P.R. (2006) Defocused orientation and position imaging (DOPI) of myosin V. *Proc. Natl Acad. Sci. USA*, **103**, 6495–6499.

14 Kishino, A. and Yanagida, T. (1988) Force measurements by micromanipulation of a single actin filament by glass needles. *Nature*, **334**, 74–76.

15 Ishijima, A., Doi, T., Sakurada, K. and Yanagida, T. (1991) Sub-piconewton force fluctuations of actomyosin *in vitro*. *Nature*, **352**, 301–306.

16 Ishijima, A., Harada, Y., Kojima, H., Funatsu, T., Higuchi, H. and Yanagida, T. (1994) Single-molecule analysis of the actomyosin motor using nano-manipulation. *Biochem. Biophys. Res. Commun.*, **199**, 1057–1063.

17 Kitamura, K., Tokunaga, M., Iwane, A.H. and Yanagida, T. (1999) A single myosin head moves along an actin filament with regular steps of 5.3 nanometres. *Nature*, **397**, 129–134.

18 Svoboda, K., Schmidt, C.F., Schnapp, B.J. and Block, S.M. (1993) Direct observation of kinesin stepping by optical trapping interferometry. *Nature*, **365**, 721–727.

19 Finer, J.T., Simmons, R.M. and Spudich, J.A. (1994) Single myosin molecule mechanics: piconewton forces and nanometre steps. *Nature*, **368**, 113–119.

20 Ashkin, A., Dziedzic, J.M. and Yamane, T. (1987) Optical trapping and manipulation of single cells using infrared laser beams. *Nature*, **330**, 769–771.

21 Bagshow, C.R. (1984) *Muscle Contraction*, Chapman & Hall.

22 Iwane, A.H., Kitamura, K., Tokunaga, M. and Yanagida, T. (1997) Myosin

subfragment-1 is fully equipped with factors essential for motor function. *Biochem. Biophys. Res. Commun.*, **230**, 76–80.

23 Yanagida, T., Nakase, M., Nishiyama, K. and Oosawa, F. (1984) Direct observation of motion of single F-actin filaments in the presence of myosin. *Nature*, **307**, 58–60.

24 Ishijima, A., Kojima, H., Funatsu, T., Tokunaga, M., Higuchi, H., Tanaka, H. and Yanagida, T. (1998) Simultaneous measurement of chemical and mechanical reactions. *Cell*, **92**, 161–171.

25 Kitamura, K., Tokunaga, M., Esaki, S., Iwane, A.H. and Yanagida, T. (2005) Mechanism of muscle contraction based on stochastic properties of single actomyosin motors observed *in vitro*. *Biophysics*, **1**, 1–19.

26 Huxley, A.F. and Tideswell, S. (1996) Filament compliance and tension transients in muscle. *J. Muscle Res. Cell Motil.*, **17**, 507–511.

27 Lymn, R.W. and Taylor, E.W. (1971) Mechanism of adenosine triphosphate hydrolysis by actomyosin. *Biochemistry*, **10**, 4617–4624.

28 Tanaka, H., Ishijima, A., Honda, M., Saito, K. and Yanagida, T. (1998) Orientation dependence of displacements by a single one-headed myosin relative to the actin filament. *Biophys. J.*, **75**, 1886–1894.

29 Woledge, R.C., Curtin, N.A. and Homsher, E. (1985) Energetic aspects of muscle contraction. *Monogr. Physiol. Soc.*, **41**, 1–357.

30 Howard, J. (2001) *Mechanics of Motor Proteins and the Cytoskeleton*, Sinauer Press.

31 Cheney, R.E., Oshea, M.K., Heuser, J.E., Chelho, M.V., Wolenski, J.S., Espreafico, E.M., Forscher, P., Larson, R.E. and Mooseker, M.S. (1993) Brain myosin-V is a two-headed unconventional myosin with motor activity. *Cell*, **75**, 13–23.

32 Mehta, A.D., Rock, R.S., Rief, M., Spudich, J.A., Mooseker, M.S. and Cheney, R.D. (1999) Myosin-V is a processive actin-based motor. *Nature*, **400**, 590–593.

33 Walker, M.L., Burgess, S.A., Sellers, J.R., Wang, F., Hammer, J.A., III Trinick, J. and Knight, P.J. (2000) Two-headed binding of a processive myosin to F-actin. *Nature*, **405**, 804–807.

34 Yildiz, A., Forkey, J.N., McKinney, S.A., Ha, T., Goldman, Y.E. and Selvin, P.R. (2003) Myosin-V walks hand over hand: single fluorophore imaging with 1.5 nm localization. *Science*, **300**, 2061–2065.

35 Okada, T., Tanaka, H., Iwane, A.H., Kitamura, K., Ikebe, M. and Yanagida, T. (2007) The diffusive search mechanism of processive myosin class-V motor involves directional steps along actin subunits. *Biochem. Biophys. Res. Commun.*, **354**, 379–384.

36 Uemura, S., Higuchi, H., Olivares, A.O., De La Cruz, E.M. and Ishiwata, S. (2004) Mechanical coupling of two substeps in a single myosin V motor. *Nature Struct. Biol.*, **11**, 833–877.

37 Nishiyama, M., Higuchi, H. and Yanagida, T. (2002) Chemomechanical coupling of the forward and backward steps of single kinesin molecules. *Nature Cell Biol.*, **4**, 790–797.

38 Taniguchi, Y., Nishiyama, M., Ishii, Y. and Yanagida, T. (2005) Entropy rectifies the Brownian steps of kinesin. *Nature Chem. Biol.*, **1**, 346–351.

39 Irving, M. and Goldman, Y.E. (1999) Motor proteins: Another step ahead for myosin. *Nature*, **398**, 463–465.

40 Goldman, Y.E. (1987) Kinetics of the actomyosin ATPase in muscle fibers. *Ann. Rev. Physiol.*, **49**, 637–654.

41 Esaki, S., Ishii, Y. and Yanagida, T. (2003) Model describing the biased Brownian movement of myosin. *Proc. Japan Acad.*, **79**, 9–14.

42 Yanagida, T., Esaki, S., Iwane, A.H., Inoue, Y., Ishijima, A., Kitamura, K., Tanaka, H. and Tokunaga, M. (2000) Single-motor mechanics and models of the myosin motor. *Philos. Trans. R. Soc. Lond. B Biol. Sci.*, **355**, 441–447.

43 Terada, T.P., Sasai, M. and Yomo, T. (2002) Conformational change of the actomyosin

complex drives the multiple stepping movement. *Proc. Natl Acad. Sci. USA*, **99**, 9202–9206.

44 Watanabe, T. M., Tanaka, H., Iwane, A. H., Yonekura, S. M., Homma, K., Inoue, A., Ikebe, R., Yanagida, T. and Ikebe, M. (2004) A one-head myosin-V molecule develops multiple large (32 nm) steps successively. *Proc. Natl. Acad. Sci. USA*, **101**, 9630–9635.

45 Iwaki, M., Tanaka, H., Iwane, A.H., Katayama, E., Ikebe, M. and Yanagida, T. (2006) Cargo-binding makes a wild-type single-headed myosin-VI move processively. *Biophys. J.*, **90**, 3643–3652.

46 Takezawa, Y., Sugimoto, Y. and Wakabayashi, K. (1998) Extensibility of the actin and myosin filaments in various states of skeletal muscle as studied by X-ray diffraction. *Adv. Exp. Med. Biol.*, **453**, 309–317.

47 Wakabayashi, K., Ueno, Y., Takezawa, Y. and Sugimoto, Y. (2001) Muscle contraction mechanism: Use of X-ray Synchrotron Radiation. *Nature Encyclopedia of Life Sciences* (nature Publishing Group/www.els.net), pp. 1–11.

48 Huxley, H.E. and Hanson, J. (1954) Changes in the cross-striations of muscle during contraction and stretch and their structural interpretation. *Nature*, **173**, 973–976.

49 Huxley, A.F. and Niedergerke, R. (1954) Structural changes in muscle during contraction; interference microscopy of living muscle fibres. *Nature*, **173**, 971–973.

50 Huxley, H.E. (1969) The mechanism of muscular contraction. *Science*, **164**, 1356–1365.

51 Huxley, A.F. (1957) Muscle structure and theories of contraction. *Prog. Biophys. Biophys. Chem.*, **7**, 255–318.

52 Yanagida, T. (1981) Angles of nucleotides bound to cross-bridges in glycerinated muscle fiber at various concentrations ε-ATP, ε-ADP and ε-AMPPNP detected by polarized fluorescence. *J. Mol. Biol.*, **146**, 539–560.

53 Cooke, R., Crowder, M.S. and Thomas, D.D. (1982) Orientation of spin labels attached to cross-bridges in contracting muscle fibres. *Nature*, **300**, 776–778.

54 Toyoshima, C. and Wakabayashi, T. (1979) Three-dimensional image analysis of the complex of thin filaments and myosin molecules from skeletal muscle. I. Tilt angle of myosin subfragment-1 in the rigor complex. *J. Biochem. (Tokyo)*, **86**, 1887–1890.

55 Cooke, R., Crowder, M.S., Wendt, C.H., Barnett, V.A. and Thomas, D.D. (1984) Muscle cross-bridges: do they rotate? *Adv. Exp. Med. Biol.*, **170**, 413–427.

56 Vibert, P. and Cohen, C. (1988) Domains, motions and regulation in the myosin head. *J. Muscle Res. Cell Motil.*, **9**, 296–305.

57 Uyeda, T.Q. and Spudich, J.A. (1993) A functional recombinant myosin II lacking a regulatory light chain-binding site. *Science*, **262**, 1867–1870.

58 Rayment, I., Rypniewski, W.R., Schmidt-Base, K., Smith, R., Tomchick, D.R., Benning, M.M., Winkelmann, D.A., Wesenberg, G. and Holden, H.M. (1993) Three-dimensional structure of myosin subfragment-1: a molecular motor. *Science*, **261**, 50–58.

59 Spudich, J.A. (2001) The Myosin Swinging Cross-Bridge Model. *Nature Rev.*, **2**, 387–392.

60 Goldman, Y.E. (1998) Wag the tail: structural dynamics of actomyosin. *Cell*, **93**, 1–4.

61 Corrie, J.E.T., Brandmeier, B.D., Ferguson, R.E., Trentham, D.R., Kendrick-Jones, J., Hopkins, S.C., van der Heide, U.A., Goldman, Y.E., Sabido-David, C., Dale, R.E., Criddle, S. and Irving, M. (1999) Dynamic measurement of myosin light-chain-domain tilt and twist in muscle contraction. *Nature*, **400**, 425–430.

62 Warshaw, D.M., Hayes, E., Gaffney, D., Lauzon, A.M., Wu, J., Kennedy, G., Trybus, K. and Berger, C. (1998) Myosin conformational states determined by single fluorophore polarization. *Proc. Natl Acad. Sci. USA*, **95**, 8034–8039.

63 Katayama, E. (1998) Quick-freeze deep-etch electron microscopy of the actin-heavy meromyosin complex during the in vitro motility assay. *J. Mol. Biol.*, **278**, 349–367.

64 Warshaw, D.M., Guilford, W.H., Freyzon, Y., Krementsova, E., Palmiter, K.A., Tyska, M.J., Baker, J.E. and Trybus, K.M. (2000) The light chain binding domain of expressed smooth muscle heavy meromyosin acts as a mechanical lever. *J. Biol. Chem.*, **275**, 37167–37172.

65 Purcell, T.J., Morris, C., Spudich, J.A. and Sweeney, H.L. (2002) Role of the lever arm in the processive stepping of myosin V. *Proc. Natl Acad. Sci. USA*, **99**, 14159–14164.

66 Sakamoto, T., Wang, F., Schmitz, S., Xu, Y., Molloy, J.E., Veigel, C. and Sellers, J.R. (2003) Neck length and processivity of myosin V. *J. Biol. Chem.*, **278**, 29201–29207.

67 Tsuda, Y., Yasutake, H., Ishijima, A. and Yanagida, T. (1996) Torsional rigidity of single actin filaments and actin-actin breaking force under torsion measurement directly by *in vitro* micromanipulation. *Proc. Natl Acad. Sci. USA*, **93**, 12937–12942.

68 Tsiavaliaris, G., Fujita-Becker, S. and Manstein, D.J. (2004) Molecular engineering of a backwards-moving myosin motor. *Nature*, **427**, 558–561.

69 Tanaka, Y., Ishijima, A. and Ishiwata, S. (1992) Super helix formation of actin filaments in an in vitro motile system. *Biochem. Biophys. Acta*, **1159**, 94–98.

70 Nishizaka, T., Yagi, T., Tanaka, Y. and Ishiwata, S. (1993) Right-handed rotation of an actin filament in an *in vitro* motile system. *Nature*, **361**, 269–271.

71 Sase, I., Miyata, H., Ishiwata, S. and Kinosita, K., Jr. (1997) Axial rotation of sliding actin filaments revealed by single-fluorophore imaging. *Proc. Natl Acad. Sci. USA*, **94**, 5646–5650.

72 Yanagida, T., Arata, T. and Oosawa, F. (1985) Sliding distance of actin filament induced by a myosin crossbridge during one ATP hydrolysis cycle. *Nature*, **316**, 366–369.

73 Higuchi, H. and Goldman, Y.E. (1991) Sliding distance between actin and myosin filaments per ATP molecule hydrolysed in skinned muscle fibres. *Nature*, **352**, 352–354.

74 Harada, Y., Sakurada, K., Aoki, T., Thomas, D.D. and Yanagida, T. (1990) Mechanochemical coupling in actomyosin energy transduction studied by *in vitro* movement assay. *J. Mol. Biol.*, **216**, 49–68.

75 Julicher, F. and Prost, J. (1997) Spontaneous oscillations of collective molecular motors. *Phys. Rev. Lett.*, **78**, 4510–4513.

76 Julicher, F., Ajdari, A. and Prost, J. (1997) Modeling molecular motors. *Rev. Mod. Phys.*, **69**, 1269–1281.

77 Feynman, R.P. (1998) *Feynman Lectures on Computation*, Addison-Wesley Longman Publishing Co., Inc.

78 Brillouin, L. (1964) *Scientific Uncertainty and Information*, Academic Press.

79 Yanagida, T., Esaki, S., Iwane, A.H., Inoue, Y., Ishijima, A., Kitamura, K., Tanaka, H. and Tokunaga, M. (2000) Single-motor mechanics and models of the myosin motor. *Philos. Trans. R. Soc. Lond. B Biol. Sci.*, **355**, 441–447.

80 Huxley, H.E. and Brown, W. (1967) The low-angle X-ray diagram of vertebrate striated muscle and its behaviour during contraction and rigor. *J. Mol. Biol.*, **30**, 383–434.

3
Imaging and Molecular Motors

Yale E. Goldman

3.1
Introduction

Molecular motors, the enzymes that power cellular motions and determine cell shape, have long been productive subjects for development of new techniques in biophysics research [1, 2]. Many of the techniques described in this book, examining forces, localization and structure of single molecules were first described in studies of molecular motors. For instance, one of the early reports of detection of single fluorescent molecules was made by T. Yanagida and his colleagues using myosin and a fluorescent analog of ATP binding to its active site [3]. The first biological application of infrared optical traps by Block, Berg, and their colleagues was for studies of the bacterial flagellar rotary motor [4]. Consequently, our understanding of these cellular nano-machines has been markedly accelerated by the development and availability of single-molecule mechanical and imaging methods.

There are several reasons for the synergy between these techniques and the molecular motor field. Firstly, the primary functional outputs of active molecular motors are forces and displacements, signals particularly amenable to detection using optical traps, scanning probes, and fluorescence microscopy. The forces, motions and dynamics are all within ranges accessible using the single-molecule techniques summarized in this book. Transduction of metabolic energy into mechanical work is accomplished in molecular motors by internal structural changes, such as rotational motions and sliding, that are particularly large and detectable. Conversely, the output of molecular motors, such as local positions, spatial orientations and fluctuations, are obscure in isotropic suspensions. Many of the motor systems operate virtually on their own, making it difficult for them to be synchronized for ensemble studies. Motor research thus needs single-molecule techniques, and single-molecule biophysics has benefited from progress on molecular motors.

In this chapter, single molecule fluorescence imaging techniques that have been applied to the three classical molecular motors, myosin, kinesin and dynein, are described and the information garnered is briefly summarized. Reconstitution of

Single Molecule Dynamics in Life Science. Edited by T. Yanagida and Y. Ishii
Copyright © 2009 WILEY-VCH Verlag GmbH & Co. KGaA, Weinheim
ISBN: 978-3-527-31288-7

motility with purified proteins has been a very powerful approach toward understanding the necessary components and their interactions. With molecular motors attached to the microscope slide, the appropriate cytoskeletal filament can be shown to translate in semblance to their relative motion in a cell. This approach is termed the "gliding filament assay" and is a common method of documenting direction, velocity and regulation of all three molecular motors [5, 6]. Strongly labeled fluorescent actin filaments or microtubules can be visualized and their motions tracked with a standard epi-fluorescence microscope. If the filament is attached to the microscope slide or otherwise immobilized, then the motor can be manipulated by attaching it to a small polymer (e.g. lucite or polystyrene) bead that serves as an easily visualized marker for motion and can also be used as a handle for manipulating the motor or applying a force. Bead assays on laser traps are discussed in detail elsewhere in this book.

The present chapter emphasizes methods to visualize single molecular motors in action without attaching large cargos. The most common strategy is to label the motor with a fluorescent probe and detect the individual fluorophore *in vitro* in a geometry that reconstitutes the motor activity. Methods for labeling protein components by extrinsic fluorophores or expressing them coupled to green fluorescent protein or its derivatives have received enormous attention because observing ensembles of these fluorescent markers, their locations, dynamics and interactions is a predominant approach in cell biology [7]. Techniques for attaching fluorescent tags and the essential step of testing for functionality after labeling have been described in detail in many contexts [8, 9]. Small organic fluorophores, auto-fluorescent proteins, such as GFP, and semiconductor quantum dots have been used as markers. For motor research on individual molecules, the labeled component is usually visualized in a gliding filament assay or while it is translating on a cytoskeletal filament track that has been immobilized in the microscope. Techniques to detect single molecular motors or motions of their cargos in live cells are discussed towards the end of the chapter.

Single fluorescent probes have been visualized at much higher precision than the diffraction limit of classical optical microscopy and their functionally relevant translational and rotational motions have been recorded *in vitro*. Sub-pixel localization of single probes has been generalized into methods to obtain images of non-motor cellular structures at much finer resolution than previously thought possible. In this way, the single-molecule imaging techniques first described with regard to molecular motors are having broader impact in cell biology.

3.2
Methods

3.2.1
Detection of Single Fluorophores

A macromolecule is labeled with a fluorescent probe, checked that it still functions normally, and placed in an assay that demonstrates its activity. An example is a molecular motor propelling a cytoskeletal filament on the surface of a microscope slide. In order to image single fluorescent probes and to measure parameters that are relevant to the functional mechanism of the target molecule, the detector needs to be sensitive enough to respond to the limited fluorescence emission from the probe and the intensity of fluorescence from non-specific sources and other sources of noise must be minimized.

To keep the intensity of extraneous background below the fluorescence of the target molecule, several approaches are used. Samples prepared for *in vitro* work are assembled onto very clean microscope slides [10] in a clean environment and with reagents filtered to remove dust and aggregates. Once the sample is placed into a glass microscope slide flow cell for microscopy and sealed, the cleanliness of the microscope itself is not as critical. The sample volume in which fluorescence is detected is minimized to reduce background fluorescence from outside that region. A common way of achieving a shallow excitation volume at the sample surface is termed Total Internal Reflection Fluorescence (TIRF) microscopy [11]. When the fluorescence excitation, typically a laser beam, is directed at a very glancing angle toward the interface between the aqueous medium and a glass or fused silica microscope slide, all of the light is reflected there and no energy is propagated into the sample chamber. Due to the boundary conditions in Maxwell's equations for electromagnetism that require continuity of the electric field (more properly, the electrical displacement vector), a non-propagating, oscillating electromagnetic field is generated at the interface and only extends a few hundred nanometers into the aqueous medium. This so-called "evanescent wave" will excite fluorescent probes that are at or near the surface. Its limited reach into the rest of the flow chamber, however, prevents excitation of any non-bound sample molecules or contaminants, thus keeping the background intensity very low. The first report of the optical detection of single myosin molecules used TIRF microscopy [3].

The fluorescent detection volume is also highly restricted in confocal microscopy [12], and confocal imaging of single molecules diffusing in solution has been reported [13–16]. For molecules that are attached to a substrate and for those that exhibit lateral motions (motor proteins), the evanescent wave at the slide surface is ideal. Further characteristics of evanescent waves are described later.

Figure 3.1 shows two optical arrangements used for imaging single molecules by TIRF microscopy. In an objective-type total internal reflection fluorescence microscope (Figure 3.1A), a carefully collimated laser beam is projected by lens L1 and mirror M1 onto the back focal plane of the microscope objective. L1 is located one focal length away from the back focal plane and generates a beam waist there. The microscope objective is designed with a high converging power. Its numerical aperture, $NA_o = n_1 \sin(\theta_{max})$, where n_1 is the index of refraction of the cover slip (typically $n_g = 1.515$ for glass) and θ_{max} is the angle, relative to the optical axis, that the objective lens refracts rays entering at the margin of its aperture. Microscope objectives, designed for this purpose, with numerical apertures from 1.45 to 1.63 are produced by several manufacturers. The input light is recollimated by the objective so that the rays of the exciting beam are parallel to each other at the sample plane. When the mirror M1 is translated in the direction shown by the double-headed

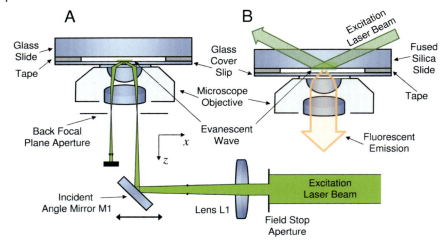

Figure 3.1 Optical arrangements used for imaging single molecules by total internal reflection fluorescence microscope microscopy. A, objective-type TIRF microscope. B. prism-type TIRF microscope. The clear regions between the glass slides and the tapes are the flow chambers. The sample is located on the bottom cover-slip in A and the top cover-slip in B.

arrow, the laser beam in the microscope objective moves laterally maintaining its direction of propagation parallel to the optic axis. When the beam at the back focal plane reaches a critical radius (explained further below), then the incident angle is large enough to reflect all of the energy at the interface, and the non-propagating evanescent wave extends a few hundred nanometers into the aqueous sample compartment. Commercial objective-type TIRF illuminators are also available that use a movable fiber optic source instead of L1 and M1.

In prism-type TIRF microscopy (Figure 3.1B), the evanescent wave is generated at the opposite side of the sample compartment from the objective. The collimated excitation laser beam is projected through a fused silica slide or prism to the aqueous interface. Quartz or fused silica is used for this purpose because its intrinsic auto-fluorescence is lower than that of glass. Polarization of the evanescent field is easier to predict with this arrangement, but it adds complexity. The imaging objective in this case is meant to operate at a longer working distance through the water of the sample chamber, reducing the maximum NA_o and brightness.

In either arrangement, fluorescent material located at the reflecting surface is excited by the oscillating electromagnetic field and emits fluorescence. The emission is collected by the objective lens and is projected onto the photo-detector, either a camera or a photo-diode. Fluorescence from material suspended in the medium but within the 100–200-nm extent of the evanescent wave also contributes to the image. This effect poses an upper limit of approximately 50 to 100 nM on the concentration of fluorescent substrates or other fluorescent ligands that bind to the protein immobilized on the surface. An approach to circumvent this limitation is to use microfabricated metal barriers to further restrict the fluorescence excitation volume [17].

Imaging cameras and spot detectors with the sensitivity to record single molecule emission are commercially available. A type of imaging detector that has come into wide use for single molecule fluorescence microscopy is the Electronic Multiplying Charge-Coupled Device (EMCCD). Thinned chips, illuminated from the side of the silicon wafer opposite to the detector junctions (back-illumination), result in increasing the quantum efficiency of converting photons into charge carriers to above 90%. The electron-multiplying shift register increases the gain (output signal/input light) several hundred-fold while introducing very little noise [18], and thereby effectively reduces the contribution to noise associated with reading the photo-electron count from the detector. Cooling the detector chip from $-30°$ to $-70°$ C reduces the dark current. These features have the result that the main source of noise in the image is the probabilistic nature of detecting limited numbers of photons (shot noise). Several manufacturers supply EMCCD cameras with 512×512 pixel arrays reading out up to 30 full frames per second.

Single spot photon counting detectors, such as avalanche photodiodes (APDs), have much higher time resolution (>1 MHz) than imaging cameras and high quantum yield extending to longer wavelengths. They do not need to be cooled, simplifying construction and reducing their cost. Background count rates as low as 10 counts per second allow counting of photons arriving significantly above this rate [19]. Limitation to a single position in the microscopic field is a disadvantage because typical experiments result in many individual molecules worth recording per microscopic field. A spot detector also requires very accurate motion of the microscope stage to position an individual molecule conjugate toward the detector [20]. For molecular motor research, the molecules of interest move during the experiment, and localizing them in line with a spot detector is only a temporary condition. Thus cameras, mostly EMCCDs, have dominated the experiments. For other research areas in which the molecules of interest can diffuse into a limited observation region, spot detectors are used [16]. Arrays of photodiodes are becoming available that retain high time resolution with single-photon counting sensitivity.

Snell's law for refraction at an interface is $n_1 \sin \theta_1 = n_2 \sin \theta_2$, where n_1 and θ_1 are the refractive index ($n_g = 1.515$) and incident angle in the glass, and n_2 and θ_2 are the refractive index ($n_w = 1.33$) and refracted angle in the aqueous medium. Angles are relative to the optical axis. Figure 3.2A shows the directions for a series of incident and refracted rays at various angles depicted by different colors. Because $n_2 < n_1$, $\theta_2 > \theta_1$, the refracted rays are bent away from the optical axis. At a certain critical angle of incidence, $\theta_c = \sin^{-1}(n_2/n_1)$, where $\sin \theta_2 = 1$, the refracted ray would be parallel to the interface (a ray at a slightly higher angle than the blue ray in Figure 3.2A). At $\theta_1 > \theta_c$, there is no real solution for θ_2, resulting in total internal reflection (purple ray). For cover slip glass and water the critical angle is $\theta_c = \sin^{-1}(1.33/1.515) = 61.4°$. The numerical aperture of the objective required to obtain higher illumination angles is greater than a critical value, $NA_c = n_g \sin \theta_c = n_g \sin (\sin^{-1}(n_w/n_g)) = n_w$. Thus, TIRF objectives must have NA_o substantially greater than 1.33, the index of refraction of water. Another point that follows from this relationship is that the coupling fluid

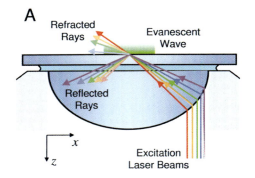

Figure 3.2 Optics near the microscope objective in TIRF microscopy. A, light rays approaching the objective at various offsets from the optical axis (different colors) are projected by the microscope objective onto the glass/water interface at different angles. Beyond the critical angle (blue ray) the light is totally reflected (purple ray). B, view of back focal plane from below. The black circle is the clear aperture of the 1.45 NA water immersion objective. The dashed circle shows the radius corresponding to the critical angle for total reflection. C, enlarged region of B. The green circle and fringes depict the Airy disk of light intensity produced by lens L1 in Figure 3.1. Panels B and C are drawn approximately to scale.

between the objective lens and the cover glass needs to have a higher refractive index than water, for instance immersion oil matching the index of the glass or glycerol, $n = 1.47$. A water immersion objective will not work.

Placement of the illuminating beam waist near the NA-limiting aperture at the back focal plane of a 1.45-NA_o objective, to obtain total internal reflection, is shown in Figure 3.2B, drawn approximately to scale. The view is along the optical (z) axis toward the back focal plane aperture (upward in Figures 3.1 and 3.2A). The black region is the clear aperture. For a lens that is well corrected for spherical aberration, an incident ray, traveling parallel to the optical axis into the objective, and refracted by the objective to an angle θ_1 at the sample, has a radial position, $h = f_o \sin \theta_1$ where f_o is the focal length of the objective. Contrary to intuition, this sine relationship (not tangent) is valid even at large θ_1 [21]. Combining the sine condition with the definition of numerical aperture leads to $NA_e = hn_1/f_0$. The radial position away from the optic axis is thus linearly proportional to the effective NA_e of any incident ray.

As a practical example, consider a 100× microscope objective with $f_o \cong 2$ mm. The dashed lines in Figure 3.2B and C indicate the minimum radius, $h_c = 1.33 \times 2\,mm/1.515 = 1.756\,mm$, for illumination at the critical angle for TIRF. The limiting aperture of the 1.45-NA_o objective at the back focal plane is $h_{max} = 1.914$ mm, giving a 158 mm band for the incident light. Using a converging lens (L1 in Figure 3.1A) with 500 mm focal length and a 30-µm sample field illuminated at $\lambda = 532$ nm, L1 and the objective lens form a Keplerian telescope. The beam at the field stop (Figure 3.1A) is $d_{L1} = 500 \cdot 30\,\mu m/2 = 7.5$ mm in diameter. At the beam waist inside the objective, the intensity distribution is termed an Airy disk, which is a centro-symmetric peaked function, the circular equivalent of the function

$\sin^2(x)/x^2$ [22]. It is similar to a two-dimensional Gaussian shape with small fringes added around the periphery. The minimum between the main peak and the first fringe has a diameter of $d_0 = 3.28\lambda f_{L1}/d_{L1} = 116\,\mu m$, which fits comfortably within the NA_e limits of 1.33 and 1.45 as shown in Figure 3.2B and C. The portion of the outer fringes at NA_e lower than 1.33, resulting in light propagated into the sample compartment contains approximately 2% of the total energy. Similarly the outer fringe scattered by the aperture is ~2%. In this example, then, 96% of the energy is totally reflected as intended. Objective lenses with NA_o even higher than 1.45 are desirable to bring the incident beam farther away from the 1.33 NA_c critical radius, but some of them require special coupling fluids and special materials for the specimen slides. Lenses from Olympus, Inc. and the Nikon, Inc, with NA_o values 1.45–1.49 have come into broad use for single molecule TIRF experiments.

The component of the incident radiation that scatters off of optical surfaces in the microscope toward the detector is several orders of magnitude greater than the single-probe fluorescence emission. Care must be taken to block this back-scattered excitation from reaching the detector. The dichroic mirrors and interference filters commonly used in standard epi-fluorescence microscopy discriminate wavelengths sufficiently well that background radiation at the excitation wavelength is usually not a difficulty. Fluorescence from the microscope objective, glass slides and other optics in the excitation path, can sometimes give significant background intensity. The optical train should be designed with these factors in mind.

The intensity of the evanescent wave decays exponentially with distance, z, from the reflecting interface, $I = I_0 \exp(z/d)$, where I_0 is the intensity at the slide surface and the spatial decay constant d, also termed the penetration depth, is given by [11]

$$d = \lambda \Big/ \left(4\pi \sqrt{n_1^2 \sin^2\theta_1 - n_2^2} \right) = \lambda \Big/ \left(4\pi \sqrt{NA_e^2 - n_2^2} \right) \tag{3.1}$$

where λ is the free-space wavelength and NA_e is the effective numerical aperture of the excitation rays. The decay parameter is plotted versus NA_e in Figure 3.3A,

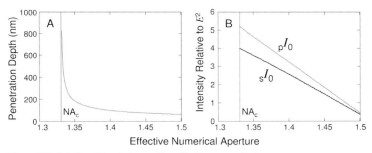

Figure 3.3 Relationships between the effective numerical aperture (NA_e) of the incident light and the penetration depth (A) and the light intensity in the evanescent wave at $z = 0$ (B). The upper and lower lines in B are plotted for p- and s-polarized illumination.

showing that very near $NA_c = 1.33$, the penetration depth of the evanescent wave is several hundred nanometers, but in the usable range of incident directions, $NA_e = 1.4 - 1.49$, d is approximately 100 nm.

The intensity of the evanescent wave at the interface, I_0, depends on polarization of the incident illumination. The plane containing the incident and reflected beams is termed the "scattering plane", the x–z plane in Figures 3.1 and 3.2. s-Polarization is defined as excitation linearly polarized (with its oscillating electric field) perpendicular to the scattering plane. p-Polarization is excitation linearly polarized parallel to the scattering plane. The intensity of the evanescent waves at the interface for the two polarizations are

$$_sI_0 = {}_sE^2 4 \frac{n_1^2 \cos^2(\theta_1)}{n_1^2 - n_2^2} = {}_sE^2 4 \frac{n_1^2 - NA_e^2}{n_1^2 - n_2^2} \tag{3.2}$$

$$_pI_0 = {}_pE^2 4 \frac{n_1^2 \cos^2(\theta_1)(2n_1^2 \sin^2(\theta_1) - n_2^2)}{n_2^4 \cos^2(\theta_1) + n_1^4 \sin^2(\theta_1) - n_1^2 n_2^2} = {}_pE^2 4 \frac{n_1^2(n_1^2 - NA_e^2)(2NA_e^2 - n_2^2)}{n_2^4 n_1^2 - n_2^4 NA_e^2 + n_1^4 NA_e^2 - n_2^2 n_1^4} \tag{3.3}$$

where $_sE$ and $_pE$ are the electric field magnitudes for the two polarizations. The expressions for p-polarization are more complex than for s-polarization because p-polarized excitation generates an elliptically polarized evanescent wave, explained further in the Section 3.2.7. Plots of these functions (Figure 3.3B) show that the $_pI_0$ is approximately 30% greater than $_sI_0$ (when $_sE = {}_pE$), they are greatest near NA_c, and they both decrease nearly linearly as NA_e increases. In the range of $NA_e = 1.4$–1.49, the evanescent wave intensities are approximately 1.5–2.5-fold greater than $_sE^2$ and $_pE^2$.

For general imaging and localization of individual fluorophores, the polarization of the evanescent wave should not select particular molecules over others. Therefore, the incident beam is often circularly polarized. If the positions of lens L1 and mirror M1 are reversed, and M1 and the objective back focal plane are both one focal length, f_{L1}, away from L1, then instead of moving L1, rotating M1 causes parallel translation of the beam inside the microscope objective. This arrangement provides a means for rapidly altering the incident angle which, in turn, alters the intensity and penetration depth.

Figures 3.7, 3.9 and 3.14 show images of single fluorophores labeling various molecular motor proteins. Each bright spot of fluorescence in the images is determined to be an individual fluorophore or several fluorophores on one molecule based on their surface density as the quantity loaded on the surface is varied, the distribution of fluorescence intensities (single component, two components, etc.) and sudden decrease of the fluorescence to a background level due to photobleaching. The location, spatial orientation, interactions of these spots with other components of the system and response to mechanical forces are parameters that can elucidate the functional mechanisms of the molecules under study. The quality of these signals depends critically on the brightness, steadiness and maintenance of the fluorescence emission.

Bright fluorophores, efficient, high-aperture collection optics, high transparency emission filters, the quantum efficiency of generating charge carriers per photon captured by the detector and low noise amplification or counting of the photoelectrons all contribute to increasing the signal. Removal of scattered excitation and spurious fluorescence by spectral filtering and by limitation of the detection volume, as described above, reduce the background. Interactions of the probe with the optical excitation and the chemical environment also determine the variability and longevity of the emission. Organic fluorescent probes, semiconductor nanocrystals (quantum dots) and auto-fluorescent proteins (GFP and its variants) all exhibit photophysical processes in the electronically excited state that lead to reversible and irreversible dark states, termed blinking, flickering and photobleaching [23]. Studies of the mechanisms of these phenomena are outside the scope of this chapter, but their consideration is an important aspect of designing all single-molecule imaging experiments. Photoisomerizations, long-lived triplet states, redox reactions, protonation and breaking of covalent bonds contribute to these fluorescence changes for different probes. Various schemes for minimizing the photobleaching rate are addition of enzymatic oxygen scavenging systems, such as glucose oxidase, catalase and glucose [24, 25], reducing agents, such as β-mercaptoethanol or dithiothreitol and triplet state quenchers, such as Trolox [26]. These agents also strongly modulate the dynamics of fluorescence blinking and flickering, making the choice of anti-fade reagents and even their supplier [27] important considerations.

At a given frame rate and chemical environment of the fluorescent probe, the only other parameter that modulates the signal magnitude is the power of the illumination, which directly controls the emission rate. In a typical fluorescence microscopy experiment, the fluorophores are far from optical saturation (the condition where a substantial proportion (e.g. $\sim$20%) of the fluorophores are excited at any instant), so increasing the excitation intensity proportionally increases the emission flux. The quantum yields for photobleaching and for some of the flickering mechanisms are also approximately constant, causing the photobleaching rate to increase proportionally with incident light intensity as well. An input laser power is used that achieves a compromise between the magnitude of the fluorescence signal, kinetics of flickering and duration of recording before photobleaching that ends the experiment.

The emission intensity per camera frame from a single rhodamine fluorophore under constant TIRF excitation and with effective deoxygenation (Figure 3.4A) is fairly constant until it decreases suddenly to a much lower background value (at 50 s in this experiment). The $\sim$10% frame-to-frame variation of the signal is due to non-steady excitation rate due to laser noise and mechanical vibrations that shift fringes in the TIRF illumination, fluorophore blinking, and the statistics of counting the photons. The sudden decrease represents the irreversible photobleaching of the fluorophore to a non-fluorescent state, presumably with altered chemical structure. Approximately 10^5 total photons were collected and registered by the camera before the photobleaching event. Given the likely collection efficiency of the optics (10%), quantum yield of the camera (0.9) and fluorescence quantum yield of the rhodamine itself (0.5), this number of detected photons represents >10^6 electronic excitation/de-excitation cycles of the fluorophore. The constant emission rate, occupying a relatively narrow distribution and the sudden

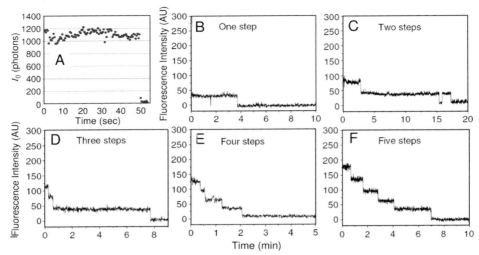

Figure 3.4 Discrete photobleaching of small numbers of fluorophores. A, single bifunctional rhodamine molecule lasts 50 s under illumination in oxygen-depleted buffer (from ref. [27]). B-F, intensity traces for bleaching of one to five Cy3 pRNA moleculae in a viral packaging motor (from Ref. [28]).

photobleaching in one step, indicate that the emission is from an individual probe.

When two or more probes occupy the same picture element, the intensity is higher (not always by an integer multiple, due to quenching by energy transfer) and the photobleaching occurs in multiple steps (Figure 3.4B–F). For oligomeric complexes, it is possible to estimate the number of components from the number of bleaching steps [28–30]. Detection of individual fluorescent molecules or small groups has become straightforward with commercially available or readily constructed instrumentation. The research effort now has shifted toward obtaining useful, mechanistically relevant, biological information.

3.2.2
Sub-Diffraction Localization of Fluorescent Molecules

Single molecule imaging has overcome the limit of spatial resolution in optical microscopy that was formerly thought to be insurmountable. The location of individual fluorophores can be determined with an uncertainty of a few nanometers, a range readily suited to studying the 8–36-nm stepping motions of molecular motors. In a fluorescence microscope, the emission from a point source is broadened at the detector due to diffraction and aberrations in the microscope optics. The point spread function (PSF) of the microscope is the apparent spatial distribution of the origin of the photons emitted from a source with essentially infinitesimal size, such as an individual fluorescent probe. Generally, the PSF is a three-dimensional function but in TIRF microscopy at the focal plane, variations along the optical axis are not relevant. The theoretical PSF for a well-focused individual fluorophore is an Airy disk, similar to a two-dimensional Gaussian shape, as described earlier. At visible optical wavelengths and with a high-aperture, well-corrected objective lens, the full width at half-intensity of the PSF is ~ 250 nm. This value sets the classical lower limit of spatial resolution in optical microscopy for resolving nearby objects. When two objects are closer than this distance, their emission distributions overlap and they appear fused.

An additional piece of information about the object, however, can enable its location to be determined much more precisely than 250 nm. If the object is known to be a single sub-resolution emitter, such as a single fluorescent probe, then its lateral (*x–y*) location can be determined from the detected distribution of emission intensity with precision considerably less than classical resolution and less than the pixel size. This idea has become known as Fluorescence Imaging to One Nanometer Accuracy (FIONA). For an optically well-corrected image of an in-focus or unpolarized point emitter, the center of the recorded intensity distribution corresponds to the physical position of the probe at sub-pixel resolution. Several groups utilized this idea to localize fluorescent spots to within tens of nm [31, 32] and P. R. Selvin and colleagues first achieved precision of 1–2 nm [27]. FIONA and its derivatives have been reviewed in [33].

The theoretical uncertainty, σ_μ, of position in a FIONA experiment is given approximately by [32]

$$\sigma_{\mu_i} = \sqrt{\frac{s_i^2 + a^2/12}{N} + \frac{8\pi s_i^4 b^2}{a^2 N^2}} \tag{3.4}$$

where the index *i* refers to the *x* or *y* direction, s_i is the width (standard deviation) of the image intensity distribution in direction *i*, *N* is the number of collected photons, *b* is the standard deviation of the background, including photon counting statistics and camera readout noise, and *a* is the pixel size of the imaging detector. The first term (s_i^2/N) is the photon noise; the second term is the effect of the finite detector pixel size of the detector; and the last term is the effect of the background. This relation assumes that other sources of error, such as mechanical vibrations and thermal drift in the microscope are negligible. As can be seen from this expression, when the pixel size, *a*, becomes smaller than s_x and s_y, the effect of pixelation on position uncertainty is small. Practical values of the pixel size for s_x, $s_y = 120$ nm are 75–150 nm referred to as the sample plane. For a camera with 13-μm square pixels, this situation corresponds to overall magnification of 80–170. Figure 3.5A shows how the expected uncertainty (standard deviation, σ_μ) in determination of position decreases as the number of photons is increased in a practical situation with 0.5-s image integration, s.d. of the background $b = 11$–33 per pixel, and effective pixel size of $a = 86$ or 43 nm [27]. With the lower background, a precision of $\sigma_\mu = 1.5$ nm is reached when the image accumulates 11 000 photoelectron counts. In this condition, 80% of the uncer-

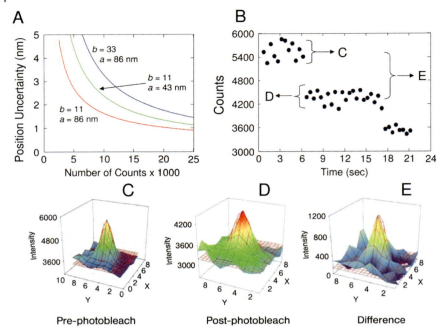

Figure 3.5 Characteristics of FIONA and SHRImP. A, theoretical positional uncertainty (standard deviation) of a single fluorophore vs. number of photons counted per imaging period (text equation 3.4). a = pixel size, b = standard deviation of the background intensity. B, peak fluorescence intensity from two nearby GFPs labeling the two heads of a myosin VI molecule, and displaying two-step photobleaching. C point spread function (PSF) of the fluorophore pair. D, PSF of the GFP remaining after the first photobleaching event. E, the difference between C and D gives the PSF of the GFP that bleached first. Positions of the peaks in D and E give the distance between the fluorophores. Panels B-E are from Ref. [111].

tainty is due to the first term under the radical in Equation 3.1, 2% is due to the second term, and the remainder is due to the third term, the background noise. Several laboratories have achieved this value of stability in localization at nearly the theoretical numbers of collected photons [27, 34]. If b increases three-fold, then the effect of the background becomes more significant (blue curve in Figure 3.5A). Higher magnification and reducing the pixel size further, also increases σ_μ, because, as the last term in Equation 3.3 shows, the broader spread of the image over more pixels causes background from those extra pixels to contribute uncertainty.

The FIONA method has contributed toward understanding the stepping mechanism of the molecular motors as described later in this chapter (Figures 3.10 and 3.12). Several new methods with entertaining acronyms which have been derived from FIONA, and have been useful for studying molecular motors are explained here. Those for general cellular imaging are mentioned at the end of the chapter.

3.2.3
Darkfield Imaging with One Nanometer Accuracy (DIONA)

Darkfield microscopy is similar to the optical arrangement shown in Figure 3.1B using a high numerical aperture condenser to illuminate the specimen with high-angle glancing rays, but at lower angles relative to the optical axis so that the incident light propagates obliquely into the sample compartment. The viewing objective has a lower NA_o than the condenser, so the oblique illumination is not collected unless the light is scattered by the sample. Several groups have placed gold nano-particles on their macromolecules which scatter light effectively, producing bright spots on the dark background, similar to a fluorescent emitter, but not limited by excited state lifetime, photobleaching or blinking [35, 36]. The number of collected photons is limited only by intensity of the illumination and the scattering efficiency of the particle. With 40-nm gold aggregates, $\sigma_\mu = 6$ nm noise was obtained at 300-μs sampling intervals. With 200-nm plastic beads a time resolution of 20 μs was achieved [37]. An application of this method to the events during myosin stepping is described in conjunction with Figure 3.12.

3.2.4
Single-molecule High Resolution Imaging with Photobleaching (SHRImP)

Photobleaching can be made into an advantage for measuring the distance between two fluorophores. When a sample molecule contains several identical fluorescent probes that are located too close together to resolve directly, they produce a PSF similar to a single fluorophore, but more intense (Figures Figure 3.3B–F and Figure 3.5C). The two fluorophores bleach sequentially (Figure 3.5B) and the point spread function from the longer-lasting probe (panel D) can be subtracted from the original two-fluorophore distribution (panel C) to generate the PSF from the shorter-lived probe (E). Fitting the two single-fluorophore distributions to Gaussians locates them with the usual FIONA precision, and the difference in positions gives the original distance between the two probes. This "SHRImP" method complements another more commonly used technique, to be described later, Fluorescence Resonance Energy Transfer (FRET) for quantifying single molecule distances. Whereas FRET between two spectrally matched fluorophores is ideal in the distance range 2.5–7.5 nm, SHRImP allows distances to be estimated at any separation, approximately 10 nm and above [38]. Application of this method to myosin VI is given later in this chapter. As described, the technique is limited to static measurements because photobleaching is irreversible. This limitation might be avoided by applying SHRImP to probes that blink.

3.2.5
Single Molecule Fluorescence Resonance Energy Transfer (smFRET)

When two fluorescent probes, such as Cy3 and Cy5, with overlapping emission and absorption spectra are near to each other, the energy of optical excitation of the shorter wavelength probe (Cy3) can be transferred to excite the longer wavelength probe (Cy5) without radiation of an intermediate photon. The FRET technique is well developed at the single molecule level, especially for nucleic acid processing enzymes. It is explained in more detail in Chapter 11 of this book. For example, when Cy3, termed the donor, is illuminated with 514-nm laser light, instead of fluorescing at its 570 nm emission wavelength, its excitation energy can sometimes be transferred to nearby Cy5 probes (the acceptors). The energy transfer is monitored by Cy5 fluorescence at 670 nm. After correction for various spectral properties and cross-over between the excitation and detector channels, a quantity termed FRET efficiency is calculated, $E = I_A/(I_D + I_A)$, where I_D and I_A are the donor and acceptor fluorescence intensities. Among other factors, E depends on the distance, r, between the two fluorophores in the useful range of 2.5–7.5 nm according to $E = r_0^6/(r^6 + r_0^6)$ where r_0 is the distance at which $E = 0.5$, which is $r_0 = 5.3$ nm for the Cy3–Cy5 pair. An example single molecule FRET experiment with kinesin is given in Figure 3.13 and explained later in the chapter.

3.2.6
Orientation of Single Molecules

Tilting and rotations of molecular domains are crucial motions in the mechanisms of many protein and RNA enzymes [39, 40]. Spatial orientation of protein domains are especially important in molecular motor research because the key structural changes leading to movement are commonly attributed to tilting of protein domains, as described below for myosin and kinesin. Quantifying angle changes in ensemble systems, though, can be very difficult. For example, during contraction of muscle fibers, most of the myosin heads are not attached to actin and have nearly random angular distributions [41]. Concerted rotational motions among the small proportion that are attached to actin are difficult to detect against the disordered background. Single-molecule fluorescence microscopy, on the other hand, provides sufficient sensitivity to detect absolute orientation and tilting on the millisecond time scale relevant to the function of the molecular motor. A bright extrinsic fluorophore is inserted into the structure by site-specific labeling and changes in the orientation of the fluorophore are considered to signal the rotational motions of the macromolecular domain. When the orientation of the probe is known relative to the attached domain, the absolute orientation in space can be estimated [42, 43]. The relative orientation between the probe and the protein can be preprogrammed into the structure by the placement of the labeling sites [42, 43] or it can be determined by crystallography [44], NMR spectroscopy [45] or molecular dynamics calculations [46]. Single molecule measurements of orientation and tilting motions have been reviewed in [39, 40, 47, 48].

Three related physical properties of a fluorescent dipole enable determination of its spatial orientation: (1) the relative absorption of light polarized in various directions, (2) the angular directions of its emitted photons, and (3) their polarization. The likelihood that a chromophore will absorb a photon is proportional to $\cos^2 \theta_a$, where θ_a is the angle between the photon's polarization (direction of the oscillating electric field) and the absorption dipole moment. When a polarizer is introduced before a narrow aperture detector, the likelihood of registering the emitted photon is given by $\cos^2 \theta_e$, where θ_e is the angle between the fluorophore emission dipole and the detector polarization. In xanthene-derivative dyes, such as rhodamine, and cy-dyes, such as Cy3, the absorption and emission dipoles are aligned with the long axis of the chromophore.

Several groups have used excitation light circularly polarized in the plane of the microscope slide (the x–y plane) and separated the fluorescence emission polarized along the x and y axes [15, 49]. With this optical arrangement, the ratio of intensities between the two detectors, corrected for any differences in their sensitivity, gives a signal, independent of total intensity, but sensitive to the angle (ϕ) of the emission dipole projected onto the x–y plane. At fluorophore angles giving ϕ of $45°$ or $-45°$, the two recorded intensities are equal; angular discrimination away from $45°$ or $-45°$ depends on differences between the two intensities. For a given average angle, rotational mobility (wobble) of the probe or of the macromolecule causes the two recorded signals to become more equal, so the signals are mixtures of the mean angle and extent of mobility. This optical arrangement does not provide information regarding the fluorophore angle (θ_p) relative to the optical axis, because both the exciting and detection polarizations are in the x–y plane. Nevertheless, interesting and useful information about motions within molecular motors were obtained [15, 49].

3.2.7
Polarized Total Internal Reflection Fluorescence Microscopy (polTIRF)

In polTIRF, the orientation of the fluorophore is detected by illuminating the sample with linearly polarized light at various known angles and resolving the polarization of the fluorescence emission. In our implementation of this scheme, Pockels cells are used to rapidly modulate the path and polarization of a laser beam in an objective-type or prism-type TIRF microscope (Figure 3.1). Four different combinations of scattering plane (either x–z, or y–z) and polarization (either s- or p-) are applied during successive 5–10-ms intervals. In Figure 3.6, the x–z direction is indicated as Path 1 and y–z direction is labeled Path 2. For each of these input directions, the s-polarized excitation beams produce x- or y-polarized evanescent waves. The p-polarized excitation beams produce strong z-polarized evanescent waves with small, out-of-phase components (5–10% intensity) along the x or y directions [11]. These four combinations of optical path and polarization give evanescent waves polarized along all three Cartesian coordinates (Figure 3.6). In a typical experiment on molecular motors, cycles of the four combinations are completed every 20–40 ms by alternating the voltages on the Pockels cells.

A bright water-immersion objective, a polarizing prism, and two avalanche photodiodes (APDs) collect the x-polarized and y-polarized fluorescence emission. The photon counts from the two APDs during each of the four intervals of input path and polarization are binned into eight temporal traces, $_{s1}I_x$, $_{s1}I_y$, $_{p1}I_x$, $_{p1}I_y$, $_{s2}I_x$, $_{s2}I_y$, $_{p2}I_x$, and $_{p2}I_y$, that contain the angular information of the probe with the 20–40-ms

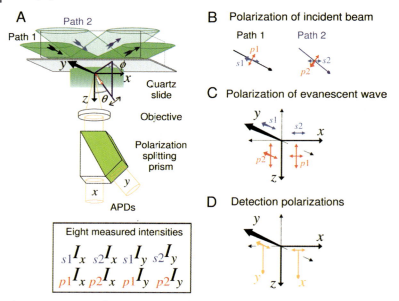

Figure 3.6 Geometry for a prism-type polarized TIRF microscope. A, Paths of incident, reflected and fluorescent emission beams and definitions of detected probe angles. APDs are avalanche photo diodes. B, definitions of incident beam polarization. s-Polarized illumination is polarized in the horizontal plane. p-Polarized i lumination is perpendicular to s-polarization. C, polarization of the evanescent wave for each of the input paths and polarizations. Note that p-polarization leads to both transverse (downward) and longitudinal (horizontal) components. D. Polarizations of the detectors. The downward colored arrows show the direction of fluorescent light propagation. From Ref. [40].

time resolution. The raw photocount traces are corrected for the different intensities of the illuminating conditions (partly given by Equations 3.2 and 3.3 above), different sensitivities of the two APDs, the elliptical components of the p-generated evanescent waves, and intermixing of the x- and y-polarization emissions by the objective lens [50]. Corrected intensities are then fitted by a model that predicts the polarized fluorescence intensities based on the angular dependence of probe fluorescence emission (Figure 3.7A), θ, the axial angle between dipole axis and optical axis, ϕ, the azimuthal angle around optical axis, and limited rotational mobility of the probe on the subnanosecond time scale and, separately, on the microsecond time scale [20]. Also taken into account are the collinear absorption and emission dipoles (e.g. for rhodamine or Cy3) and background intensity measured after photobleaching. Snapshots of these angular parameters are estimated every 40 ms. A few hundred photocounts per 10-ms time gate are typically recorded, leading to standard deviations of the angular measurements of 5–15° [20]. The main source of this uncertainty is photon counting shot noise, making much higher time resolutions feasible at higher laser intensities.

Because the dipole moments of the probe and the excitation and emission light all exhibit C2 symmetry (rotating 180° about a central axis does not alter the structure),

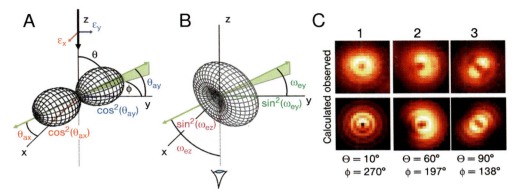

Figure 3.7 Relative probabilities for fluorophore absorption (A, $\cos^2 \theta_a$) and collection of its emission (B, $\sin^2 \omega_e$). In the microscope coordinate frame (x, y, z), the optical axis is z. For axial illumination (heavy arrow in A), ε_x, and ε_y are excitation polarizations. The probe absorption and emission dipole moments (considered to be parallel) in the (x, y, z) frame are defined by θ (axial angle) and ϕ (azimuth: angle between the projection of the dipole onto the x–y plane and the positive x axis), θ_{ax} and θ_{ay} are angles between the probe dipole and detector polarizations along the x- and y-axes. ω_{ey} and ω_{ez} are angles between the probe dipole and detector optical paths in the y- and z-axes. From Ref. [40] C, defocused images of quantum dots (frozen in 1% polyvinyl alcohol) showing examples of vertical, inclined, and horizontal emission dipoles (upper row) and corresponding calculated patterns (lower row). From Ref. [56].

polarized fluorescence cannot discriminate an angle (θ, ϕ) from the corresponding angle $(180° - \theta, \phi + 180°)$ in the opposite hemisphere. With excitation and emission polarizations symmetrical about the x, y, and z planes, as in Figure 3.6, ambiguities caused by the additional symmetries restrict the range of unambiguous probe angular discrimination to 1/4 of a hemisphere, such as $0 < \theta < 90°$ and $0 < \phi < 90°$. Additional incident polarizations in polTIRF (e.g. $\pm 45°$ angles) expand this range to a full hemisphere [51]. This feature has been utilized to determine the handedness of actin motion powered by myosin *in vitro* [51] and the path of myosin motors along actin [52], as described later.

3.2.8
Defocused Orientational and Positional Imaging (DOPI)

As mentioned, along with the polarization, the paths of emitted photons carry information about the orientation of a fluorescent dipole. The radiation pattern can be detected by collecting an image slightly away from the sharpest focus [53, 54]. Some intuition into the cause of this behavior can be gained by considering the distribution of photon paths from the fluorescent dipole. Figure 3.7A shows a contour surface representing the probability of photons absorbed or detected with polarizations at angles θ_a relative to the dipole axis, the $\cos^2\theta_a$ distribution. Photons with polarization parallel to the dipole are preferentially absorbed and those polarized perpendicular to the dipole are not absorbed. The distribution of radiation directions of photons originating from a dipole emitter is given by $\sin^2\omega_e$, where ω_e is the angle between the path of the photon and the emission dipole (Figure 3.7B). Photons are most likely to travel perpendicular to the dipole and they are not emitted along the dipole axis. An out of focus image is similar to sectioning the $\sin^2\omega_e$ contour slightly above or below the x–y plane, giving an asymmetric distribution of intensities, containing lobes and fringes and an intensity gap in the direction of the dipole.

In DOPI, images of the fluorophore are recorded approximately 0.5 µm out of focus and the resulting asymmetric intensity distributions are fitted with theoretically predicted models (Figure 3.7C) [55]. The detected image of the single fluorophore depends on θ and ϕ as in polTIRF, and the amount of defocusing δz from the focal plane. DOPI enables relatively straightforward determination of the full hemisphere of dipole orientations with useful angular and temporal resolution. Compared with polTIRF, the instrumentation in DOPI is simpler to implement and can also simultaneously determine the probe's spatial position, a feature that has only been partially implemented with polTIRF [73]. The temporal resolution of DOPI ($\sim$0.5–1 s) is lower than that of polTIRF (40–80 ms), but in many cases the biological reaction can be slowed down to match the time resolution of the method, for instance by reducing the substrate concentration [56]. The angular resolutions of the two techniques are comparable ($\sim$10–20°). PolTIRF, however, is sensitive to the rate and amplitude of rotational wobbling motions of the probe and the labeled macromolecule, whereas DOPI and similar techniques provide the average angle over the recording gate time. DOPI has been applied to the stepping of myosin V, as described later.

3.3
Molecular Motors

The single-molecule imaging techniques described above are readily applicable to the understanding of the dynamics and energetics of the transport machines of the cell. Some of the molecular motors that have been studied with single molecule imaging are shown in Figure 3.8. Myosin V is a 450-kDa isoform of the myosin superfamily having two heavy chains each with an N-terminal ATP- and actin-binding motor domain, six peptide sequences, termed IQ motifs, each of which bind a calmodulin subunit or a calmodulin-like light chain, and a tail that carries the dimerization and cargo-binding domains.

Myosin V and some 20 other the myosin isoforms translate along actin to power muscle contraction and cell shape changes, locomotion, cytokinesis, intracellular vesicle transport and other motile, architectural and signaling functions. Muscle myosin, designated myosin II, is similar to myosin V in having two globular N-terminal motor domains, and a tail that dimerizes into an α-helical coiled coil. In muscle, the tails of $\sim$300 myosin II molecules self-polymerize into bipolar filaments that align sideways in the muscle to form the so-called A-band of the sarcomere, the contractile organelle. Myosin II contains two IQ motifs that carry light chain subunits structurally related to CaM.

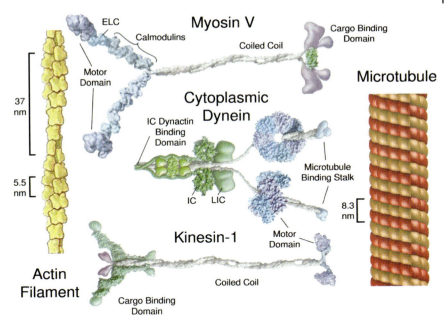

Figure 3.8 Molecular motor and cytoskeleton "toolbox". Myosin V, cytoplasmic dynein and conventional kinesin are drawn in surface rendering approximately to scale with actin and a microtutuble. Motor catalytic domains are displayed in blue, mechanical amplifiers (myosin lever, kinesin neck linker) in light blue, and cargo attachment domains are shown in purple. Light and intermediate chains are in green. Dynein is shown in mixed purple and blue shading to illustrate the distinct domains that comprise the motor head. The stalks extending from the ring bind to microtubules at their globular tips. Adapted from Vale R.D. (2003), Cell, **112**:467–480.

Kinesin and dynein are microtubule-based motors that translocate predominantly toward the plus and minus end of microtubules, respectively. Of the three classic molecular motors, kinesin is the smallest, having two 40-kDa head domains with a nucleotide binding fold very similar to the core of the myosin motor domain [57]. The tails dimerize and also carry cargo-binding domains. Like myosin, the large family of kinesin isoforms are all related by amino acid sequence homology in the motor domain, but otherwise have diverse structures.

Dynein, at ~2 MDa, has a more complex head structure containing a ring of protein domains carrying at least four ATP-binding sites. The motor domain belongs to the family of ATPases Associated with various cellular Activities (AAA proteins), including the F1 ATP synthase, heat shock proteins and proteinases [58]. The other members of this family are oligomers of ~6 AAA domain peptides, whereas the dynein ring contains one large polypeptide with seven domains. A stalk extending from the ring contains the microtubule binding domain. The N-terminal tails dimerize and contain complex intermediate and light chains and the cargo-binding functionality. Flagellar dyneins have one to three heads, whereas cytoplasmic dynein contains two heads. An additional large, multi-subunit 1.2-MDa complex, termed dynactin (not shown), is required for most functions of cytoplasmic dynein [59].

3.3.1
Myosin V

Myosin V (Figure 3.8) is involved in transport of vacuoles and RNA–protein particles in yeast vesicular cargoes in neurons, and pigment granules in melanocytes. Due to its cellular roles, deletion or misexpression of myosin V in mice and humans causes a developmental neurological deficit, with immune suppression and light pigmentation [60]. The term processivity refers to the feature of many DNA processing enzymes and some molecular motors that take many productive mechanical steps upon each diffusional encounter with their track. Myosin V exhibits very high processivity (20–60 steps [61–63]), a very large step size (36 nm [27, 61]) and is abundant enough to be isolated from brain tissue [64]. These characteristics have caused myosin V to be a very popular subject for optical trapping and single molecule imaging studies, making it one of the best understood molecular motors.

In the actin-activated ATPase cycle of myosin V, the dissociation of ADP from the acto–myosin–ADP complex is very slow ($15 \, s^{-1}$) compared to that step in conventional muscle myosin II ($>200 \, s^{-1}$). This slow rate implies that myosin V spends a large proportion of the ATPase cycle attached to actin, leading investigators to suspect that myosin V might exhibit processive motility [65]. This supposition was confirmed in a single-molecule imaging study by Ando and colleagues [62]. Fluorescent labeled CaM was introduced into the myosin V molecule by dissociating some of the endogenous calmodulin (CaM) subunits at high Ca^{2+} concentration. Actin filaments were attached to the microscope slide surface using sparse biotin–streptavidin linkages, and fluorescent myosin V molecules added to the sample chamber were observed by TIRF microscopy (Figure 3.9). Individual fluorescent spots, representing single myosin V molecules, were seen to localize with actin and move along the filament for several μm. The myosin V molecules did not readily dissociate from the actin and fluorescence built up at the barbed ends of the actin filaments as the myosin molecules accumulated there. This direct demonstration of processive motility confirmed optical trapping experiments at nearly the same time [66]. Further insightful experiments using single molecule fluorescence imaging and optical trapping have elucidated many details of myosin V's stepping mechanism [67–72].

The initial application of FIONA was an investigation of myosin V [27]. Myosin V molecules were labeled with rhodamine or Cy3 on one of the CaM subunits at low stoichiometry to minimize the number of multiply labeled molecules. The position of the CaM was resolved to within a few nanometers during processive motility as described above in Section 3.2.2. The labeled CaM was found to move stepwise along the actin filament, and several classes of motion were detected: molecules with fluorophores that took 74-nm steps, those that alternated between 52- and 23-nm steps and those with alternating 42- and 33-nm steps. Such alternating large and small steps are expected if the molecule translocates along actin using a

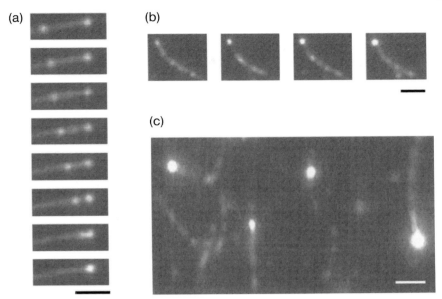

Figure 3.9 Processive movement of individual brain myosin V molecules along actin filaments. A, successive video images at 0.5-s intervals. A light spot at the left-hand side is moving toward the stationary light spot at the actin-filament end. Scale bar: 2 μm. b, video images at successive 2-s intervals. The barbed end of the actin filament becomes brighter with time due to accumulated motors. Single myosin V molecules produce moving spots on the filament. Scale bar: 2 μm. c, a video image showing bright spots at the barbed end of filaments after 1 min of incubation with the myosin V. From Ref. [62].

hand-over-hand mechanism in which both heads can bind to actin and a step is accomplished by the trailing head moving past its partner to become the new leading head (Figure 3.10A). When the arm containing the fluorescent probe is the stepping one, its displacement is the larger value and when the other arm moves, the probe is displaced by a smaller value.

In later studies adding angular detection to the lateral motion detection of FIONA, the three classes of molecule were determined to give alternating 64-10-64, 50-25-50 and 44-30-44 nm step displacements of the probe (DOPI: [56]; partial polTIRF: [73]). These values are just what would be expected if the molecule moves 36–37 nm per step and the labeled CaM exchanges into the second, fourth and sixth positions away from the motor domain (Figure 3.10A). A distance of 36 nm is the half-pitch of actin (Figure 3.8), allowing the molecule to walk relatively straight along the actin axis, rather than twirl in the tight helical path defined by the actin monomers.

When a GFP fluorophore was engineered into the head of myosin V, the steps alternated between 74 nm and virtually zero, also as expected [74]. With two different colored probes simultaneously exchanged into various pairs of locations in myosin V, the two arms are shown directly to swap between leading and trailing positions along the actin axis (Figure 3.10B, also [75, 76]). These FIONA experiments provide strong

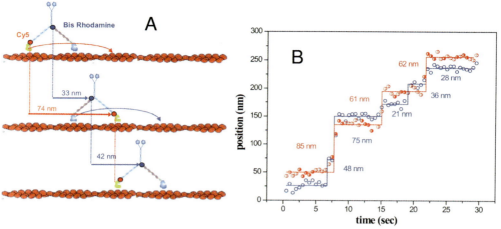

Figure 3.10 Dual color FIONA, A, cartoon showing the expected steps of a myosin V molecule labeled with bifunctional rhodamine and Cy5 in different positions. B, FIONA recording of rhodamine and Cy5 swapping the leading position, as expected. A. Yildiz, P.R. Selvin and Y.E. Goldman, unpublished observations.

evidence for the hand-over-hand mechanism and rule out hypotheses in which the two heads do not exchange places during the step.

What causes the motion? The crystal structures of various myosin conformations suggest that the CaM/light chain region can tilt relative to the motor domain. This data leads to a natural hypothesis that the tilting is a working stroke and the light chain domain acts as a lever arm to amplify smaller structural changes in the motor domain into the nanometer translocation of the step. Angular measurements of the myosin V CaM subunits using polTIRF [42, 73] or DOPI [56] showed that they do rotate approximately 70° in the plane of the actin filament on each mechanical step (Figure 3.11). Note the repeated increase and decrease of the axial angle β (bottom red trace) as the myosin V lever arm adopts the leading and trailing positions during stepping. This motion is often straight along the actin axis (in Figure 3.11, the azimuthal angle around actin, α, the green trace at the bottom, is relatively constant). In some cases, tipping around the filament axis is also possible [56, 73, 77]. With engineered constructs having various lengths above and below the native number of six CaM/light chain sites, the step size of the molecule varies directly with the length of the lever arm [67, 71, 78, 79], confirming that the tilting of the CaM lever arm determines the distance moved.

In native myosin V, the lever arm is approximately 24 nm long (six CaM subunits, each 3.6 nm long plus 2.4 nm for the tilting part of the motor domain). A 70° tilt along actin would translate the lever arm–tail junction by 27.5 nm ($2 \cdot 24 \cdot \sin(70°/2)$), which is considerably less than the 36-nm step measured by FIONA, polTIRF, and the optical trap [27, 42, 61]. Mechanical measurements on a single-headed construct of myosin V gave ~24 nm for the stroke size due to tilting of the lever arm [69].

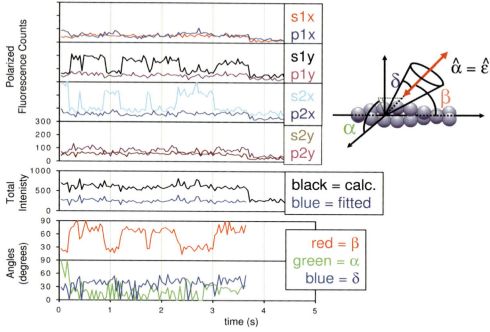

Figure 3.11 PolTIRF recordings from a bifunctional rhodamine labeled calmodulin exchanged into myosin V. The motor is walking processively along actin at 5 µM MgATP. Polarized fluorescence intensities are given as photocounts per 10-ms gate. Polarized intensity subscripts indicate excitation/detection polarizations as defined in the Figure 3.6. The sum of the polarized intensities is plotted as "calc"; fitted total intensity, less background, is blue. The single discrete decrease of all intensities to background levels at 3.7 s is due to photo-bleaching of the fluorophore. From Ref. [42].

Those authors hypothesized that the remainder of the 36-nm step of the whole, double-headed molecule is accomplished by thermal fluctuations of a singly-bound intermediate randomly carrying the free head to the actin monomer 36 nm in front of the stationary head.

Using the darkfield adaptation of FIONA explained above (DIONA) to make a major improvement to the time resolution, Dunn and Spudich [36] were able to detect the singly-bound state expected from the diffusional search model. A 40-nm gold nano-particle was attached to a CaM subunit and viewed by scattered light under highly oblique, propagating illumination (Figure 3.12). A transient intermediate position with fluctuations was evident (horizontal arrows in panel B), confirming the random search hypothesis for completion of the step.

This remarkable behavior implies that the molecule is harnessing random thermal fluctuations to extend its reach. No thermodynamic laws are violated here in making fluctuations useful because the energy liberated in the ATPase cycle enables the overall process to proceed in the forward direction [80]. When the wobbling head in

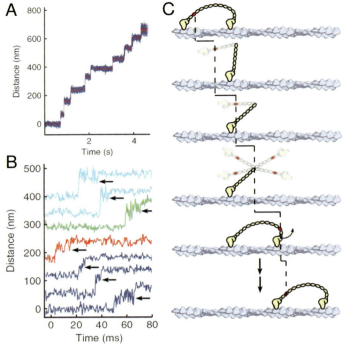

Figure 3.12 A myosin V dimer is labeled with a gold nanoparticle on one of its lever arms and is located by darkfield microscopy. A, data trace with a 40 nm gold particle, 3 μM ATP. Frame interval: 0.32 ms. The red trace represents low-pass filtered DIONA. B, 49-nm substeps. The ends of the substeps are indicated by arrows. Note the increase in variance during the substep, suggesting single-headed attachment. C, cartoon model that explains the results; increased variance during the step is the "thermal search". From Ref. [36].

the myosin-ADP-P_i state binds to actin, the P_i is released quickly and nearly irreversibly [81], capturing the distortion in the molecule produced by the random process.

Why does the trailing head detach before the leading head? The high processivity of the motor requires that the biochemical cycles of the two heads be kept out of synchrony. This feature ensures that the trailing head is the one to step forward and it also prevents both heads from detaching at the same time. The difference between the stroke size (24 nm) and the step size (36 nm), and the extra reach delivered by the diffusional search imply that when both heads are bound to actin, the molecule is under internal mechanical strain. Each head pulls the other one toward the center of the molecule. Due to the slow kinetics of ADP release, both heads have ADP bound in the waiting state between steps. If the backward strain on the leading head suppresses ADP release and/or the forward strain on the trailing head accelerates ADP release, then the trailing head will release its ADP first, ATP will bind rapidly to the trailing head inducing detachment and initiating a forward step [63, 68, 70, 82]. Thus the intramolecular strain generated by the thermal fluctuations is trapped by the biochemical transitions and used for communication between the two heads. These mechanisms require two-headed motors to maintain contact with the filament and to communicate intramolecular strain. Thus, processive motility of a single-headed myosin V construct, reported in an optical trap assay [72], should be confirmed by single fluorophore imaging.

There are several levels of myosin V regulation necessary to determine activity, cargo binding, and cooperation with other motor proteins. CaM is involved in many cellular signaling cascades in which Ca^{2+} binding to CaM enables it to interact with downstream effector proteins, such as kinases [83]. The CaM binding IQ motifs in myosins are unusual because they serve a structural/mechanical role as the lever arm and they interact strongly with CaM in the Ca-free state. The CaM subunits may also be involved in regulation of motility. In the absence of Ca^{2+}, the myosin V molecule folds into a triangular shape with the globular tail domain interacting with the head, thus preventing it from associating with actin [84–87]. Moderate Ca^{2+} (1 µM) or cargo binding causes the molecule to open into the active form. At higher [Ca^{2+}], one or more CaM subunits dissociate from their IQ domains, again suppressing motility, either by diminishing actin binding or by structurally weakening or kinking the lever arm [88, 89].

Dissociation of the CaM subunits is measured biochemically in bulk samples [86, 88], but the ensemble experiments cannot address whether each molecule homogeneously loses the same number of molecules or there is a distribution of lost subunits. Single molecule imaging resolves this type of question by allowing the number of dissociating subunits to be counted for individual myosin V molecules. When CaM subunits were dissociated at high [Ca^{2+}] and then replaced with fluorescent labeled CaM [90], most of the molecules lost two of the 12 CaM light chains. Processivity measurements on single fluorescent labeled molecules were consistent with a model in which dissociation of one CaM subunit is sufficient to terminate a processive series of steps [91]. Whether these effects of Ca^{2+} are part of the normal cellular regulatory pathways is not known, although other myosins are dynamically modulated by Ca^{2+} [92, 93]. Other modes of myosin V regulation have been found in ensemble cell biological studies [94, 95], but they have not been approached yet by single molecule imaging.

3.3.2
Myosin II

The absence of processivity in muscle myosin has caused single molecule imaging studies of myosin II to lag behind those of myosin V. In the presence of ATP, myosin II dissociates rapidly from actin, and the proportion of the ATPase cycle occupied in actin-bound states is much lower than for myosin V. In normal activity, only one of the myosin heads binds to actin at a time. Thus it has been difficult to image single myosin II molecules during normal function. Cy3-labeled ATP molecules were visualized in TIRF microscopy when they bound to Cy5-labeled myosin molecules [3]. Free fluorescent nucleotide molecules near the reflecting surface of the TIRF microscope diffuse rapidly in the aqueous medium causing delocalization of their intensity. Binding to myosin was recognized as fixed localization identifying individual ATPase turnovers.

Combining TIRF microscopy with optical trap measurements of actin displacements by myosin II enabled correlation of the mechanical working stroke with binding of ATP and dissociation of ADP [96]. As expected from solution biochemistry, ATP caused rapid dissociation of myosin from actin. Surprisingly, ADP often dissociated from the myosin–ADP complex before it rebound to actin, leading to the suggestion that the energy liberated by ATP hydrolysis can be stored in the protein even after the nucleotide is released. This apparent protein "hysteresis" has been detected in other enzymes [97, 98].

The angle of the myosin II light chain region has been measured during the working stroke in bulk assays within muscle fibers using fluorescence polarization [41, 99] and low angle X-ray diffraction [100]. This structural change takes place on the 1–5-ms timescale. Whether the average structural signals from these bulk assays faithfully represent the behavior of the individual molecules is an important application for single molecule imaging. Angular motions have been made by single molecule fluorescence polarization [15, 101]. The limited lifetime of the actomyosin complex, however, and the rapid kinetics of the myosin II working stroke have so far challenged single molecule detection of the angular change during its working stroke.

3.3.3
Myosin VI

This molecular motor, involved in cargo transport, endocytosis and anchoring in neural development and in sensory cells, exhibits several interesting and puzzling features [102]. Unlike the other characterized myosins which translate toward the barbed end of the actin filament, myosin VI moves toward the pointed end [103, 104] due to a peptide insert in the motor domain that orients the lever arm opposite to the other myosins [105]. The directionality can be determined using actin filaments whose polarity is marked by a bright region of fluorescent actin or a specific fluorescent actin-binding protein [103, 105]. Polarity marked filaments show the direction of the active motion in both the single-molecule TIRF processivity assay (myosin walks on actin filaments attached to the substrate) and the gliding filament assay (myosin is attached to the surface and moves the actin).

Double-headed molecules of myosin VI are processive with highly variable step sizes ranging from 20 to 40 nm and display many more backward steps than myosin V [106]. The distance myosin VI can traverse per mechanical step is surprisingly large because it contains only two CaMs per putative lever arm, compared to six per lever arm in myosin V. Whether the motor is a dimer or monomer within cells is controversial [107]. There is evidence that cargos or actin can facilitate dimerization [108, 109] and that even a single head can move processively when it is attached to a bead cargo [109]. As with myosin V, processive motility by single-headed constructs has not been reported using single-molecule fluorophore imaging, possibly indicating that an attached bead enhances processivity by restricting diffusion away from the filament.

FIONA experiments showed that the individual heads of the myosin VI dimer translate by 70 nm on each step [34, 110], implying a hand-over-hand mechanism, as with myosin V. As the lever arms are too short for the molecule to span over the corresponding 35-nm distance between target actin sites, asymmetric models of the motion were considered in which the motor or dimerization domains would partially unfold to enhance the molecule's reach. Such a difference between the two heads might result in two sets of distances between the heads corresponding to the unfolded one leading or trailing. Measurement of the distance between two GFP-labeled myosin VI heads by SHRImP showed this not to be the case (Figure 3.5) [111]. A strongly dimerizing construct of myosin VI was expressed with a GFP tag on each head. When this species was bound to actin and imaged using TIRF microscopy, sequential bleaching of the fluorophores on each of the heads, and subtraction of the one- and two-fluorophore PSF distributions (Figure 3.5) gave the inter-probe distance. For myosin VI, the distance between the two heads bound to actin was found to be 29 nm with an upper limit of ∼14 nm for any supposed asymmetry (e.g. two distances 22 nm and 36 nm). These data are consistent with a symmetrical hand-over-hand stepping mechanism.

Optical trap experiments [112] and X-ray crystallography [113] have suggested that the lever arm rotates nearly 180° during the working stroke of myosin VI. With a 10-nm long lever arm, the maximum rotational working stroke is 20 nm, necessitating a diffusional search as in myosin V [110, 112]. A polTIRF study of myosin VI dimers detected large rotational motions of the CaM during processive stepping [52], with a highly variable angle when the head binds actin in the leading position. Taken together with the high variance of mechanical step distance [106] and leading position [110], these experiments suggest that the thermal search reaches actin monomers having a large range of axial distances and azimuthal angles around the actin axis. Due to the structure of the actin filament, binding of the leading head to actins 6, 7, 9, 11 and 13 monomer units away from the attached head require sideways tilting of various amounts up to nearly a ±90° azimuth [52]. This wide choice of target actin monomers may enable myosin VI to navigate around obstacles and to switch actin filaments in order to carry out cargo transport functions in the crowded cellular environment. Investigations into the kinetic coupling between the two heads suggest that intramolecular strain modulates both ADP release and ATP binding kinetics [114, 115], another novel mechanism in this unusual motor.

The description here of myosin experiments has emphasized points that are reasonably agreed upon in order to illustrate the advances made by imaging single fluorescent probes. But there are many other issues that are not settled and will be approached by single molecule biophysics. How is binding of the myosin head to actin coupled to the working stroke and dissociation of P_i and ADP? How is the path along actin determined? How often do backward steps occur and what is their significance and biochemical pathway? Is there a torque associated with the working stroke? Does the tilting of the lever arm power the motion or just sense it? Do single-headed myosins work processively? What novel mechanisms will be revealed by study of the almost 20 myosin family members that have hardly been studied by single molecule techniques? Do multiple motors cooperate during motion of a cargo? The methods described here are poised to address these questions.

3.3.4
Conventional Kinesin

The founding member of the kinesin family (Figure 3.8, kinesin-1) transports a variety of cargos in cells, including membranous organelles, mRNA, and signaling molecules. It is essential for mitosis and is the predominant molecular motor that conducts antegrade transport of vesicular cargos in nerve axons. Mutations have been linked to neurological diseases [116]. Kinesin-1 a highly processive, dimeric motor, whose 8-nm steps toward the plus (faster polymerizing) end of microtubules match the spacing of the tubulin dimers (Figure 3.8). An early TIRF microscopic assay showed that Cy3-labeled dimeric kinesins moved an average of 600 nm along microtubules attached to the microscope slide [117], confirming the high processivity of kinesin's mechanism found earlier in gliding filament and optical trap studies [118, 119]. A 600-nm processive run requires completion of 75 8-nm steps before the molecule dissociates. Monomeric constructs are not processive, indicating that the two heads in the native molecule cooperate to maintain contact with the microtubule during stepping FIONA experiments, using kinesin labeled on the motor domain with Cy3, gave alternating 17-, 0-, 17-nm steps indicating hand-over-hand motility [120 4] in which the trailing head detaches, passes by the attached head moving forward by a distance twice the step size, and becomes the leading head.

The motor domains of a dimeric kinesin molecule are too close to the junction with the coiled-coil tail to accommodate a stiff lever arm that could produce an 8-nm stroke (Figure 3.8). In crystal structures of the dimeric molecule, the two heads are only 5 nm apart raising the issue of how both motor domains can bind to tubulin dimers separated along the microtubule by 8 nm [121]. There is a 15-amino acid sequence at the C-terminus of the motor domain, just before the coiled-coil tail, that was found to dock and undock from the motor domain in various steps of the ATPase cycle [122]. Ensemble distance measurements between fluorescent probes in this "neck linker" and the motor domain, as well as cryo-EM and EPR studies supported its role in generating the working stroke and enabling two-headed attachment of kinesin to successive tubulin dimers.

A single molecule measurement, using FRET to measure intramolecular distance [123] (Figure 3.13), enabled the structural changes in the neck linker to be correlated with active processive motility. Engineered cysteine residues in the motor domain (amino acid 215) and in the tail (amino acid 342) were labeled with Cy3 and Cy5. The distance between them was estimated using smFRET as described above. The FRET efficiency oscillated between $E = 0.4$ and 0.9, values consistent with the 2-nm and 6-nm distance between the probes expected on the basis of neck linker docking and undocking (Figure 3.13A). The docked and undocked configurations could be assigned to trailing and leading positions and the rate of switching between the high and low E states was consistent with one structural change per 8-nm step of motility. These results link changes of the motor domain–neck linker interaction to

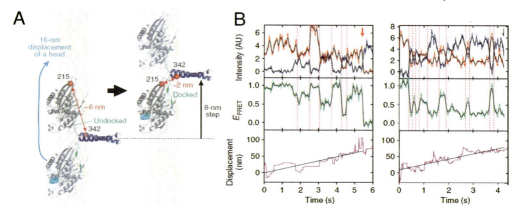

Figure 3.13 Neck linker positions in dimeric kinesin-1 with both heads bound to a microtubule. A, In this model, the neck linker in the leading head points rearward and the one in the trailing head, docked onto the motor domain, points forward. Cys^{215} and Cys^{342}, labeled in one chain of the dimer (red spheres), are expected to be farther from each other in the leading head than in the trailing head. B, time traces of donor (Cy3, blue) and acceptor (Cy5, red) fluorescence intensities (arbitrary units), calculated FRET efficiency (E_{FRET}, green) and axial displacement of the fluorophore (purple), moving along axonemes at 0.5 or 1 µM ATP. Image sampling rate: 15 frames s^{-1}. Black lines show running averages or linear fits. Vertical dotted lines mark anticorrelated FRET changes. The red arrow indicates photobleaching of the acceptor dye; black arrow indicates photobleaching of the donor dye or detachment from the axoneme. From Ref. [123].

processive motility. PolTIRF measurements of the angle of the neck linker and its rotational mobility also showed variations during motility consistent with the neck linker model [124]. The free energy liberated when this short peptide binds to the rest of the head, however, is insufficient to account for the ~48 zJ (8 nm × 6 pN) mechanical work kinesin can perform per ATPase cycle [125] and other steps such as binding of the free head must contribute to generating force and motion. Evidence for another interpretation of the data, that ATP binding causes increased mobility of the attached head rather than docking of the neck linker, is summarized in [126].

As discussed for myosins V and VI, the high processivity of kinesin requires a mechanism to gate the biochemical reactions of the two heads and maintain them out of phase. As in myosin, intramolecular strain between the two heads of kinesin bound to the microtubule alters the affinity of the heads for ADP [127] and prevents premature ATP binding to the leading head [128–131]. These characteristics are similar to the strain-dependent gating of myosin VI [115].

3.3.5
Other Kinesins

The other kinesin family members have diverse and interesting mechanisms and cellular roles. As mentioned, in conventional kinesins, the motor domain is at the N-terminus, whereas some classes of kinesin have C-terminal or central motor domains. The C-terminal motors translate in the opposite direction, toward the slow growing (minus) end of microtubules [132]. Gliding filament and processivity assays with engineered chimeras of plus-end and minus-end directed kinesin classes have shown that the sequences of the linkers and their interaction with the motor domain control the direction of motion [133].

Kinesins and dyneins are essential for formation of the mitotic spindle and its motions during chromosome segregation. A tetrameric, processive plus-end directed motor, kinesin-5, also termed Eg5, crosslinks microtubules and causes relative microtubule sliding [134, 135]. Another mitotic motor, kinesin-13 also termed MCAK (mitotic centromere-associated kinesin), targets the ends of microtubules and facilitates their depolymerization, presumably helping to power chromosome separation. MCAK oligomerizes into rings that encircle the microtubule [136]. A single-molecule TIRF assay [137] showed that MCAK reaches the ends of the microtubule by a one-dimensional, ATP-independent, diffusional process that is 20–50-fold faster than direct binding to the microtubule ends from the solution. The interactions of many proteins with microtubules are sensitive to the ionic strength suggesting that they are mediated by ionic bonds.

For kinesin-1, the evidence that both heads are required for processivity came from gliding filament assays [138, 139], optical trap mechanics [118], single-molecule fluorescence imaging [117] and FIONA [108]. Therefore it was a surprise when a monomeric isoform, kinesin-3 also termed KIF1A, was shown to exhibit processive motility [140]. Like myosin VI, whether this motor is monomeric or dimeric in cells was called into question [141] and the motions have an extensive diffusive component, like MCAK, as well as an ATP-driven directionality. Still, the intriguing question remains: how can a single motor domain dissociate from tubulin to move along the microtubule and not diffuse away? KIF1A contains an unusual loop containing six closely spaced lysine residues (the K-loop) forming a highly positively charged region that seems to interact with a C-terminal segment of tubulin containing negatively charged glutamate residues (the E-hook). Deleting the K-loop or proteolyzing the E-hook eliminated processive motility by single-headed KIF1A [140]. Thus, a secondary interaction between the KIF1A motor domain and tubulin might hold the motor near the microtubule while the ATP-dependent tubulin binding domain is weakly bound or detached. Crystal structures of the KIF1A motor domain showed that the position of the K-loop is dependent on the identity of the bound nucleotide [142]. This result suggests that the KIF1A motor domain has two alternating, ATP-dependent sites of microtubule interaction, thereby obviating the requirement for a second head. Whether this concept applies to any other kinesins or other molecular motors is unknown.

Many questions remain on the mechanism of kinesin motility [143]. Does kinesin take substeps? How does kinesin track a protofilament? Which tubulin dimers are targeted during a step? How are the biochemical and mechanical cycles related? Is the neck linker hypothesis compatible with the energy transduction and thermodynamic efficiency of kinesin? What is the structure and biochemical complement of the rate-limiting step during processive motility? How often do backward steps occur and what is their significance and biochemical pathway? Do single-headed kinesins move processively *in vitro* or in a cell? Do multiple kinesin motors cooperate during motion of a cargo? Some of these questions may be answered by single-molecule imaging.

3.3.6
Dyneins

Axonemal dynein is the molecular motor that bends and waves eukaryotic cilia and flagella. Cytoplasmic dynein helps to position the Golgi complex and other organelles in the cell, transports vesicular cargos derived from endoplasmic reticulum, endosomes, and lysosomes, and takes part in movement of the chromosomes and the mitotic spindle in mitosis [144]. Dynein powers retrograde axonal transport and its disruption causes neurological disease [145]. Biophysical and molecular biological studies of dynein lag behind those of myosin and kinesin because the large modular nature of dynein complicates its function. Purifying dynein from cells while retaining full activity and manufacturing it within heterologous expression systems are difficult. But progress is being made on these fronts [146–148].

Questions arise directly from the structure of cytoplasmic dynein shown in Figure 3.8: how is the nucleotide state in the active site (AAA domain 1) communicated to the microtubule-binding domain on the stalk, 15 nm away? What structural change accomplishes the working stroke? Cryo-electron micrographs [149] and ensemble FRET studies [150] have shown that the stalk and stem adopt different dispositions relative to the nucleotide-binding ring at various ATPase intermediates, simulated with nucleotide analogs. These results suggest a hypothetical rotation of the ring relative to the stalk or stem which could drive microtubule motion.

Processivity of cytoplasmic dynein was suggested from motility experiments with latex beads coated with dynein at very low motor densities [151]. When beads were placed on a microtubule using an optical trap, the likelihood of binding and of moving was linearly proportional to the probability that one active motor was located on the bead within interaction range of the microtubule [151]. Processivity of dynein molecules requires at least two heads [148, 152], but compared to kinesin, the motility of the dynein dimer is more flexible, varying in step distance, and direction, including sideways and backward steps. Dynein also shows strong thermally-driven diffusive motion along microtubules even when the normal ATPase activity is blocked [153], suggesting an electrostatic interaction with the microtubule like that of MCAK and KIF1A.

The step size for cytoplasmic dynein was reported to vary with mechanical load from 8 nm (the spacing of the tubulin dimers) at a resisting force of $\sim$1 pN up to 24 or 32 nm at lower loads [154]. Another study, however, found mostly 8-nm mechanical steps up to forces of 7–8 pN [155]. FIONA experiments with quantum dots bound to the dynein tail gave mostly 8-nm steps [148, 155], but when quantum dots were bound to the head at the C-terminus, they showed predominantly 16-nm and larger steps [148]. These results suggest an alternating leading–trailing mechanism as with the other motors. The dynein rings (13–15 nm diameter) are larger than the tubulin dimers (8.3 nm) however, making direct binding of the two heads to adjacent dimers along a protofilament unlikely. Flexibility of the stalk and the tendency to take variable sideways and backward steps resemble the motions of myosin VI.

Organelle motion in cells is bidirectional with single particles often reversing their direction. The destinations of vesicular cargos vary with physiological stimuli. Thus the relationship between molecular motors directed toward the cell periphery, e.g. kinesin and myosin V, and those mainly moving toward the center, such as dynein and myosin VI is an important question in cell biology. Inhibition of dynein in cells disrupts both directions of transport [59, 156, 157], suggesting a close cooperation or interaction between dynein activity and plus-end directed kinesin transport. This interaction might be mediated by the dynactin accessory complex [158].

When the dynactin complex was made fluorescent in a transgenic mouse, the purified dynein–dynactin complex showed strong ATP-dependent bidirectional motion along microtubules (Figure 3.14). In an equivalent processivity assay, kinesin moved unidirectionally with a regular stepping rate (Figure 3.14). Thus,

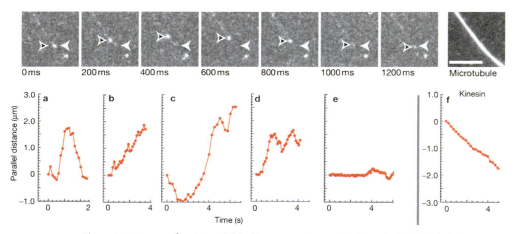

Figure 3.14 Traces of motion exhibited by dynein—dynactin—GFP as visualized by TIRF. Top row: video images at 200 ms intervals, 10 mM ATP. The filled arrowhead indicates the starting position of the complex. The open arrowhead indicates the location at each time point. The far right image is an epifluorescence micrograph of the rhodamine-labeled microtubule. Scale bar: 4 µm. Lower row a-e, kymographs from dynein—dynactin—GFP time-series show reversal of direction with long processive motion in each direction. f, kinesin-GFP motility is much steadier. From Ref. [159].

the dynein–dynactin complex may participate directly in bidirectional transport of cargos within cells. Whether this activity requires dynactin and how the directionality and reversals are controlled are not yet known. Higher resolution studies of dynein constructs and dual color imaging, such as shown for myosin V in Figures 3.10 and 3.12, may lead to an understanding of the stepping and control mechanisms of this complex molecular motor. An approach toward studying the interactions among the different molecular motor families is to construct arrays of cytoskeletal filaments *in vitro* that introduce some of the complexity of cells. A few reports of this type of study have been published [159, 160].

3.3.7
Single Molecule Intracellular Imaging

Dramatic technical advances in optical microscopy over the past 50–75 years, including phase contrast, differential interference contrast, epifluorescence, and confocal microscopy, the commercial availability of highly sensitive cameras, and the improvement in specific markers of cellular structures using organic fluorescent dyes and auto-fluorescent proteins have made optical microscopy one of the core research tools in cell biology [7, 9]. Within the last few years, the techniques developed for imaging single molecules have been applied to fluorescent structures within live cells both to study how molecular motors operate in their native environment and to improve spatial resolution of images in many other cell biological contexts. These evolving methods have great potential for further improving light microscopic imaging to understand molecular processes.

With brightly fluorescent vesicles containing dozens of GFP tags [161], quantum dots endocytosed into vesicles in cells [162, 163], or optically dense melanin transporting vesicles [164], the brightness or contrast of the particle relative to the surrounding cytoplasm is high enough to resolve nanometer movements within milliseconds. These vesicles display 8- or 35-nm discrete stepwise motions when they are transported by several microtubule-based or actin-based molecular motors. Transport velocities up to 10-fold higher than those measured *in vitro*, were observed in some of the experiments without alteration of the 8-nm stepwise character. These results are puzzling, because increasing the number of motors carrying a cargo *in vitro* does not generally increase the velocity of motility above that of the unloaded single molecule [139] and it tends to reduce the observed step size due to the smoothing effect of multiple unsynchronized working strokes [165]. Understanding the high velocities and quantized step size of cargos observed *in vivo* will await further detailed imaging studies.

Direct imaging of single molecules within cells, rather than their cargo, is limited by autofluorescence of cellular components that reduces the contrast between the target molecule and its surroundings. Several routes to increasing the signal above background have been reported. A kinesin fused to three tandem yellow fluorescent proteins is bright enough to observe within cultured cells [166]. TIRF illumination reduces the volume illuminated within the cell, but then imaging is restricted to the area of substrate contact. Reducing the incident angle (measured relative to the optical axis) to slightly below the critical angle for total internal reflection produces oblique illumination propagating into the cell (Figure 3.2A, blue light ray near the critical angle). For favorable cases with low autofluorescence, this arrangement has been used to image individual signaling molecules [167].

Fast diffusion or active translocation of a labeled macromolecule, can spread its fluorescence emission over many pixels, causing reduced intensity above the cellular autofluorescence. This problem has been tackled for markers of gene expression by immobilization in the cell membrane [168], but that route would not be appropriate for functioning molecular motors. Techniques that have been used with other systems to capture images of moving targets might be promising for molecular motor studies. A very brief pulse of exciting light acts like a stroboscope and localizes the fluorescence to the spatial region occupied by the fluorophore during the pulse, thereby improving the signal-to-noise ratio [169]. Lastly, fluorescence lifetime imaging microscopy (FLIM) is a method using repeated sub-nanosecond excitation pulses and time-gated photon counting to resolve and form an image from the nanosecond decay rates of the excited fluorescent species. The emission wavelength and lifetime of the fluorophore of interest may differ from those of the non-specific background emission, enabling separation of the probe photons from the background luminescence [170].

The sub-diffraction localization of individual fluorophores, developed for single-molecule studies can also be used to obtain images with dramatically improved spatial resolution of multiple cellular structures. A promising route to achieve this goal is to use fluorophores that can be repeatedly switched on and off [171, 172]. Both organic fluorophores and modified fluorescent protein analogs have been developed that can be repeatedly transferred by pulses of light from fluorescent and non-fluorescent states, or between two different-color states [173–177]. These light-controlled fluorophores are useful in studying the dynamics of cell processes, transfer between compartments, and mobility [178].

Examples of using photoactivatable fluorescent proteins in sub-diffraction microscopy are shown in Figure 3.15 [179]. The authors used blue (405 nm) laser pulses to switch on sparse populations of a photoactivatable GFP and then imaged and photobleached them under green (561 nm) light. The individual fluorescent GFP molecules were located within ∼20 nm by fitting two-dimensional Gaussian functions, as in FIONA. After photobleaching, a new set of fluorophores were activated by the 405-nm laser pulse and imaged at 561 nm. After several thousand photo-activation, imaging, and photobleaching cycles, the summed information gave an image representing the likelihood of a GFP molecule being located at each position, but with spatial resolution given by the nanometer-scale uncertainty of individual molecule localization (Figure 3.15 lower panels).

Several other approaches using switchable fluorophores [171, 172] and excitation fields "structured" by optical interference [180] or stimulated emission [181] have been described. This area of hyper-resolution microscopy is rapidly evolving for imaging molecular structures in cells.

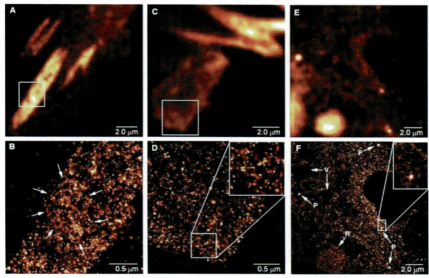

Figure 3.15 Photo Activated Localization Microscopy (PALM). A, a summed-molecule (equivalent to standard) TIRF image of focal adhesions for a cell expressing vinculin tagged with the photoactivatable fluorescent protein, dEos. B, magnified inset of adhesion region in A. C, summed-molecule TIRF image near the periphery of a cell expressing tdEos-tagged actin. D, magnified PALM view of the actin distribution within the box outline in C. Inset, further magnified view. E and F, summed-molecule TIRF and PALM images, respectively, of a cell expressing dEos-tagged, retroviral protein Gag. Considerable detail in the PALM image is not resolved in the standard image. From Ref [179].

3.4
Conclusions

Studying individual molecules using novel types of optical microscopy is helping to reveal the functional mechanisms of the molecular motors, myosin, kinesin and dynein in ways not thought feasible only a few years ago. Single molecule biophysics avoids the inevitable loss of information in classical ensemble experiments when a signal is averaged over the members of the sample population. Fluctuations, reversals, rare events and other heterogeneities are the necessary consequence of the stochastic nature of chemical and physical reactions. This "noise" is interesting! Many of the signals that are the primary functional output of molecular motors, such as piconewton forces and nanometer motions, are not accessible in a collection of molecules suspended in a cuvette. Techniques have evolved for localization of molecules at far better precision than the classical resolution of light microscopy and for detecting structural dynamics such as rotational motions and intramolecular distances. The experiments are both resolving questions and raising further questions about these "nano-machines": how they produce force and motion, how they transduce metabolic energy into work and how they are regulated. Although our level of understanding is increasing rapidly, none of the molecular motors have been "solved".

Further advances in fluorescent probes, instrumentation and biological assays will accelerate progress. Techniques for introducing some of the complexity of the cell into *in vitro* studies and for collecting high resolution signals within live cells are beginning to emerge. Single-molecule imaging is likely to impact over a broad segment of cell biology as novel methods for collecting hyper-resolution images become more practical and images that give truly molecular insight become available. Watching and understanding every important molecule in a cell do its job is farfetched, but the tools are being assembled to enable us to consider what would be required and how close we can get to that goal.

Acknowledgements

I thank the members of my laboratory for the productive and enjoyable environment, Drs. P.R. Selvin and A. Yildiz for permission to use unpublished data, Yujie Sun, John Lewis and John Beausang for helpful comments on the manuscript, and the NIAMS, NIGMS and NSF for research support.

References

1 Sellers, J.R. (2004) Fifty years of contractility research post sliding filament hypothesis. *J. Muscle Res. Cell. Motil.*, **25**, 475–482.

2 Huxley, H.E. and Holmes, K.C. (1997) Development of synchrotron radiation as a high-intensity source for X-ray diffraction. *J. Synchrotron Radiat.*, **4**, 366–379.

3 Funastsu, T., Harada, Y., Tokunaga, M., Saito, K. and Yanagida, T. (1995) Imaging of single fluorescent molecules and individual ATP turnovers by single myosin molecules in aqueous solution. *Nature*, **374**, 555–559.

4 Block, S.M., Blair, D.F. and Berg, H.C. (1989) Compliance of bacterial flagella measured with optical tweezers. *Nature*, **338**, 514–518.

5 Paschal, B.M. and Vallee, R.B. (1993) Microtubule and axoneme gliding assays for force production by microtubule motor proteins. *Methods Cell Biol.*, **39**, 65–74.

6 Rock, R.S., Rief, M., Mehta, A.D. and Spudich, J.A. (2000) *In vitro* assays of processive myosin motors. *Methods*, **22**, 373–381.

7 Allan, V.J. (2000) *Protein Localization by Fluorescence Microscopy: A Practical Approach*, Oxford University Press, New York.

8 Zhang, J., Campbell, R.E., Ting, A.Y. and Tsien, R.Y. (2002) Creating new fluorescent probes for cell biology. *Nat. Rev. Mol. Cell. Biol.*, **3**, 906–918.

9 Giepmans, B.N.G., Adams, S.R., Ellisman, M.H. and Tsien, R.Y. (2006) The fluorescent toolbox for assessing protein location and function. *Science*, **312**, 217–224.

10 Howard, J., Hunt, A.J. and Baek, S. (1993) Assay of Microtubule Movement Driven by Single Kinesin Molecules, *Methods Cell Biol.*, **39**, 137–147.

11 Axelrod, D., Burghardt, T.P. and Thompson, N.L. (1984) Total internal reflection fluorescence. *Annu. Rev. Biophys. Bioeng.*, **13**, 247–268.

12 Yoo, H., Song, I. and Gweon, D.G. (2006) Measurement and restoration of the point spread function of fluorescence confocal microscopy. *J. Microsc.*, **221**, 172–176.

13 Edman, L., Mets, Ü. and Rigler, R. (1996) Conformational transitions monitored for single molecules in solution. *PNAS*, **93**, 6710–6715.

14 Nie, S., Chiu, D.T. and Zare, R.N. (1994) Probing individual molecules with confocal fluorescence microscopy. *Science*, **266**, 1018–1021.

15 Warshaw, D.M., Hayes, E., Gaffney, D., Lauzon, A.M., Wu, J., Kennedy, G., Trybus, K., Lowey, S. and Berger, C. (1998) Myosin conformational states determined by single fluorophore polarization. *Proc. Natl Acad. Sci. USA*, **95**, 8034–8039.

16 Kudryavtsev, V., Felekyan, S., Woźniak, A.K., König, M., Sandhagen, C., Kühnemuth, R., Seidel, C.A.M. and Oesterhelt, F. (2007) Monitoring dynamic systems with multiparameter fluorescence imaging. *Anal. Bioanal. Chem.*, **387**, 71–82.

17 Levene, M.J., Korlach, J., Turner, S.W., Foquet, M., Craighead, H.G. and Webb, W.W. (2003) Zero-mode waveguides for single-molecule analysis at high concentrations. *Science*, **299**, 682–686.

18 Robbins, M.S. and Hadwen, B.J. (2003) The noise performance of electron multiplying charge-coupled devices. *IEEE Trans. Electron Devices*, **50**, 1227–1232.

19 Akiba, M., Fujiwara, M. and Sasaki, M. (2005) Ultrahigh-sensitivity high-linearity photodetection system using a low-gain avalanche photodiode with an ultralow-noise readout circuit. *Optics Lett.* **30**, 123–125.

20 Forkey, J.N., Quinlan, M.E. and Goldman, Y.E. (2005) Measurement of single macromolecule orientation by total internal reflection fluorescence polarization microscopy. *Biophys. J.*, **89**, 1261–1271.

21 Born, M. and Wolf, E. (1975) *Principles of Optics: Electromagnetic Theory of Propagation, Interference and Diffraction of Light*, Pergamon Press, New York.

22 Hecht, E. (2002) *Optics*, Addison-Wesley, Reading, MA.

23 Bagshaw, C.R. and Cherny, D. (2006) Blinking fluorophores: what do they tell us about protein dynamics? *Biochem. Soc. Trans.*, **34**, 979–982.

24 Benesch, R.E. and Benesch, R. (1953) Enzymatic removal of oxygen for polarography and related methods. *Science*, **118**, 447–448.

25 Harada, Y., Sakurada, K., Aoki, T., Thomas, D.D. and Yanagida, T. (1990) Mechanochemical coupling in acto-myosin energy transduction studied by *in vitro* movement assay. *J. Mol. Biol.*, **216**, 49–68.

26 Rasnik, I., McKinney, S.A. and Ha, T. (2006) Nonblinking and long-lasting single-molecule fluorescence imaging. *Nat. Methods.*, **3**, 891–893.

27 Yildiz, A., Forkey, J.N., McKinney, S.A., Ha, T., Goldman, Y.E. and Selvin, P.R. (2003) Myosin V walks hand-over-hand: single fluorophore imaging with 1.5-nm localization. *Science*, **300**, 2061–2065.

28 Shu, D., Zhang, H., Jin, J. and Guo, P. (2007) Counting of six pRNAs of phi29 DNA-packaging motor with customized single-molecule dual-view system. *EMBO J.* **26**, 527–537.

29 Leake, M.C., Chandler, J.H., Wadhams, G.H., Bai, F., Berry, R.M. and Armitage, J.P. (2006) Stoichiometry and turnover in single, functioning membrane protein complexes. *Nature*, **443**, 355–358.

30 Ross, J.L., Shuman, H., Holzbaur, E.L. and Goldman, Y.E. (2008) Kinesin and Dynein-dynactin at intersecting microtubules: Motor density affects dynein function. *Biophys. J.*, **94**, 3115–3125.

31 Michalet, X., Lacoste, T.D. and Weiss, S. (2001) Ultrahigh-resolution colocalization of spectrally separable point-like fluorescent probes. *Methods*, **25**, 87–102.

32 Thompson, R.E., Larson, D.R. and Webb, W.W. (2002) Precise nanometer localization analysis for individual fluorescent probes. *Biophys. J.*, **82**, 2775–2783.

33 Toprak, E. and Selvin, P.R. (2007) New fluorescent tools for watching nanometer-scale conformational changes of single molecules. *Annu. Rev. Biophys. Biomol. Struct.*, **36**, 349–369.

34 Ökten, Z., Churchman, L.S., Rock, R.S. and Spudich, J.A. (2004) Myosin VI walks hand-over-hand along actin. *Nat. Struct. Mol. Biol.*, **11**, 884–887.

35 Yasuda, R., Noji, H., Yoshida, M., Kinosita, K., Jr. and Itoh, H. (2001) Resolution of distinct rotational substeps by submillisecond kinetic analysis of F_1-ATPase. *Nature*, **410**, 898–904.

36 Dunn, A.R. and Spudich, J.A. (2007) Dynamics of the unbound head during myosin V processive translocation. *Nat. Struct. Mol. Biol.*, **14**, 246–248.

37 Nishiyama, M., Muto, E., Inoue, Y., Yanagida, T. and Higuchi, H. (2001) Substeps within the 8-nm step of the ATPase cycle of single kinesin molecules. *Nat. Cell. Biol.*, **3**, 425–428.

38 Gordon, M.P., Ha, T. and Selvin, P.R. (2004) Single-molecule high-resolution imaging with photobleaching. *Proc. Natl Acad. Sci. USA*, **101**, 6462–6465.

39 Forkey, J.N., Quinlan, M.E. and Goldman, Y.E. (2000) Protein structural dynamics by single-molecule fluorescence polarization. *Prog. Biophys. Mol. Biol.*, **74**, 1–35.

40 Rosenberg, S.A., Quinlan, M.E., Forkey, J.N. and Goldman, Y.E. (2005) Rotational motions of macro-molecules by single-molecule fluorescence microscopy. *Acc. Chem. Res.*, **38**, 583–593.

41 Irving, M., St Claire Allen, T., Sabido-David, C., Craik, J.S., Brandmeier, B., Kendrick-Jones, J., Corrie, J.E.T., Trentham, D.R. and Goldman, Y.E. (1995) Tilting of the light-chain region of myosin during step length changes and active force generation in skeletal muscle. *Nature*, **375**, 688–691.

42 Forkey, J.N., Quinlan, M.E., Shaw, M.A., Corrie, J.E.T. and Goldman, Y.E. (2003) Three-dimensional structural dynamics of myosin V by single-molecule fluorescence polarization. *Nature*, **422**, 399–404.

43 Asenjo, A.B., Krohn, N. and Sosa, H. (2003) Configuration of the two kinesin motor domains during ATP hydrolysis. *Nat. Struct. Biol.*, **10**, 836–842.

44 Otterbein, L.R., Graceffa, P. and Dominguez, R. (2001) The crystal structure of uncomplexed actin in the ADP state. *Science*, **293**, 708–711.

45 Mercier, P., Ferguson, R.E., Irving, M., Corrie, J.E.T., Trentham, D.R. and Sykes, B.D. (2003) NMR structure of a bifunctional rhodamine labeled N-domain of troponin C complexed with the regulatory "switch" peptide from troponin I: implications for in situ fluorescence studies in muscle fibers. *Biochemistry*, **42**, 4333–4348.

46 Sale, K., Sár, C., Sharp, K.A., Hideg, K. and Fajer, P.G. (2002) Structural determination of spin label immobilization and orientation: a Monte Carlo minimization approach. *J. Magn. Reson.*, **156**, 104–112.

47 Ha, T., Laurence, T.A., Chemla, D.S. and Weiss, S. (1999) Polarization spectroscopy of single fluorescent molecules. *J. Phys. Chem. B.*, **103**, 6839–6850.

48 Peterman, E.J.G., Sosa, H. and Moerner, W.E. (2004) Single-molecule fluorescence spectroscopy and microscopy of biomolecular motors. *Annu. Rev. Phys. Chem.*, **55**, 79–96.

49 Sase, I., Miyata, H., Ishiwata, S. and Kinosita, K., Jr. (1997) Axial rotation of sliding actin filaments revealed by single-fluorphore imaging. *Proc. Natl Acad. Sci USA*, **94**, 5646–5650.

50 Axelrod, D. (1989) Fluorescence polarization microscopy. *Methods Cell. Biol.*, **30**, 333–352.

51 Beausang, J.F., Schroeder, H.W., III, Gilmour, J.A. and Goldman, Y.E. (2006) Twirling of actin by myosins II and V. *Biophys. J.*, **90**, 587a.

52 Sun, Y., Schroeder, H.W., Beausang, J.F., Homma, K., Ikebe, M. and Goldman, Y.E. (2006) Single molecule fluorescence polarization of calmodulin in myosin VI. *Biophys. J.*, **90**, 431a.

53 Bartko, A.P. and Dickson, R.M. (1999) Imaging three-dimensional single molecule orientations. *J. Phys. Chem. B.*, **103**, 11237–11241.

54 Böhmer, M. and Enderlein, J. (2003) Orientation imaging of single molecules by wide-field epifluorescence microscopy. *J. Opt. Soc. Am. B.*, **20**, 554–559.

55 Patra, D., Gregor, I. and Enderlein, J. (2004) Image analysis of defocused single-molecule images for three-dimensional molecule orientation studies. *J. Phys. Chem. A.*, **108**, 6836–6841.

56 Toprak, E., Enderlein, J., Syed, S., McKinney, S.A., Petschek, R.G., Ha, T., Goldman, Y.E. and Selvin, P.R. (2006) Defocused orientation and position imaging (DOPI) of myosin V. *Proc. Natl Acad. Sci. USA*, **103**, 6495–6499.

57 Kull, F.J., Sablin, E.P., Lau, R., Fletterick, R.J. and Vale, R.D. (1996) Crystal structure of the kinesin motor domain reveals a structural similarity to myosin. *Nature*, **380**, 550–555.

58 Erzberger, J.P. and Berger, J.M. (2006) Evolutionary relationships and structural mechanisms of AAA + proteins. *Annu. Rev. Biophys. Biomol. Struct.*, **35**, 93–114.

59 Waterman-Storer, C.M., Karki, S.B., Kuznetsov, S.A., Tabb, J.S., Weiss, D.G., Langford, G.M. and Holzbaur, E.L.F. (1997) The interaction between cytoplasmic dynein and dynactin is required for fast axonal transport. *Proc. Natl Acad. Sci. USA*, **94**, 12180–12185.

60 Langford, G.M. and Molyneaux, B.J. (1998) Myosin V in the brain: mutations lead to neurological defects. *Brain. Res. Brain. Res. Rev.*, **28**, 1–8.

61 Rief, M., Rock, R.S., Mehta, A.D., Mooseker, M.S., Cheney, R.E. and Spudich, J.A. (2000) Myosin-V stepping kinetics: a molecular model for processivity. *Proc. Natl Acad. Sci. USA*, **97**, 9482–9486.

62 Sakamoto, T., Amitani, I., Yokota, E. and Ando, T. (2000) Direct observation of processive movement by individual myosin V molecules. *Biochem. Biophys. Res. Commun.*, **272**, 586–590.

63 Baker, J.E., Krementsova, E.B., Kennedy, G.G., Armstrong, A., Trybus, K.M. and Warshaw, D.M. (2004) Myosin V processivity: multiple kinetic pathways for head-to-head coordination. *Proc. Natl Acad. Sci. USA*, **101**, 5542–5546.

64 Cheney, R.E. O'Shea, M.K., Heuser, J.E., Coelho, M.V., Wolenski, J.S., Espreafico, E.M., Forscher, P., Larson, R.E. and Mooseker, M.S. (1993) Brain myosin-V is a two-headed unconventional myosin with motor activity. *Cell*, **75**, 13–23.

65 De La Cruz, E.M., Wells, A.L., Rosenfeld, S.S., Ostap, E.M. and Sweeney, H.L. (1999) The kinetic mechanism of myosin V. *Proc. Natl Acad. Sci. USA*, **96**, 13726–13731.

66 Mehta, A.D., Rock, R.S., Rief, M., Spudich, J.A., Mooseker, M.S. and Cheney, R.E. (1999) Myosin-V is a processive actin-based motor. *Nature*, **400**, 590–593.

67 Purcell, T.J., Morris, C., Spudich, J.A. and Sweeney, H.L. (2002) Role of the lever arm in the processive stepping of myosin V. *Proc. Natl Acad. Sci. USA*, **99**, 14159–14164.

68 Purcell, T.J., Sweeney, H.L. and Spudich, J.A. (2005) A force-dependent state controls the coordination of processive myosin V. *Proc. Natl Acad. Sci. USA*, **102**, 13873–13878.

69 Veigel, C., Wang, F., Bartoo, M.L., Sellers, J.R. and Molloy, J.E. (2002) The gated gait of the processive molecular motor, myosin V. *Nat. Cell Biol.*, **4**, 59–65.

70 Veigel, C., Schmitz, S., Wang, F. and Sellers, J.R. (2005) Load-dependent kinetics of myosin-V can explain its high processivity. *Nat. Cell Biol.*, **7**, 861–869.

71 Moore, J.R., Krementsova, E.B., Trybus, K.M. and Warshaw, D.M. (2004) Does the myosin V neck region act as a lever? *J. Muscle. Res. Cell. Motil.*, **25**, 29–35.

72 Watanabe, T.M., Tanaka, H., Iwane, A.H., Maki-Yonekura, S., Homma, K., Inoue, A., Ikebe, R., Yanagida, T. and Ikebe, M. (2004) A one-headed class V myosin molecule develops multiple large ($\approx$32-nm) steps successively. *Proc. Natl Acad. Sci. USA*, **101**, 9630–9635.

73 Syed, S., Snyder, G.E., Franzini-Armstrong, C., Selvin, P.R. and Goldman, Y.E. (2006) Adaptability of myosin V by simultaneous detection of position and orientation. *EMBO J.* **25**, 1795–1803.

74 Snyder, G.E., Sakamoto, T., Hammer, J.A., 3rd, Sellers, J.R. and Selvin, P.R. (2004) Nanometer localization of single green fluorescent proteins: evidence that myosin V walks hand-over-hand via telemark configuration. *Biophys. J.*, **87**, 1776–1783.

75 Churchman, L.S., Ökten, Z., Rock, R.S., Dawson, J.F. and Spudich, J.A. (2005) Single molecule high-resolution colocalization of Cy3 and Cy5 attached to macromolecules measures intramolecular distances through time. *Proc. Natl Acad. Sci. USA*, **102**, 1419–1423.

76 Warshaw, D.M., Kennedy, G.G., Work, S.S., Krementsova, E.B., Beck, S. and Trybus, K.M. (2005) Differential labeling of myosin V heads with quantum dots allows direct visualization of hand-over-hand processivity. *Biophys. J.*, **88**, L30–L32.

77 Ali, M.Y., Uemura, S., Adachi, K., Itoh, H., Kinosita, K., Jr. and Ishiwata, S. (2002) Myosin V is a left-handed spiral motor on the right-handed actin helix. *Nat. Struct. Biol.*, **9**, 464–467.

78 Sakamoto, T., Wang, F., Schmitz, S., Xu, Y., Xu, Q., Molloy, J.E., Veigel, C. and Sellers, J.R. (2003) Neck length and processivity of myosin V. *J. Biol. Chem.*, **278**, 29201–29207.

79 Sakamoto, T., Yildiz, A., Selvin, P.R. and Sellers, J.R. (2005) Step-size is determined by neck length in myosin V. *Biochemistry*, **44**, 16203–16210.

80 Astumian, R.D. (2001) Making molecules into motors. *Sci. Am.*, **285**, 56–64.

81 Rosenfeld, S.S. and Sweeney, H.L. (2004) A model of myosin V processivity. *J. Biol. Chem.*, **279**, 40100–40111.

82 Uemura, S., Higuchi, H., Olivares, A.O., De La Cruz, E.M. and Ishiwata, S. (2004) Mechanochemical coupling of two substeps in a single myosin V motor. *Nat. Struct. Mol. Biol.*, **11**, 877–883.

83 Griffith, L.C. (2004) Regulation of calcium/calmodulin-dependent protein kinase II activation by intramolecular and intermolecular interactions. *J. Neurosci.*, 24, 8394–8398.

84 Wang, F., Thirumurugan, K., Stafford, W.F., Hammer, J.A. 3rd, Knight, P.J. and Sellers, J.R. (2004) Regulated conformation of myosin V. *J. Biol. Chem.*, **279**, 2333–2336.

85 Li, X.D., Mabuchi, K., Ikebe, R. and Ikebe, M. (2004) Ca^{2+}-induced activation of ATPase activity of myosin Va is accompanied with a large conformational change. *Biochem. Biophys. Res. Commun.*, **315**, 538–545.

86 Krementsov, D.N., Krementsova, E.B. and Trybus, K.M. (2004) Myosin V: regulation by calcium, calmodulin, and the tail domain. *J. Cell. Biol.*, **164**, 877–886.

87 Liu, J., Taylor, D.W., Krementsova, E.B., Trybus, K.M. and Taylor, K.A. (2006) Three-dimensional structure of the myosin V inhibited state by cryoelectron tomography. *Nature*, **442**, 208–211.

88 Homma, K., Saito, J., Ikebe, R. and Ikebe, M. (2000) Ca^{2+}-dependent regulation of the motor activity of myosin V. *J. Biol. Chem.*, **275**, 34766–34771.

89 Li, X.D., Jung, H.S., Mabuchi, K., Craig, R. and Ikebe, M. (2006) The globular tail domain of myosin Va functions as an inhibitor of the myosin Va motor. *J. Biol. Chem.*, **281**, 21789–21798.

90 Nguyen, H. and Higuchi, H. (2005) Motility of myosin V regulated by the dissociation of single calmodulin. *Nat. Struct. Mol. Biol.*, **12**, 127–132.

91 Lu, H., Krementsova, E.B. and Trybus, K.M. (2006) Regulation of myosin V processivity by calcium at the single molecule level. *J. Biol. Chem.*, **281**, 31987–31994.

92 Sellers, J.R., Chantler, P.D. and Szent-Györgyi, A.G. (1980) Hybrid formation between scallop myofibrils and foreign regulatory light-chains. *J. Mol. Biol.*, **144**, 223–245.

93 Batters, C., Arthur, C.P., Lin, A., Porter, J., Geeves, M.A., Milligan, R.A., Molloy, J.E. and Coluccio, L.M. (2004) Myo1c is designed for the adaptation response in the inner ear. *EMBO J.* **23**, 1433–1440.

94 Karcher, R.L., Roland, J.T., Zappacosta, F., Huddleston, M.J., Annan, R.S., Carr, S.A. and Gelfand, V.I. (2001) Cell cycle regulation of myosin-V by calcium/calmodulin-dependent protein kinase II. *Science*, **293**, 1317–1320.

95 Wu, X., Wang, F., Rao, K., Sellers, J.R. and Hammer, J.A. 3rd. (2002) Rab27a is an essential component of melanosome receptor for myosin Va. *Mol. Biol. Cell*, **13**, 1735–1749.

96 Ishijima, A., Kojima, H., Funatsu, T., Tokunaga, M., Higuchi, H., Tanaka, H. and Yanagida, T. (1998) Simultaneous observation of individual ATPase and mechanical events by a single myosin molecule during interaction with actin. *Cell*, **92**, 161–171.

97 Frieden, C. (1970) Kinetic aspects of regulation of metabolic processes. *J. Biol. Chem.*, **245**, 5788–5799.

98 Lu, H.P., Xun, L. and Xie, X.S. (1998) Single-molecule enzymatic dynamics. *Science*, **282**, 1877–1882.

99 Corrie, J.E.T., Brandmeier, B.D., Ferguson, R.E., Trentham, D.R., Kendrick-Jones, J., Hopkins, S.C., van der Heide, U.A., Goldman, Y.E., Sabido-David, C., Dale, R.E., Criddle, S. and Irving, M. (1999) Dynamic measurement of myosin light-chain-domain tilt and

twist in muscle contraction. *Nature*, **400**, 425–430.

100 Reconditi, M., Linari, M., Lucii, L., Stewart, A., Sun, Y.B., Boesecke, P., Narayanan, T., Fischetti, R.F., Irving, T., Piazzesi, G., Irving, M. and Lombardi, V. (2004) The myosin motor in muscle generates a smaller and slower working stroke at higher load. *Nature*, **428**, 518–558.

101 Quinlan, M.E., Forkey, J.N. and Goldman, Y.E. (2005) Orientation of the myosin light chain region by single molecule total internal reflection fluorescence polarization microscopy. *Biophys. J.*, **89**, 1132–1142.

102 Sweeney, H.L. and Houdusse, A. (2007) What can myosin VI do in cells? *Curr. Opin. Cell. Biol.*, **19**, 57–66.

103 Wells, A.L., Lin, A.W., Chen, L.Q., Safer, D., Cain, S.M., Hasson, T., Carragher, B.O., Milligan, R.A. and Sweeney, H.L. (1999) Myosin VI is an actin-based motor that moves backwards. *Nature*, **401**, 505–508.

104 Nishikawa, S., Homma, K., Komori, Y., Iwaki, M., Wazawa, T., Iwane, A.H., Saito, J., Ikebe, R., Katayama, E., Yanagida, T. and Ikebe, M. (2002) Class VI myosin moves processively along actin filaments backward with large steps. *Biochem. Biophys. Res. Commun.*, **290**, 311–317.

105 Park, H., Li, A., Chen, L.Q., Houdusse, A., Selvin, P.R. and Sweeney, H.L. (2007) The unique inset at the end of the myosin VI motor is the sole determinant of directionality. *Proc. Natl Acad. Sci. USA*, **104**, 778–783.

106 Rock, R.S., Rice, S.E., Wells, A.L., Purcell, T.J., Spudich, J.A. and Sweeney, H.L. (2001) Myosin VI is a processive motor with a large step size. *Proc. Natl Acad. Sci. USA*, **98**, 13655–13659.

107 Buss, F., Spudich, G. and Kendrick-Jones, J. (2004) Myosin VI: cellular functions and motor properties. *Annu. Rev. Cell. Dev. Biol.*, **20**, 649–676.

108 Park, H., Ramamurthy, B., Travaglia, M., Safer, D., Chen, L.Q., Franzini-

Armstrong, C., Selvin, P.R. and Sweeney, H.L. (2006) Full-length myosin VI dimerizes and moves processively along actin filaments upon monomer clustering. *Mol. Cell.*, **21**, 331–336.

109 Iwaki, M., Tanaka, H., Iwane, A.H., Katayama, E., Ikebe, M. and Yanagida, T. (2006) Cargo-binding makes a wild-type single-headed myosin-VI move processively. *Biophys. J.*, **90**, 3643–3652.

110 Yildiz, A., Park, H., Safer, D., Yang, Z., Chen, L.Q., Selvin, P.R. and Sweeney, H.L. (2004) Myosin VI steps via a hand-over-hand mechanism with its lever arm undergoing fluctuations when attached to actin. *J. Biol. Chem.*, **279**, 37223–37226.

111 Balci, H., Ha, T., Sweeney, H.L. and Selvin, P.R. (2005) Interhead distance measurements in myosin VI via SHRImP support a simplified hand-over-hand model. *Biophys. J.*, **89**, 413–417.

112 Bryant, Z., Altman, D. and Spudich, J.A. (2007) The power stroke of myosin VI and the basis of reverse directionality. *Proc. Natl Acad. Sci. USA*, **104**, 772–777.

113 Ménétrey, J., Bahloul, A., Wells, A.L., Yengo, C.M., Morris, C.A., Sweeney, H.L. and Houdusse, A. (2005) The structure of the myosin VI motor reveals the mechanism of directionality reversal. *Nature*, **435**, 779–785.

114 Altman, D., Sweeney, H.L. and Spudich, J.A. (2004) The mechanism of myosin VI translocation and its load-induced anchoring. *Cell*, **116**, 737–749.

115 Sweeney, H.L., Park, H., Zong, A.B., Yang, Z., Selvin, P.R. and Rosenfeld, S.S. (2007) How myosin VI coordinates its heads during processive movement. *EMBO J.* **26**, 2682–2692.

116 Hirokawa, N. and Takemura, R. (2004) Molecular motors in neuronal development, intracellular transport and diseases. *Curr. Opin. Neurobiol.*, **14**, 564–573.

117 Vale, R.D., Funatsu, T., Pierce, D.W., Romberg, L., Harada, Y. and Yanagida, T. (1996) Direct observation of single kinesin molecules moving along microtubules. *Nature*, **380**, 451–453.

118 Block, S.M., Goldstein, L.S.B., and Schnapp, B.J. (1990) Bead movement by single kinesin molecules studied with optical tweezers. *Nature*, **348**, 348–352.

119 Svoboda, K., Schmidt, C.F., Schnapp, B.J. and Block, S.M. (1993) Direct observation of kinesin stepping by optical trapping interferometry. *Nature*, **365**, 721–727.

120 Yildiz, A., Tomishige, M., Vale, R.D., and Selvin, P.R. (2004) Kinesin walks hand-over-hand. *Science*, **303**, 676–678.

121 Block, S.M. (1998) Kinesin: what gives? *Cell*, **93**, 5–8.

122 Rice, S., Lin, A.W., Safer, D., Hart, C.L., Naber, N., Carragher, B.O., Cain, S.M., Pechatnikova, E., Wilson-Kubalek, E.M., Whittaker, M., Pate, E., Cooke, R., Taylor, E.W., Milligan, R.A. and Vale, R.D. (1999) A structural change in the kinesin motor protein that drives motility. *Nature*, **402**, 778–784.

123 Tomishige, M., Stuurman, N. and Vale, R.D. (2006) Single-molecule observations of neck linker conformational changes in the kinesin motor protein. *Nat. Struct. Mol. Biol.*, **13**, 887–894.

124 Asenjo, A.B., Weinberg, Y. and Sosa, H. (2006) Nucleotide binding and hydrolysis induces disorder–order transition in the kinesin neck-linker region. *Nat. Struct. Mol. Biol.*, **13**, 648–654.

125 Rice, S., Cui, Y., Sindelar, C., Naber, N., Matuska, M., Vale, R. and Cooke, R. (2003) Thermodynamic properties of the kinesin neck-region docking to the catalytic core. *Biophys. J.*, **84**, 1844–1854.

126 Schief, W.R. and Howard, J. (2001) Conformational changes during kinesin motility. *Curr. Opin. Cell. Biol.*, **13**, 19–28.

127 Uemura, S. and Ishiwata, S. (2003) Loading direction regulates the affinity of ADP for kinesin. *Nat. Struct. Biol.*, **10**, 308–311.

128 Rosenfeld, S.S., Fordyce, P.M., Jefferson, G.M., King, P.H. and Block, S.M. (2003)

Stepping and stretching. How kinesin uses internal strain to walk processively. *J. Biol. Chem.*, **278**, 18550–18556.

129 Crevel, I.M.-T.C., Nyitrai, M., Alonso, M.C., Weiss, S., Geeves, M.A. and Cross, R.A. (2004) What kinesin does at roadblocks: the coordination mechanism for molecular walking. *EMBO J.* **23**, 23–32.

130 Klumpp, L.M., Hoenger, A. and Gilbert, S.P. (2004) Kinesin's second step. *Proc. Natl Acad. Sci. USA*, **101**, 3444–3449.

131 Guydosh, N.R. and Block, S.M. (2006) Backsteps induced by nucleotide analogs suggest the front head of kinesin is gated by strain. *Proc. Natl Acad. Sci. USA*, **103**, 8054–8059.

132 Higuchi, H. and Endow, S.A. (2002) Directionality and processivity of molecular motors. *Curr. Opin. Cell. Biol.*, **14**, 50–57.

133 Skowronek, K.J., Kocik, E. and Kasprzak, A.A. (2007) Subunits interactions in kinesin motors. *Eur. J. Cell. Biol.*, **86**, 549–568.

134 Kapitein, L.C., Peterman, E.J.G., Kwok, B.H., Kim, J.H., Kapoor, T.M. and Schmidt, C.F. (2005) The bipolar mitotic kinesin Eg5 moves on both microtubules that it crosslinks. *Nature*, **435**, 114–118.

135 Valentine, M.T., Fordyce, P.M., Krzysiak, T.C., Gilbert, S.P. and Block, S.M. (2006) Individual dimers of the mitotic kinesin motor Eg5 step processively and support substantial loads *in vitro*. *Nat. Cell Biol.*, **8**, 470–476.

136 Tan, D., Asenjo, A.B., Mennella, V., Sharp, D.J. and Sosa, H. (2006) Kinesin-13s form rings around microtubules. *J. Cell. Biol.*, **175**, 25–31.

137 Helenius, J., Brouhard, G., Kalaidzidis, Y., Diez, S. and Howard, J. (2006) The depolymerizing kinesin MCAK uses lattice diffusion to rapidly target microtubule ends. *Nature*, **441**, 115–119.

138 Berliner, E., Young, E.C., Anderson, K., Mahtani, H.K. and Gelles, J. (1995) Failure of a single-headed kinesin to track parallel to microtubule protofilaments. *Nature*, **373**, 718–721.

139 Howard, J., Hudspeth, A.J. and Vale, R.D. (1989) Movement of microtubules by single kinesin molecules. *Nature*, **342**, 154–158.

140 Okada, Y., Higuchi, H. and Hirokawa, N. (2003) Processivity of the single-headed kinesin KIF1A through biased binding to tubulin. *Nature*, **424**, 574–577.

141 Tomishige, M., Klopfenstein, D.R. and Vale, R.D. (2002) Conversion of Unc104/KIF1A kinesin into a processive motor after dimerization. *Science*, **297**, 2263–2267.

142 Nitta, R., Kikkawa, M., Okada, Y. and Hirokawa, N. (2004) KIF1A alternately uses two loops to bind microtubules. *Science*, **305**, 678–683.

143 Block, S.M. (2007) Kinesin motor mechanics: binding, stepping, tracking, gating, and limping. *Biophys. J.*, **92**, 2986–2995.

144 Höök, P. and Vallee, R.B. (2006) The dynein family at a glance. *J. Cell. Sci.*, **119**, 4369–4371.

145 Chevalier-Larsen, E. and Holzbaur, E.L. (2006) Axonal transport and neurodegenerative disease. *Biochim. Biophys. Acta.*, **1762**, 1094–1108.

146 Bingham, J.B., King, S.J. and Schroer, T.A. (1998) Purification of dynactin and dynein from brain tissue. *Methods Enzymol.* **298**, 171–184.

147 Nishiura, M., Kon, T., Shiroguchi, K., Ohkura, R., Shima, T., Toyoshima, Y.Y. and Sutoh, K. (2004) A single-headed recombinant fragment of Dictyostelium cytoplasmic dynein can drive the robust sliding of microtubules. *J. Biol. Chem.*, **279**, 22799–22802.

148 Reck-Peterson, S.L., Yildiz, A., Carter, A.P., Gennerich, A., Zhang, N. and Vale, R.D. (2006) Single-molecule analysis of dynein processivity and stepping behavior. *Cell*, **126**, 335–348.

149 Burgess, S.A., Walker, M.L., Sakakibara, H., Knight, P.J. and Oiwa, K. (2003) Dynein structure and power stroke. *Nature*, **421**, 715–718.

150 Kon, T., Mogami, T., Ohkura, R., Nishiura, M. and Sutoh, K. (2005) ATP hydrolysis cycle-dependent tail motions in cytoplasmic dynein. *Nat. Struct. Mol. Biol.,* **12**, 513–519.

151 Wang, Z., Khan, S. and Sheetz, M.P. (1995) Single cytoplasmic dynein molecule movements: characterization and comparison with kinesin. *Biophys J.* **69**, 2011–2023.

152 Shima, T., Imamula, K., Kon, T., Ohkura, R. and Sutoh, K. (2006) Head-head coordination is required for the processive motion of cytoplasmic dynein, an AAA+ molecular motor. *J. Struct. Biol.,* **156**, 182–189.

153 Vale, R.D., Soll, D.R. and Gibbons, I.R. (1989) One-dimensional diffusion of microtubules bound to flagellar dynein. *Cell,* **59**, 915–925.

154 Mallik, R., Carter, B.C., Lex, S.A., King, S.J. and Gross, S.P. (2004) Cytoplasmic dynein functions as a gear in response to load. *Nature,* **427**, 649–652.

155 Toba, S., Watanabe, T.M., Yamaguchi-Okimoto, L., Toyoshima, Y.Y. and Higuchi, H. (2006) Overlapping hand-over-hand mechanism of single molecular motility of cytoplasmic dynein. *Proc. Natl Acad. Sci. USA,* **103**, 5741–5745.

156 Ling, S.C., Fahrner, P.S., Greenough, W.T. and Gelfand, V.I. (2004) Transport of Drosophila fragile X mental retardation protein-containing ribonucleoprotein granules by kinesin-1 and cytoplasmic dynein. *Proc. Natl Acad. Sci. USA,* **101**, 17428–17433.

157 He, Y., Francis, F., Myers, K.A., Yu, W., Black, M.M. and Baas, P.W. (2005) Role of cytoplasmic dynein in the axonal transport of microtubules and neurofilaments. *J. Cell. Biol.,* **168**, 697–703.

158 Gross, S.P. (2003) Dynactin: coordinating motors with opposite inclinations. *Curr. Biol.,* **13**, R320–R322.

159 Ross, J.L., Wallace, K., Shuman, H., Goldman, Y.E. and Holzbaur, E.L.F. (2006) Processive bidirectional motion of dynein-dynactin complexes *in vitro. Nat. Cell. Biol.,* **8**, 562–570.

160 Ali, M.Y., Krementsova, E.B., Kennedy, G.G., Mahaffy, R., Pollard, T.D., Trybus, K.M. and Warshaw, D.M. (2007) Myosin Va maneuvers through actin intersections and diffuses along microtubules. *Proc. Natl Acad. Sci. USA,* **104**, 4332–4336.

161 Kural, C., Kim, H., Syed, S., Goshima, G., Gelfand, V.I. and Selvin, P.R. (2005) Kinesin and dynein move a peroxisome *in vivo*: a tug-of-war or coordinated movement? *Science,* **308**, 1469–1472.

162 Nan, X., Sims, P.A., Chen, P. and Xie, X.S. (2005) Observation of individual microtubule motor steps in living cells with endocytosed quantum dots. *J. Phys. Chem. B.,* **109**, 24220–24224.

163 Watanabe, T.M. and Higuchi, H. (2007) Stepwise movements in vesicle transport of HER2 by motor proteins in living cells. *Biophys. J.,* **92**, 4109–4120.

164 Kural, C., Serpinskaya, A.S., Chou, Y.H., Goldman, R.D., Gelfand, V.I. and Selvin, P.R. (2007) Tracking melanosomes inside a cell to study molecular motors and their interaction. *Proc. Natl Acad. Sci. USA,* **104**, 5378–5382.

165 Leduc, C., Ruhnow, F., Howard, J. and Diez, S. (2007) Detection of fractional steps in cargo movement by the collective operation of kinesin-1 motors. *Proc. Natl Acad. Sci. USA,* **104**, 10847–10852.

166 Cai, D., Verhey, K.J. and Meyhofer, E. (2007) Tracking single kinesin molecules in the cytoplasm of mammalian cells. *Biophys. J.,* **92**, 4137–4144.

167 Sako, Y. (2006) Imaging single molecules in living cells for systems biology. *Mol. Syst. Biol.,* **2**, 56.

168 Yu, J., Xiao, J., Ren, X., Lao, K. and Xie, X.S. (2006) Probing gene expression in live cells, one protein molecule at a time. *Science,* **311**, 1600–1603.

169 Xie, X.S., Yu, J. and Yang, W.Y. (2006) Living cells as test tubes. *Science,* **312**, 228–230.

170 Knemeyer, J.P., Herten, D.P. and Sauer, M. (2003) Detection and identification of single molecules in living cells using spectrally resolved fluorescence lifetime imaging microscopy. *Anal. Chem.*, **75**, 2147–2153.

171 Hofmann, M., Eggeling, C., Jakobs, S. and Hell, S.W. (2005) Breaking the diffraction barrier in fluorescence microscopy at low light intensities by using reversibly photoswitchable proteins. *Proc. Natl Acad. Sci. USA*, **102**, 17565–17569.

172 Rust, M.J., Bates, M. and Zhuang, X. (2006) Sub-diffraction-limit imaging by stochastic optical reconstruction microscopy (STORM). *Nat. Methods*, **3**, 793–795.

173 Miyawaki, A. (2003) Fluorescence imaging of physiological activity in complex systems using GFP-based probes. *Curr. Opin. Neurobiol.*, **13**, 591–596.

174 Patterson, G.H. and Lippincott-Schwartz, J. (2004) Selective photolabeling of proteins using photoactivatable GFP. *Methods*, **32**, 445–450.

175 Lukyanov, K.A., Chudakov, D.M., Lukyanov, S. and Verkhusha, V.V. (2005) Photoactivatable fluorescent proteins. *Nat. Rev. Mol. Cell. Biol.*, **6**, 885–891.

176 Bates, M., Blosser, T.R. and Zhuang, X. (2005) Short-range spectroscopic ruler based on a single-molecule optical switch. *Phys. Rev. Lett.*, **94**, 108101.

177 Sakata, T., Yan, Y. and Marriott, G. (2005) Family of site-selective molecular optical switches. *J. Org. Chem.*, **70**, 2009–2013.

178 Chudakov, D.M., Chepurnykh, T.V., Belousov, V.V., Lukyanov, S. and Lukyanov, K.A. (2006) Fast and precise protein tracking using repeated reversible photoactivation. *Traffic*, **7**, 1304–1310.

179 Betzig, E., Patterson, G.H., Sougrat, R., Lindwasser, O.W., Olenych, S., Bonifacino, J.S., Davidson, M.W., Lippincott-Schwartz, J. and Hess, H.F. (2006) Imaging intracellular fluorescent proteins at nanometer resolution. *Science*, **313**, 1642–1645.

180 Gustafsson, M.G.L. (2005) Nonlinear structured-illumination microscopy: wide-field fluorescence imaging with theoretically unlimited resolution. *Proc. Natl Acad. Sci. USA*, **102**, 13081–13086.

181 Hell, S.W. (2007) Far-field optical nanoscopy. *Science*, **316**, 1153–1158.

4
Ion Channels

Toru Ide, Minako Hirano, and Yuko Takeuchi

4.1
Introduction

Single-molecule imaging techniques are very powerful tools for investigating biomolecules. These techniques have been applied to study various types of molecules and have revealed many novel properties. Funatsu *et al.* observed, for the first time, interaction between a single myosin molecule and single ATP molecules using a TIRF (total internal reflection fluorescence) microscope [1]. They combined single molecule imaging techniques with single molecule manipulation techniques using a scanning probe or optical tweezers observing important properties of motor proteins not shown by multi-molecular experiments [2]. These techniques, which were initially developed to study motility of the acto-myosin system, have spread rapidly through many fields. Now the motion of other types of motor proteins, such as kinesin and F_0,F_1-ATPase, are routinely measured using these techniques.

In contrast to water-soluble proteins, application of these techniques to ion-channel proteins is not as advanced because it is much more difficult to image single ion-channels while recording their functions simultaneously. This is due to the fact that, unlike most water-soluble proteins, it is difficult to maintain channel function with isolated ion-channel proteins. Lipid molecules are indispensable for maintaining the channel activity. This means that recording channel function requires the addition of a large number of lipid molecules that may produce optical noise. On the other hand, experimental techniques to investigate single ion-channel function, namely the single channel recording techniques, are well established compared to the technologies applied to measure functions of other types of proteins. Measurement of motor protein motility, for example, requires highly specialized equipment, such as an apparatus to measure small displacements of beads or a special kind of atomic force microscope. These problems are minimal when investigating single ion-channel function, because of the availability of commercial equipment.

Our present purpose is to develop an experimental apparatus for simultaneous optical and electrical observation of single ligand bindings to a single channel protein, i.e. a fusion of single molecule imaging and single channel current recording. Edelstein *et al.* showed that ligand binding events to ligand-gated ion-channels are more complex than ionic events, due to multiple interconversions between different conformational states at the same degree of ligation [3]. Compared to single channel current recording, which directly measures current through a channel pore, single ligand binding events have been inferred indirectly from multi-molecule experiments. There has been a strong need for technology that allows simultaneous measurement of single binding events and single channel current fluctuations, which would enable us to establish a new field of "single molecule pharmacology". In this chapter, we will introduce some technologies developed for this purpose.

4.2
Artificial Bilayers

Lipid bilayers are indispensable to maintain channel activity and measure the ionic current. So far, most single channel imaging experiments have made use of artificial bilayer membranes into which fluorescently labeled channels are incorporated. In the 1990s single fluorescent particles in artificial bilayers were detected for the first time. Thermal diffusion of individual lipid molecules was directly observed in solid supported bilayers [4] and self-standing bilayers [5, 6]. In both types of membrane, these results indicate a rapid lateral diffusion of lipids, which is thought to relate to the function of channel proteins. Supported bilayers are durable and suitable for long-term observation but not for single channel current recording because to date it has not been possible to block leakage from the edge of the membrane and from defects in the membranes which are caused mainly by unevenness along the surface of the solid supports. In contrast, self-standing bilayers in an aqueous environment or on a hydrophilic gel are suitable for single channel recording because they have little leakage current although they are very fragile and difficult to handle.

4.2.1
Solid Supported Bilayers

As described above, we can directly see lateral motion of single fluorescent particles in the membranes if the membranes are sufficiently large. Furthermore, interactions between receptor molecules and ligands can be detected at the single molecule level using a TIRF microscope. However, individual fluorophores in solution cannot be visualized by video rate TIRF recordings but do increase background intensity because of their rapid three-dimensional thermal diffusion. They are detected as single spots only when bound to the receptor in membranes. Using this technique, we studied interaction between a single cardiac ryanodine receptor channel (RyR2) and single ryanodine (Ry) molecules. RyR2 is the calcium release channel from the cardiac sarcoplasmic reticulum membrane, which is named after its agonist, ryanodine. Several facts regarding Ry binding to RyR are known: (1) Ry binds to an open form of RyR from the cytoplasmic side; (2) RyR is a homotetramer, but contains only one site with a high affinity for Ry; (3) there is no high affinity binding at high pCa (intracellular Ca^{2+} concentration less than 10^{-7} M); (4) the binding time constant of Ry to RyR has been obtained using biochemical methods (161 min).

The cardiac sarcoplasmic reticulum vesicles containing RyR2 were incubated with anti-RyR antibodies labeled with a Cy5-dye (dye/protein = 1). A single RyR2 was labeled with at most four fluorophores because it contains four subunits. Ry was labeled with Bodipy-FL. Glass supported bilayer membranes were made by the method of vesicle fusion. Vesicle suspension containing fluorescently labeled channels was placed on the glass support. When the calcium concentration in the solution was adjusted to 1 mM, the vesicles fused and lipid bilayers were formed on the glass surface. By controlling vesicle concentration, we were able to produce bilayers in a very small area within which channel proteins were confined. Thus it was possible to neglect the effect of lateral diffusion of channels.

As shown in Figure 4.1a, we were able to directly observe the interaction between fluorescent Ry (BodipyFL-Ry) and RyR (Cy5-RyR2) at the single molecule level by adding a solution that contained fluorescent Rys to RyR immobilized on a glass. Figures 4.1b–e show the images of Cy5–RyRs immobilized on glass and Bodipy FL–Rys bound to the channels. Figure 4.1b and Figure 4.1d are images of Cy5–RyRs, and Figure 4.1c and Figure 4.1e show BodipyFL–Rys. Figure 4.1b and Figure 4.1c are fluorescence images excited by a red laser (633 nm) while Figure 4.1d and Figure 4.1e were taken during excitation with both a green (532 nm) and a red (633 nm) laser. These figures show that there were three RyRs on the glass (Figures 4.1b and 4.1d) and at least two Rys bound to two out of three channels (Figure 4.1e). We determined binding durations of individual Rys. Figure 4.1f shows the binding duration histogram measured in the presence of 10 nM Ry and 10 μM Ca^{2+}, which highly activates the channel. The histogram can be fitted by two exponential functions with the time constants of $\tau_1 = 445$ ms and $\tau_2 = 3294$ ms. Table 4.1 summarizes the binding durations at pCa5 and pCa3, which shows the single RyR2 channel and ryanodine binding dynamics for the first time by using single molecule imaging techniques. In the active state (10^{-5} M calcium) ryanodine showed both long (3–6 s) and short (300–500 μs) binding durations to RyR while it only showed short durations at pCa3. These bindings might correspond to high and low affinity binding of ryanodine. This method is very simple and has a high temporal resolution, allowing sub-millisecond resolution using faster detection systems, such as an avalanche photo-diode (APD).

4.2.2
Self-Standing Bilayers

The artificial planar bilayer technique has been used to study many types of channels and is a potential tool to study ionic channels having revealed both pharmacological and dynamic behaviors [7]. For our purposes, bilayers were formed horizontally so single fluorescent particles in the membrane could be imaged using a TIRF microscope [6, 8]. To prevent the vertical movement of the membrane and possible breakage of the membrane by touching the glass bottom of the chamber,

ryanodine receptor   ryanodine

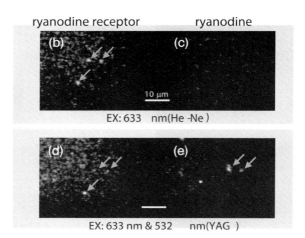

EX: 633   nm(He -Ne)

EX: 633 nm & 532   nm(YAG)

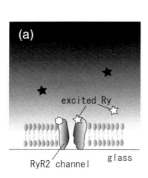

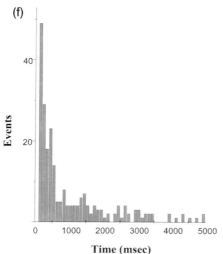

Figure 4.1 Single molecule observations in Ry–RyR2 binding. (a) Schematic representation of the experiment. The receptor channels were immobilized on the glass. Fluorescent ryanodine molecules were not excited (★) when they were distant from the glass surface. Excitation occurs at a distance of less than 200 nm. However, they cannot be visualized by video rate recording because of their rapid three-dimensional thermal motion. They were visualized as bright spots only when bound to the channels. (b–e) Fluorescence images of the same field. The upper panel (b, c) shows images recorded by excitation with a red laser (633 nm). The lower panel (d, e) show excitation with both a green (532 nm) and red (633 nm) laser. (b) and (d) show images of Cy5–RyR2 while (c) and (e) represent BodipyFL–Ry. (f) Histogram of Ry–RyR2 binding duration measured in the presence of 10 nM Ry and 10 μM Ca^{2+}.

the base of the chamber was coated with a thin layer of agarose. The bilayers were positioned so as to make contact with the agarose layer. The TIRF microscope allowed us to visualize single fluorophores in the membrane by reducing the background noise and also to detect single fluorescent molecules in the

Table 4.1 Binding duration of ryanodine to RyR.

			pCa 5			
	Liposome	pCa 3				
Ryanodine concentration	10 nM	10 nM	1nM	10nM	100nM	1M
τ1	254	468	464	445	370	
τ2			5690	3294	3657	≫120 000

solution only when they came very close to the membrane and bound to the receptors.

Figure 4.2 shows the horizontal bilayer apparatus we developed for single channel imaging [8]. Figure 4.2a shows the bilayer apparatus consisting of two chambers. The upper chamber is made from a glass tube which can be moved using a piezo micromanipulator. A thin plastic film with a small pore in the center is attached to the bottom of the upper chamber. The pore on the film was made on a projection by the same method used to make an aperture for a conventional vertical bilayer. The lower chamber consists of a glass dish with a hole in the bottom over which a coverslip is fixed with adhesive just prior to the experiment. The upper and lower solution chambers are connected to a patch-clamp amplifier through Ag-AgCl electrodes. The coverslips were washed thoroughly, coated with agarose by painting with a warmed solution of 0.2–0.5% agarose in water, and then air-dried at room temperature. There was no special attachment between the agarose and the glass surface. The dried coverslip was fixed over the hole in the bottom of the chamber with adhesive.

The membrane was formed as follows: a thick membrane was formed across the hole by adding a small amount of lipid solution (20 mg lipid/ml *n*-decane) to the

(a)

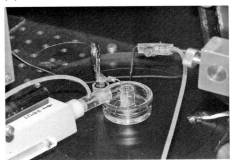

(b)

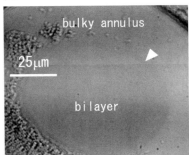

Figure 4.2 (a) The bilayer apparatus. This apparatus consists of two chambers; the upper chamber is made from a glass tube. A thin (0.1–0.2 mm thick) polypropylene film is attached to the bottom of the upper chamber. This film has a small pore in the center across which the bilayers were formed. The bilayer membrane formed on the agarose consisted of two parts: the bilayer membrane in the center and the surrounding thick annulus. (b) An aerial view of the membrane taken under a bright field microscope. The contrast was slightly enhanced with a CCD camera.

bottom of the chamber. The upper chamber was then moved in a downward direction until the membrane came into contact with the agarose-coated coverslip. By slightly increasing the pressure in the upper chamber, the membrane began to thin out and finally the center of the membrane became a bilayer (Figure 4.2b). This process of thinning was facilitated by applying membrane voltages (±100 mV) and the bilayer was then observed using a normal bright field microscope.

A bilayer membrane can also be formed at the tip of a glass pipette and single fluorophores in the membrane can be detected optically with TIRF. Because glass pipettes generate much lower auto-fluorescence and scatter excitation light much less than plastic apertures, they are suitable for single channel imaging. However, they are not convenient for measuring drug binding to channels since perfusion inside the pipettes presents difficulties.

4.3
Simultaneous Optical and Electrical Recording of the Single BK-Channels

The techniques described above were applied to achieve an image of a channel protein [9]. Figure 4.3 shows an example of simultaneous optical and electrical single channel recordings with a self-standing bilayer in which a Ca-activated K-channel was fluorescently labeled and incorporated into a bilayer by the vesicle fusion technique. Figure 4.3A illustrates the strategy used to incorporate the fluorescently labeled channel into the artificial membrane. The vesicles prepared from bovine trachea were incubated with monoclonal anti-BK-channel antibodies, which bind specifically to the BK-channel (Figure 4.3B). After removal of unreacted antibodies by ultracentrifugation, the channel in the vesicular membrane was transferred into the bilayer by vesicle-fusion. In order to induce rapid fusion of the vesicle into the limited area of the bilayer, vesicles that had been osmotically loaded were puffed through a fine glass pipette.

Figure 4.4 shows the results of the experiments. The channel was labeled with Cy5-dye molecules and applied to the upper side of the bilayer. Channel proteins cannot be reconstituted into bilayer membranes simply by adding them to the aqueous solution. This is in direct contrast to small amphipathic peptides that spontaneously become incorporated into bilayers. We utilized vesicle-fusion techniques to reconstitute channel proteins into the artificial bilayers. In a specified area of the membrane and within a short time-frame, only one channel protein is expected to be incorporated into the membrane. However, to observe the moment the channel becomes incorporated into the artificial membrane, more sophisticated techniques are required. The vesicles were added directly to the bilayers through a fine capillary tube. This was immediately followed by an instantaneous increase in ionic concentration in the vicinity of the membrane by injecting a small volume of solution of high salt concentration. The optimum conditions for successfully incorporating the channel into the membrane were not the same for all channel types. In fact, even for proteins of the same type, the conditions changed from one protein preparation to the next. Thus, the experimental conditions such as the amount of protein to be used

A

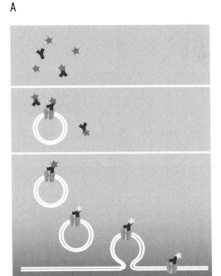

B

SDS-PAGE and Western analysis of BK-channel

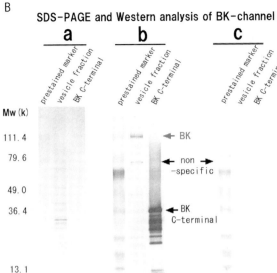

Figure 4.3 (A) Schematic diagram showing the incorporation of channel proteins into a membrane. The anti-BK channel antibodies were labeled with Cy5 dye molecules. The vesicles were incubated with labeled antibodies. After removal of unreacted antibodies by ultracentrifugation, the vesicles were added directly to the bilayer through a fine glass capillary. The channel protein was transferred into the planar bilayer by fusion between the vesicular membrane and the planar bilayer. (B) The specificity of anti-BK antibody investigated using SDS-PAGE and Western blotting. Column (a) shows CBB staining of SDS-PAGE (4~20%). Columns (b) and (c) represent Western blots produced with anti-BK antibody and secondary antibody respectively as the negative control. For the positive control, the fusion protein from *Schistosoma japonicum* glutathione-S-transferase (GST) and the C-terminal part of the mouse α subunit from Alomone labs (Jerusalem, Israel) were used in lane 3 in each column. The 37-kDa control fusion protein only stained intensely when treated with anti-BK antibody. The 70-kDa protein which is stained in both in b and c indicates that this molecule is a non-specific protein. In the sarcolemmmal vesicle fraction, the 120-kDa protein was stained intensely when exposed to the anti-BK antibody (column b, lane 2). A prote in of this size corresponds with the expected MW (125 kDa) of BK [10]. As shown in column a, the sarcolemmal vesicle fraction contained very little 125-kDa BK protein. These results confirm the specificity of this anti-BK antibody against BK in the vesicle fraction.

and the most suitable ionic concentrations had to be determined for every vesicle preparation.

A bright spot appeared just before the current across the membrane began to fluctuate as shown in Figure 4.4A. After incorporation, the spot moved rapidly within the bilayer as shown in Figure 4.4B. The diffusion constant, D, was calculated to be $3.0 \pm 1.5 \times 10^{-8} \, \text{cm}^2/\text{s}$ ($n = 5$) which agrees well with the value for a channel moving freely in a membrane. This result shows that the fluorescently labeled channel proteins in the vesicular membrane moved thermally in the solution and when in close proximity to the horizontal planar bilayer became incorporated into the bilayer through vesicle fusion. In other words, the bright spot shown in Figure 4.4

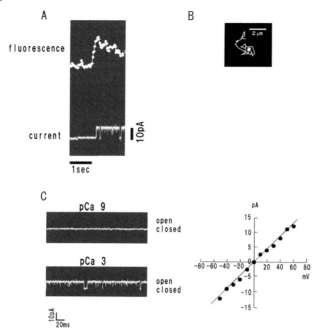

Figure 4.4 Simultaneous electrical and optical recordings of the single BK-channel. (A) The BK-channel was labeled with Cy5-dye molecules and incorporated into the bilayer using vesicle-fusion techniques. The top trace shows the fluorescence intensity and the bottom trace shows the simultaneous recording of the single channel current. The decrease in fluorescence intensity represents the photobleaching effect of the Cy5-dyes. The channel had been in the open state before it was transferred into the planar bilayer because the solution contained 1 mM $CaCl_2$. The number of Cy5-dyes attached to the channel was estimated to be approximately 25 from the fluorescence intensity. (B) Thermal motion of the channel protein in the membrane. The channel protein moved thermally in the membrane with a decrease in its fluorescence intensity as a result of photobleaching. The diffusion constant was determined to be $4.0 \times 10^{-8}\,cm^2/s$ corresponding well to the predicted values of small particles moving freely in the membrane. This indicates that there was no strong interaction between the channel and the agarose layer. (C) Single channel recording of the BK-channel incorporated into the horizontal planar bilayer. The traces show single channel fluctuations taken from the same bilayer shown in A and B (left). The free calcium concentration in the upper chamber was controlled by adding EGTA to the solution. Each trace shows the current fluctuation at pCa 9 and pCa 3, respectively. The single channel conductance was determined to be 225 pS (right).

corresponded to the channel protein itself. In this study, we did not observe that more than two channel proteins were incorporated into the membrane at the same time since the density of the BK-channel in the tracheal membrane was very low [10] and each vesicle contained either no channel protein or only one.

Typical current traces recorded with the horizontal bilayer are shown in Figure 4.4C. The open probability, P_O, of the BK-channel was <0.01 at pCa 9 and >0.95 at pCa3, respectively. The single channel conductance was determined to be $229 \pm 8\,pS$ ($n = 5$). The channel properties elucidated in this study and shown in

Figure 4.4 were consistent with the results obtained in patch-clamp or conventional planar bilayer experiments [10], showing that it is possible to record the natural properties of channel proteins with the method described herein.

We have developed a novel method for simultaneously recording the optical and electrical properties of single ion-channels. This method is so sensitive that we can image single fluorphores in the membrane and should prove applicable to a wide variety of channel proteins. It is also possible, with this method, to observe single fluorescent molecules in solution when they are very close to the membrane. This means that we can directly observe the interaction between a single channel protein and its ligand molecules labeled with a fluorescent dye as we did in Ry–RyR coupling using solid supported bilayers. Such observations will greatly increase our understanding of the dynamics of ligand–receptor interaction and the activation mechanisms at the single-molecule level in ligand-activated receptor channels such as nACh and glutamate receptor channels.

4.4
Detection of Channel Conformational Change

Brisenko *et al.* observed conformational change in single gramicidin channels using the same type of membranes [11]. They reported an approach for simultaneous fluorescence imaging and electrical recording of single gramicidin channels. Fluorescently labeled (Cy3 and Cy5) gramicidin derivatives were imaged at the single-molecule level using far-field illumination and cooled CCD camera detection. Simultaneous electrical recording detected gramicidin homodimer (Cy3/Cy3, Cy5/Cy5) and heterodimer (Cy3/Cy5) channels. Heterodimer formation was observed optically by the appearance of a fluorescence resonance energy transfer (FRET) signal (irradiation of Cy3, detection of Cy5). The number of FRET signals increased with increasing channel activity. In numerous cases the appearance of a FRET signal was observed to correlate with a channel opening event that could be detected electrically.

Lu and his colleagues reported their studies on the conformational changes in the gramicidin channel using patch-clamp fluorescence microscopy, which simultaneously combines single-molecule fluorescence spectroscopy and single-channel current recordings using artificial membranes formed at the tip of glass pipettes [12]. By measuring single-pair fluorescence resonance energy transfer and fluorescence self-quenching from dye-labeled gramicidin channels, they showed that the efficiency of single-pair fluorescence resonance energy transfer and self-quenching is widely distributed, which reflects a broad distribution of conformations.

4.5
"Optical Patch-Clamping"

Demuro and Parker recently described an optical technique known as "optical patch-clamping" which uses total internal reflection fluorescence (TIRF) microscopy to obtain simultaneous and independent recordings from ion channels via imaging of single-channel Ca^{2+} flux [13]. Acetylcholine (ACh) receptor channels were expressed in Xenopus oocytes where single channel Ca^{2+} fluorescence transients were imaged using fluo-4 as the indicator. Consistent with their passage through the opening of individual nicotinic channels, fluorescent signals were seen only when a nicotinic agonist was present in the bathing solution, and were not observed in the presence of curare, and increased in frequency roughly with the second power of ACh concentration. The rise and fall times of fluorescence were as fast as 2 ms, providing a kinetic resolution that was sufficiently adequate to characterize channel gating kinetics.

4.6
Conclusion

Single molecule imaging of ion-channel proteins is still not as well developed as that for water-soluble proteins. Nevertheless, single molecule imaging has such potential that it is worthwhile developing techniques that can be applied to ion-channel proteins. Progress, although slow, has been made as demonstrated by overcoming lateral diffusion [14]. In this chapter we have described an application of single molecule imaging techniques to simultaneously observe the optical and electrical behavior of RyR. This is a very significant advance which will produce large volumes of information related to single channels.

References

1 Funatsu, T., Harada, Y., Tokunaga, M., Saito, K. and Yanagida, T. (1995) Imaging of single fluorescent molecules and individual ATP turnovers by single myosin molecules in aqueous solution. *Nature*, **374** (6522), 555–559.

2 Ishii, Y., Ishijima, A. and Yanagida, T. (2001) Single molecule nanomanipulation of biomolecules. *Trends Biotechnol.*, **19** (6), 211–216.

3 Edelstein, S.J., Schaad, O. and Changeux, J.P. (1997) Single binding versus single channel recordings: a new approach to study ionotropic receptors. *Biochemistry*, **36** (45), 13755–13760.

4 Schmidt, T., Schutz, G.J., Baumgartner, W., Gruber, H.J. and Schindler, H. (1996) Imaging of single molecule diffusion. *Proc. Natl Acad. Sci. USA*, **93** (7), 2926–2929.

5 Schutz, G.J., Schindler, H. and Schmidt, T. (1997) Single-molecule microscopy on model membranes reveals anomalous diffusion. *Biophys. J.*, **73** (2), 1073–1080.

6 Ide, T. and Yanagida, T. (1999) An artificial lipid bilayer formed on an agarose-coated glass for simultaneous electrical and optical measurement of single ion channels. *Biochem. Biophys. Res. Commun*, **265** (2), 595–599.

7 Miller, C. (ed.) (1986) *Ion Channel Reconstitution*, Plenum Press, New York.

8 Ide, T., Takeuchi, Y. and Yanagida, T. (2002) Development of an experimental apparatus for simultaneous observation of optical and electrical signals from single ion channels. *Single Mol.*, **3**, 33–42.

9 Ide, T., Takeuchi, Y. and Yanagida, T. (2002) Simultaneous optical and electrical

recording of a single ion-channel. *Jpn J. Physiol.*, **52** (5), 429–434.

10 Latorre, R. (1986) The large calcium-activated potassium channel, in *Ion Channel Reconstitution* (ed. C. Miller), Plenum Press, New York, pp. 431–467.

11 Borisenko, V., Lougheed, T., Hesse, J., Füreder-Kitzmüller, E., Fertig, N., Behrends, J. C., Woolley, G.A. and Schütz, G.J. (2003) Simultaneous optical and electrical recording of single gramicidin channels. *Biophys. J.*, **84** (1), 612–622.

12 Harms, G.S., Orr, G., Montal, M., Thrall, B.D., Colson, S.D. and Lu, H.P. (2003) Probing conformational changes of gramicidin ion channels by single-molecule patch-clamp fluorescence microscopy. *Biophys. J.*, **85** (3), 1826–1838.

13 Demuro, A. and Parker, I. (2005) "Optical patch-clamping": single-channel recording by imaging Ca2+ flux through individual muscle acetylcholine receptor channels. *J. Gen. Physiol.*, **126** (3), 179–192.

14 Ichikawa, T., Aoki, T., Takeuchi, Y., Yanagida, T. andIde, T. (2006) Immobilizing single lipid and channel molecules in artificial lipid bilayers with annexin A5. *Langmuir*, **22** (14), 6302–6307.

5
Signal Transduction across the Plasma Membrane

Masahiro Ueda, Tatsuo Shibata, and Yasushi Sako

5.1
Introduction

The plasma membrane is the entrance for various extracellular signals into the cytoplasm. This chapter describes the single-molecule imaging approach toward understanding signal transduction across the plasma membrane. Because the plasma membrane makes a good target for single-molecule microscopy, the behavior of cell signaling molecules in the plasma membrane has been studied in living cells using single-molecule imaging [1–3]. This chapter deals with the signaling of epidermal growth factor receptor (EGFR) in mammalian cells and cAMP receptor 1 (cAR1) in *Dictyostelium* cells. EGFR and cAR1 belong to the receptor protein tyrosine kinases (RTKs) and the trimeric G-protein coupled receptor (GPCRs) superfamilies, respectively. RTKs and GPCRs are two large superfamilies of membrane receptors situated on the plasma membrane which excite complicated reaction systems inside cells. Analysis of the reaction systems requires quantitative information in both time and space. Superior quantitative data with spatiotemporal resolution in single-molecule imaging measurements is ideal to satisfy this requirement. Hence, an aim of the studies reported in this chapter is to understand the behavior of complicated reaction networks inside cells based on unitary protein reactions as visualized in single molecules.

5.2
Signal Transduction Mediated by Receptor Tyrosine Kinase

RTK is a large super family of membrane receptors found on the cell surface [4]. The role of most members of the RTKs involves signal transduction for cell proliferation and differentiation. A typical RTK consists of a single membrane-spanning protein with a ligand-binding domain at the extracellular side and a tyrosine kinase domain at the cytoplasmic side. Upon binding with the ligand, the kinase activity of RTK is

Single Molecule Dynamics in Life Science. Edited by T. Yanagida and Y. Ishii
Copyright © 2009 WILEY-VCH Verlag GmbH & Co. KGaA, Weinheim
ISBN: 978-3-527-31288-7

stimulated and several tyrosine residues of RTK in the cytoplasmic domain become phosphorylated. This tyrosine phosphorylation is critical for the signal transduction of RTKs as the phosphotyrosine residues provide the scaffolds for cytoplasmic proteins to signal downstream reactions.

Interactions between RTK molecules are indispensable for the transduction of signals across the plasma membrane. Having only one membrane-spanning domain, which is thought to form an α-helix, conformational changes of the RTK extracellular domain induced by ligand binding should have no influence on the structure of the cytoplasmic kinase domain in single molecules. In addition, it is usually thought that an RTK molecule cannot phosphorylate itself. Actually, ligand-bound RTKs form homodimers in specific structures which are known as "signaling dimmers". Signaling dimer formation changes the steric relationship between two cytoplasmic domains in the dimers to induce mutual phosphorlyation. In some cases, such as nerve growth factor (NGF), the ligand is a homodimer which crosslinks two receptor molecules to form a signaling dimer. In other cases, for example epidermal growth factor (EGF), two ligands contained in a signaling dimer do not interact with each other but stimulate an allosteric conformational change in the receptor molecules to induce formation of the signaling dimer [5].

Two types of RTKs are described in this section: EGF receptor (EGFR) and TrkA nerve growth factor receptor. Activation of EGFR is responsible for proliferation, morphological changes, chemotactic movements and carcinogenesis in various types of cells. Signals from NGF induce differentiation, chemotactic movements and survival of nerve cells. NGF recognizes two types of membrane receptors, TrkA and p75. Only TrkA belongs to the RTK superfamily.

5.3
Association between EGF and EGFR and Formation of the Signaling Dimers of EGFR

Association between EGF and EGFR and formation of the EGFR signaling dimers are the initial crucial steps in EGF signal transduction. Extensive studies have been carried out with regard to the association between EGF and EGFR showing that cells have only one type of receptor for EGF (EGFR, also called ErbB1) but, on the surface of living cells, there are multiple populations of binding sites which vary in their affinity and association and dissociation rates with EGF. On the other hand, clustering and predimerization have been suggested for at least part of the EGFR population on the cell surface. However, the relationship between such functional and structural information is largely unknown [6, 7].

We used EGF conjugated with tetramethylrhodamine (Rh-EGF) at the amino terminus to trace association processes between EGF and EGFR on living HeLa cells [8]. Modifications at the amino terminus do not inhibit the biological activities of EGF. HeLa cells were cultured on coverslips and Rh-EGF was applied under a fluorescent microscope. In order to observe the cells' apical surface, oblique angle illumination fluorescence microscopy was used. The numbers and intensities of individual fluorescent spots of Rh-EGF that appeared on the cell surface were

A

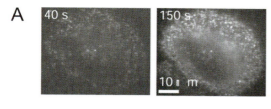

B

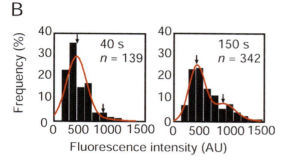

C

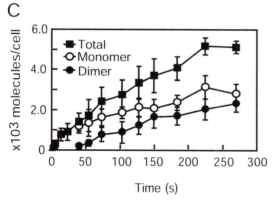

Figure 5.1 Single-molecule visualization of Rh-EGF bound on the surface of a living cell. (A) Rh-EGF was added to the extracellular medium of cultured HeLa cells. Images taken 40 and 150 s after the addition of Rh-EGF (0.5 nM final concentration) are shown. (B) Distribution of the fluorescence intensity of Rh-EGF spots bound to the cell shown in (A) at 40 and 150 s after the addition of Rh-EGF. The distributions were fitted to a sum of two Gaussian functions (red line). Arrows indicate the mean of the fractions containing one and two Rh-EGF molecules. $n =$ total number of spots. AU, arbitrary unit. (C) The total number (closed squares), monomers (open circles) and dimers (closed circles) of Rh-EGF bound to the cells were counted at the indicated times. The average and standard deviation of 10 cells are shown.

examined after the addition of Rh-EGF to the culture medium (Figure 5.1). The fluorescence intensity of single Rh-EGF molecules was determined from the step size of photobleaching observed under the same conditions. Early on, most of the fluorescent spots contained single EGF monomers, but the fraction of EGF dimers increased gradually with time. Thus single molecule visualization allowed

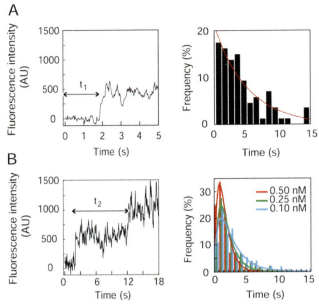

Figure 5.2 Kinetic analysis of Rh-EGF binding. (A) A typical change in the fluorescence intensity of Rh-EGF at the binding site is shown (left). At time 0, 0.5 nM Rh-EGF was added to the medium. A sudden step-like increase in the fluorescence intensity indicates binding of a single Rh-EGF molecule. A single exponential function (a red line; right) was fitted to the histogram showing the duration of binding (τ_1). The total number of events was 102. (B) A typical change in the fluorescence intensity at the site of the formation of an EGFR dimer is shown (left). Each step represents the first and second binding event of Rh-EGF to a dimeric binding site. Histograms of the duration (τ_2) are shown (right). The total number of events was 147–163. A function of tandem reaction (red, green and blue lines) was fitted to the histograms. AU, arbitrary unit.

quantification of the total number of EGF molecules bound on the cell surface as well as dividing the total number of molecules into monomeric or dimeric groups.

Unitary binding processes can be observed directly in single-molecule experiments (Figure 5.2). The sudden appearance of a fluorescent spot on the cell surface indicates that Rh-EGF from the solution has bound to a vacant EGFR molecule. When the same binding site contains two receptors, a second ligand from the solution will bind at a certain time after the first to produce a signaling dimer. Kinetic information concerning EGF/EGFR association can be derived from the lengths of time required for the first binding to take place ($\tau1$) and between the binding of the first and second ligand ($\tau2$) to the EGFR molecule. The time required for the first binding to take place can be described by a single exponential function which provides an association rate constant of $4.0 \times 10^8 \, \text{M}^{-1} \, \text{s}^{-1}$. On the other hand, the waiting time before the second binding can take place showed a peak suggesting that there is a kinetic intermediate. Conducting similar observations using different EGF concentrations, it was concluded that the first process was achieved by some type of structural change with a rate constant of $1.5 \, \text{s}^{-1}$ and that the second process involved the binding of the EGF molecule with a rate constant $2.0 \times 10^9 \, \text{M}^{-1}\text{s}^{-1}$. Practically all of the dimeric binding sites were formed by binding Rh-EGF from the solution. The association of two binding sites by lateral diffusion and collision along the plasma membrane was rarely observed [9].

A simple reaction network for the formation of EGFR signaling dimers from monomers and predimers of EGFR was constructed including the newly found kinetic intermediate (Figure 5.3A). In this network, the intermediate is formed by a conformational change of dimeric receptors after the binding of the first EGF molecule. Solutions of the coupled differential equations for this reaction network was obtained analytically and used to fit the experimentally observed time course for

A

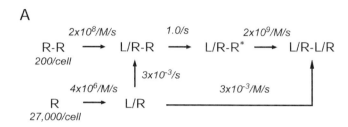

B

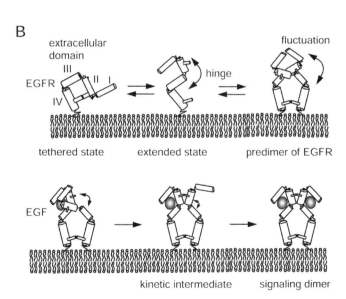

Figure 5.3 A model for the formation of signaling dimers of EGFR. (A) The simplest schemes for the formation of a signaling dimmer to explain the experimental results. L and R represent ligand (EGF) and receptor (EGFR), respectively. The model includes the novel intermediate L/R-R* found using single-molecule analysis. The best fit parameters obtained by this model are shown. (B) A dynamic conformational change in the predimer facilitates the formation of the signaling dimer composed of EGF/EGFR complexes. See text for details.

the monomeric and dimeric bindings of EGF (Figure 5.1C). Best fit values of the reaction parameters agreed very well with the results from direct observation (Figure 5.2) and were consistent with previous studies. Since this is the simplest model which takes into account both the presence of the reaction intermediate and is consistent with experimentally obtained kinetic parameters, it was concluded that this model is appropriate for the formation of signaling dimers of EGFR.

Our result can be fitted to recent X-ray crystallographic studies by assuming dynamic structural changes in EGFR. In crystal form, EGFR has two conformations, a tethered and an extended state. Only the extended state is thought to bind firmly to EGF and form signaling dimers [5]. Single-molecule experiments suggest a model for the formation of the EGFR signaling dimer as follows (Figure 5.3B). Most of the receptor molecules are in the monomeric tethered state-like structure, which is the slow binding site for EGF. However, a small percentage of receptor molecules form predimers in an extended-like structure that binds with EGF more rapidly than the monomer by a factor of about 100. EGF molecules selectively bind to the predimers of EGFR especially when their concentration is low. Binding of the first EGF molecule to the predimer induces an allosteric conformational change in the vacant binding site to make binding of the second EGF even more rapid. In this model, signal transduction of EGF is facilitated through receptor predimerization and positive cooperative ligand binding.

5.4
Amplification and Propagation of EGFR Activation

Phosphorylation (activation) of EGFR after formation of the signaling dimers was examined in single molecules [10]. Cells were incubated with Rh-EGF for 1 min and any unbound Rh-EGF was then washed out. After incubation for various periods of time, cells were fixed and activated EGFR was detected using the Fab' fragment of a monoclonal antibody which recognizes the active conformation of the cytoplasmic domain of EGFR after phosphorylation. The Fab' fragment was labeled with a green fluorophore Alexa 488 (Alx-Fab').

Rh-EGF and Alx-Fab' on the plasma membrane were visualized in single molecules. Since, unfortunately, Rh-EGF dissociated from the cell surface during fixation, binding of Rh-EGF before the fixation and Alx-Fab' after the fixation were compared between similarly treated but different cells. Both numbers of molecules and binding sites of Alx-Fab' increased with time after stimulation with Rh-EGF. At the peak of activation, the density of molecules and binding sites for Alx-Fab' were greater than those for Rh-EGF by a factor of 3.0 and 2.3, respectively, i.e. the EGF signal was amplified during the activation process of EGFR.

A semi-intact cell technique [11] was used for the simultaneous imaging of Rh-EGF and Alx-Fab' to study the process of amplification. Semi-intact cells were prepared by perforating the plasma membrane using the antibiotic streptolysin-O. Rh-EGF was applied to the semi-intact cells. Since most of the cytoplasm leaked through the pores, activation of EGFR did not take place at this stage. The cells were then washed and loaded with Alx-Fab' through the pores. Finally, ATP was added to initiate the reaction. Three types of the fluorescent spots were observed on the cell surface; 20% were Rh-EGF binding sites that were not colocalized with Alx-Fab', another 10% were Rh-EGF binding sites that were colocalized with Alx-Fab', and the other were attributable to Alx-Fab' without colocalization of Rh-EGF. The last population is responsible for amplifying EGF signals. Overall, the time-course of amplification of EGFR activation observed in semi-intact cells was similar to that observed in intact cells (Figure 5.4A).

The fluorescence intensities of EGF, activated receptors, and activated receptors co-localized with EGF were examined at different time points after the addition of ATP (Figure 5.4B). Cluster size distributions of EGF were not changed with time while activated receptors formed clusters whose size increased with time. This increase was not caused by the clustering of receptors bound with EGF since the distribution of EGF did not change. From the histograms representing activated receptors colocalized with EGF, it can be seen that clustering was more rapid and evident at the colocalization. This is probably because the binding sites of EGF are the leading spots for activation. Thus, secondary activated receptors without EGF binding formed clusters around the receptors primarily activated by EGF. Some of the secondarily activated receptors diffused out of the clusters.

Figure 5.4C shows a model for signal amplification using dynamic clustering and lateral mobility of receptors. Receptors primarily activated by EGF binding exchange the pair of dimers and activate other receptors unoccupied by EGF. Activation of the receptor was propagated on the cell surface by reorganization of the receptor clusters and lateral mobility. Fusion and splitting of the receptor clusters moving around the cell surface by thermal diffusion were actually observed using single-molecule visualization.

5.5
Dynamics of the NGF/NGFR Complex

NGF induces morphological and functional changes of the rat pheonochromocytoma cell line PC12 to a neuron-like cell [12]. NGF (2.5S NGF) labeled with a single Cy3 or Cy3.5 dye per molecule was prepared to investigate movements of NGF/NGFR complexes on the PC12 cell surface in single molecules [13].

Typical trajectories of the lateral diffusion movements of Cy3-NGF/NGFR complexes showed periods of mobility and immobility with abrupt switching between the two (Figure 5.5A). The mobile and immobile phases were separated according to the method described by Simson *et al.* [14]. In brief, for every segment of the trajectory, the local diffusion coefficient and the maximum radius of the segment were obtained, and the probability of a molecule diffusing randomly with the local diffusion coefficient to remain within the maximum radius was calculated. In the case of NGF on PC12, when a segment longer than 270 ms showed probability less than 10^{-4}, the segment was defined to be immobile. These parameters were determined from a comparison between experiments and simulations of random walks.

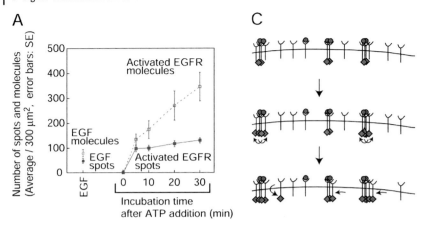

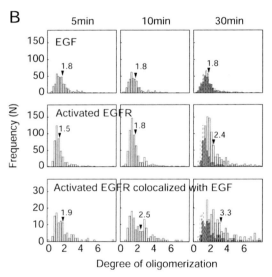

Figure 5.4 Single-molecule analysis of EGF activation. (A) Number of spots and molecules of bound EGF and activated EGFR produced after stimulation in semi-intact cells. Average and SE for 11 cells are shown. See text for details of the experiment. (B) Fluorescence intensity distribution of the spots of EGF (top row), activated EGFR (bottom row), and activated EGFR colocalized with EGF (bottom row) at indicated time points after stimulation. Fluorescence intensity (horizontal axis) was normalized with respect to the intensity of single molecules. Numbers above the arrowheads indicate the average cluster size. (C) A schematic model of the amplification process of EGFR activation through the dynamic reorganization of EGFR.

The diffusion coefficients for mobile phases, immobile phases and Cy3-NGF fixed on a glass surface were 0.18, 0.02 and 0.01 $\mu m\,s^{-1}$, respectively. Thus, even in the immobile phase, molecules were still moving but more slowly than those in the mobile phase by a factor of 9. The distribution of the durations of both mobile and immobile periods fit well to a single exponential function with decay times of

A

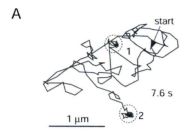

start

1

7.6 s

1 μm   2

B

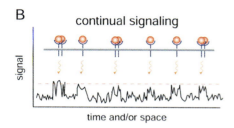

continual signaling

signal

time and/or space

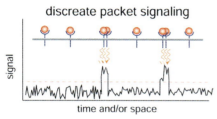

discreate packet signaling

signal

time and/or space

Figure 5.5 Single-molecule dynamics of the movements of NGF/NGFR complexes. (A) A representative trajectory of a Cy3–NGF receptor complex over 7.6 s showing reversible transitions in mobile and immobile behavior. Immobile segments are circled. Scale bar, 1 μm. (B) Signals from an NGF/NGFR complex do not flow to Raf1 continually but at discrete times and space. See text for details.

1 and 1.5 s, respectively. Single exponential distributions suggest that the motional modes switch in single rate-limiting steps.

During the immobilization period, the NGF/NGFR complex often formed clusters. The clustering continued throughout the immobile phase. Since the formation of simple clusters of this size cannot reduce the membrane protein diffusion coefficient significantly, the complex is more likely to consist of a transient molecular complex including cytoplasmic cell signaling proteins. It is also possible that some of the elements in the complex associated with the membrane skeleton.

TrkA is an NGF receptor that belongs to the RTK superfamily. Treatment of cells with a trkA-specific kinase inhibitor, k252a, decreased the population of molecules which showed immobilization. However, for the remaining population of molecules which showed immobilization, the lifetimes of both the mobile and immobile phase did not change, suggesting that immobilization depended on the phosphorylation of individual receptor molecules, but that k252a did not change the global membrane structure. Phosphorylation of TrkA induces activation of Ras, which in turn induces translocation of Raf1 from the cytoplasm to the plasma membrane. GFP-Raf1 and Cy3.5-NGF were simultaneously observed in single molecules. Colocalization of NGF and Raf1 suggests the formation of signaling complexes including Shc, Grb2, and Sos. Measurement of the diffusion movement has revealed that colocalization takes place only during the immobile periods.

Thus, single-molecule tracking of NGF/NGFR dynamics has revealed that the complex repeats random states of diffusion and immobilization and that the signaling complexes are formed during the immobilization period. NGF signaling does not occur continually but occasionally in discrete time periods and positions (Figure 5.5B). Although the importance of this phenomenon is not fully understood, this type of discrete signaling seems to be more efficient under conditions of high background noise. If a protein continually emits small signals, all of the signals will be hidden under the high background noise. However, signals emitted in larger packets can be distinguished from the high background noise even if the level of the individual signals is not increased.

5.6
Stochastic Signal Processing and Transduction in Living Cells

Intracellular signal transduction depends on stochastic processes such as association/dissociation, enzymatic catalysis, chemical modification, conformational changes and diffusion of signaling molecules, and thus intracellular signals are inevitably accompanied by random noise. How stochastic signaling systems in living cells operate reliably to receive, process and transduce signals under the strong influence of thermal and stochastic fluctuations is an open question. Chemotactic signaling systems in eukaryotic cells are an ideal model system for elucidating mechanisms of stochastic signal processing and transduction in living cells.

Chemotaxis is a directional motile response in living cells, in which cells move in a preferential direction in response to a chemical gradient. Chemotactic cells are extremely sensitive to chemical gradients. In eukaryotic cells, a difference of only a few percent in the concentration of the chemoattractant across cells is sufficient to trigger chemotactic movements in a wide range of background concentrations [15–17]. Because ligand binding to specific receptors is a stochastic process, receptor occupancy should fluctuate with time and space, and thereby signal inputs for chemotaxis should become noisy. The noise resulting from the measurement of chemical concentrations has been studied theoretically, and has revealed physical limits to a cell's sensing ability [18–21]. Theoretical estimations of the signal inputs suggest that cells receive a faint signal under the strong influence of stochastic noise generated during ligand-binding reactions. Thus, how cells obtain reliable information regarding the gradient is a critical question for directional sensing in chemotaxis.

To gain insights into mechanisms of stochastic signal processing and transduction, it is important to elucidate experimentally the stochastic nature of signaling molecules. For this purpose, we have developed single-molecule imaging techniques in living cells, and have applied these techniques to chemotactic signaling systems in *Dictyostelium* cells, which have successfully revealed the stochastic nature of the signaling molecules responsible for chemotaxis [22–24]. We have also developed a theoretical framework to analyze the impact of noise associated with the signal transduction processes [25]. We begin with a brief introduction to chemotactic signaling systems in *Dictyostelium* cells.

5.7
Chemotactic Signaling System of Eukaryotic Cells

Molecular mechanisms of chemotactic response are highly conserved among many eukaryotic cells including human leukocytes and *Dictyostelium* cells. In *Dictyostelium* cells, extracellular adenosine $3',5'$-monophosphate (cAMP) molecules trigger chemotactic signaling by binding to G protein-coupled cAMP receptor (cARs), which is mediated through G protein-liked signaling pathways including heterotrimeric G protein, Ras, PI3K, PTEN, PH-domain-containing proteins, guanylyl cyclase, PLCγ and PLA$_2$ [26, 27]. One of the key reactions in this signaling system is a distinctive localization of phosphatidylinositol 3,4,5-trisphosphates (PI(3,4,5)P$_3$) on the membrane exposed to a higher concentration of cAMP, which can be monitored by the binding of PH-domain-containing proteins to PI(3,4,5)P$_3$ on the membrane. Similar localizations of PH-domain-containing proteins have been observed in mammalian leukocytes and fibroblasts. The localization of PI(3,4,5)P$_3$ depends on dynamic signaling processes by PI3K and PTEN because they control the production and degradation of PI(3,4,5)P$_3$, respectively, in which reciprocal distributions of PI3K and PTEN leads to the accumulation of PI(3,4,5)P$_3$ at the leading edge of chemotaxing cells. PI(3,4,5)P$_3$ has been suggested to produce pseudopod extensions by recruiting PH-domain-containing proteins to the side of membrane exposed to a higher concentration of chemoattractants. Thus, chemotactic signaling systems can convert small differences in extracellular signals into localized signals for promoting the preferential formation of pseudopods. The localization of PI(3,4,5)P$_3$ takes place in an all-or-none manner, suggesting that noisy input signals are somehow processed to generate a clear signal reflecting the direction of the gradient of extracellular cAMP within cells.

5.8
Stochastic Nature of Chemotactic Signaling Molecules

To monitor input signals for chemotactic response, we prepared a fluorescent analog of cAMP (Cy3-cAMP) and used a total internal reflection fluorescence microscope (TIRFM) with which Cy3–cAMP binding to the receptors can be observed on the

A

Cy3-cAMP

laser

fluorescence

B

C

D

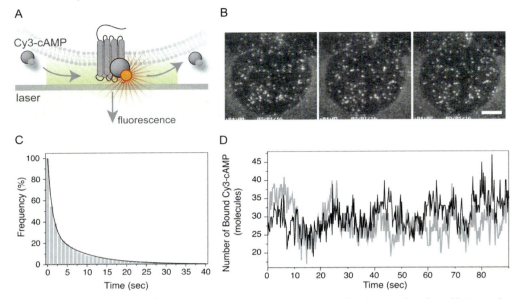

Figure 5.6 Stochastic properties of chemoattractant receptor. (A) Schematic drawing of single-molecule imaging of Cy3–cAMP bound to cAMP receptors. The basal membrane of *Dictyostelium* cells was observed by total internal reflection fluorescence microscopy. Cy3–cAMP molecules are visible as fluorescent spots when they bind to the receptors. (B) Single-molecule imaging of Cy3–cAMP bound to *Dictyostelium* cells. Scale bar, 1 µm. (C) Cumulative frequency histogram of the lifetime of Cy3–cAMP spots. The lifetimes of individual Cy3–cAMP molecules were obtained by measuring the time that elapsed between the appearance and the disappearance of the fluorescent spots. The line represents the fit of the data to the sum of two exponential functions with dissociation rates of 1.0 and 0.13 s^{-1}. (D) Time course of receptor occupancy in chemotaxing cells. The number of Cy3–cAMP spots bound to the basal surface of the cells was counted. The numbers of Cy3–cAMP bound to the anterior (black line) and the posterior (gray line) halves of the cells exhibited fluctuations in the receptor occupancy, as shown by the noisy input signals.

membrane of living *Dictyostelium* cells at the single molecule level (Figure 5.6A and B) [22]. The stochastic properties of the receptors were examined by determining the lifetime of the cAMP–receptor complex and receptor occupancy in the living cells. The lifetimes of the cAMP–receptor complexes exhibit an exponential distribution (Figure 5.6C), meaning that ligand dissociation from the receptor is a random process. The time series of receptor occupancy exhibits fluctuations with exponential time correlations, revealing that ligand binding follows a Poisson distribution. Thus, ligand binding takes place in a random manner and the input signals are therefore noisy.

Figure 5.6D shows receptor occupancy in the cells moving toward the higher concentration of Cy3–cAMP. Receptor occupancy was sometimes inversely proportional to the gradient of the chemoattractant. In our experimental system, only the basal surface of the cells could be visualized, and hence the receptor occupancy shown here does not represent the total input of chemotactic signals to the entire surface of the cells. However, assuming that ligand binding follows a Poisson distribution, receptor occupancy on the whole surface of the cells can be calculated by numerical simulations based on parameter values obtained experimentally for *Dictyostelium* cells. Such simulations confirm that input signals for chemotaxis are noisy.

Single-molecule imaging analysis has also revealed stochastic behaviors for other signaling molecules responsible for chemotactic response [23, 24]. Crac, which is one of the PH-domain-containing proteins, is stably localized at the pseudopod of chemotaxing cells. We observed GFP-tagged Crac (Crac-GFP) and examined the membrane-binding properties of Crac-GFP in cells undergoing chemotaxis. At the leading edge of the pseudopod of chemotaxing cells, individual molecules of Crac-GFP bind to the membrane in the order of ∼100 ms, while populations of Crac-GFP molecules appear to be localized in a stable manner on the membrane. Thus, Crac localization at the pseudopod is maintained dynamically by rapid exchanges of the individual Crac-GFP molecules. Such rapid exchanges of individual Crac molecules cause inevitable fluctuations in the ensemble concentrations of Crac molecules at the pseudopod, which is the molecular basis of signal noise. However, at the same time, the rapid exchange of molecules provides a molecular basis for rapid reorientation in response to directional changes in the chemical gradients, which can contribute to an accurate and sensitive chemotactic response. These dynamic properties have also been found in PTEN molecules [24]. Overall, this may be a general process for signaling molecules involved in chemotaxis.

5.9
Stochastic Model of Transmembrane Signaling by Chemoattractant Receptors

How can signal and noise propagation during transmembrane signaling by the receptors be characterized? What properties of the receptors are important for signal and noise propagation? We consider a simple but general scheme for transmembrane signaling in which receptors receive ligands stochastically as signal inputs. These activated receptors generate second messengers stochastically as outputs (Scheme 1; Figure 5.7A) [25].

$$R + L \underset{k_{off}}{\overset{k_{on}}{\rightleftharpoons}} R^*, \quad R^* + X \overset{k_p}{\longrightarrow} R^* + X^*, \quad X^* \overset{k_d}{\longrightarrow} X \tag{5.1}$$

where R, R^* and L represent inactive receptors, active receptors and the ligand, respectively. X and X^* are inactive and active second messengers. X can be regarded as the G protein for chemotactic signaling in *Dictyostelium* cells. According to this scheme, the average number of active receptors $\overline{R^*}$ and second messengers $\overline{X^*}$ per cell can be calculated using Michaelis–Menten kinetics,

$$\overline{R^*} = R_{total} \cdot L \cdot (K_R + L)^{-1}, \quad \overline{X^*} = X_{total} \cdot \overline{R^*} \cdot (K_X + \overline{R^*})^{-1} \tag{5.2}$$

where R_{total} is the total molecular number of receptors per single cell, $K_R = k_{off}/k_{on}$ is the affinity for the ligand with association and dissociation rate constants k_{on} and k_{off},

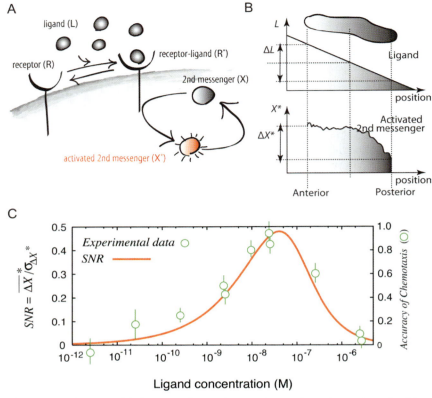

Figure 5.7 Stochastic model of transmembrane signaling by chemoattractant receptors. (A) Signal transduction reactions by chemoattractant receptors. The ligand (L) binds to the inactive receptor (R) leading to the formation of an active receptor (R*), which produces the active second messenger (X*) from the inactive precursor (X). The active X* is switched off to the inactive state X in due time. (B) The cell is placed under a ligand concentration gradient, which leads to variations in the concentration of the second messenger between the anterior and posterior ends of the cell. See text for details. (C) Dependence of the SNR of chemotactic signals on ligand concentration obtained theoretically using Equation 5.5. Parameter values used for the calculation are summarized in our previous report [25]. The theoretically-obtained SNR (red line) was overlaid on the experimental data to verify the chemotactic accuracy of *Dictyostelium* cells (green circles) as reported by Fisher *et al.* [16].

X_{total} is the total number of second messenger molecules per cell, $K_X = k_d/k_p$ is the concentration of active receptors where the activation of the second messenger reaches half-maximum with production and degradation rates of k_p and k_d for the second messenger. Note that Equation 5.2 represents the input–output relationship between the average number of active receptors and the average number of active second messengers. To describe the signal and noise propagation during transmembrane signaling by the receptors, the input–output relationship between temporal noise in the concentration of the active receptor (σ_R^2) and noise in the concentration of the active second messenger (σ_X^2) should be taken into account.

The noise of the active receptor (σ_R^2) and the active second messenger (σ_X^2) are given by the *gain fluctuation relationship* [28] which is shown below,

$$\frac{\sigma_R^2}{\overline{R^*}^2} = g_R \frac{1}{\overline{R^*}} \tag{5.3}$$

$$\frac{\sigma_X^2}{\overline{X^*}^2} = g_X \frac{1}{\overline{X^*}} + g_X^2 \frac{\tau_R}{\tau_X + \tau_R} \frac{\sigma_R^2}{\overline{R^*}^2} \tag{5.4}$$

where τ_R and τ_X are the characteristic time constants for the ligand binding reaction and the reaction to produce the second messenger, respectively. In Equation 5.1, the time constants of the reactions are given by the reaction rates as $\tau_R = (k_{on}L + k_{off})^{-1}$ and $\tau_X = (k_p R^* + k_d)^{-1}$. The gains, g_R and g_X, quantify the amplification rate of the output response to small changes in input. These gains are given by $g_R = K_R \cdot (K_R + L)^{-1}$ and $g_X = K_X \cdot (K_X + \overline{R^*})^{-1}$. The first and second terms on the right-hand side of Equation 5.4 are known as the intrinsic and the extrinsic noise, respectively. The intrinsic noise represents inherently-generated noise due to the stochastic nature of the reactions produced by the second messenger, while the extrinsic noise represents the noise propagated from ligand–receptor binding reactions. Thus, Equation 5.4 describes the input–output relationship between noise in the active receptor concentration and noise in the active second messenger concentration.

The *gain-fluctuation relationship* tells us that signal and noise propagation along the signaling cascade can be characterized by the gain and the characteristic duration of the signaling reactions (Equations 5.3 and 5.4). The reactions with higher gains generate more noise. Also, the propagation of the noise from the active receptor to the second messenger concentration depends on the factor $\tau_R/(\tau_X + \tau_R)$ in the second term. As $\tau_R/(\tau_X + \tau_R)$ decreases with a decrease in τ_R and/or increase in τ_X, the extrinsic noise decreases due to an increase in time-averaging effects. In order to reveal how signal and noise are propagated along the signaling cascade, it is important to determine experimentally the gains and the time constants, which is possible by using single-molecule imaging detection and other techniques.

To evaluate the effects of the noise on gradient sensing, we studied the signal-to-noise ratio (SNR) of the chemotactic signals. As shown in Figure 5.7B, the difference in concentration of the ligand (ΔL) may produce the difference in receptor occupancy (ΔR^*), which may in turn produce the difference in second messenger concentration (ΔX^*) between the anterior and posterior regions of chemotactic cells. ΔR^* and ΔX^* should include the noise $\sigma_{\Delta R}^2$ and $\sigma_{\Delta X}^2$ around the average values $\overline{\Delta R^*}$ and $\overline{\Delta X^*}$, respectively. Based on the *gain-fluctuation relationship* (Equation 5.4), the relationship between the relative noise intensities $\sigma_{\Delta R}/\overline{\Delta R^*}$ and $\sigma_{\Delta X}/\overline{\Delta X^*}$ is given by

$$\frac{\sigma_{\Delta X}^2}{\overline{\Delta X^*}^2} = \frac{1}{g_X \overline{X^*}} \left(\frac{\overline{R^*}}{\overline{\Delta R^*}}\right)^2 + \frac{\tau_R}{\tau_X + \tau_R} \left(\frac{\sigma_{\Delta R}}{\overline{\Delta R^*}}\right)^2 \tag{5.5}$$

where the first and second terms on the right-hand side are derived from the intrinsic and extrinsic noise, respectively. We defined $\overline{\Delta X^*}/\sigma_{\Delta X}$ as the SNR of the chemotactic signals, which can be calculated by using the appropriate parameter values obtained experimentally for *Dictyostelium* cells. Figure 5.7C shows the dependence of the SNR on the average ligand concentration with a 2% gradient. The SNR dependence resembles the dependence of chemotactic accuracy on ligand concentration measured experimentally for *Dictyostelium* cells [16]. This agreement between the SNR and the accuracy of chemotaxis indicates that the ability of directional sensing is limited by the stochastic noise generated inherently during the transmembrane signaling of receptors. Furthermore, it suggests that chemotactic accuracy is determined primarily at the most upstream reactions of the chemotactic signaling system. Our stochastic model can be further applied to other chemotactic cells. Similar dependence of chemotactic accuracy has been also observed in mammalian leukocytes and neurons [15, 29].

Equation 5.5 suggests how the SNR of chemotactic signals is enhanced or diminished by the properties of the receptors and the downstream second messenger. The time constants and the gains of the signaling reactions determine the signal and noise propagation and hence the SNR of chemotactic signals. For example, a longer lifetime of the second messenger leads to more effective noise reduction by time-averaging the extrinsic noise because the time constant, τ_X, becomes larger, suggesting that the regulatory mechanism for second messenger inactivation may play a pivotal role in signal enhancement during chemotaxis. The GTPase-activating proteins such as regulators of G protein signaling (RGS) can regulate the quality of the signal by modulating the inactivation rates of the G protein. Modulation of the time constant, τ_R, also has an effect on the SNR of chemotactic signals. Acceleration of the on-rate (k_{on}) and the off-rate (k_{off}) in the ligand-binding reaction would cause a decrease in τ_R leading to an enhancement in the SNR. Polarity in receptor kinetic states along the length of chemotactic cells has been observed by single-molecule measurements [22], suggesting a polarity in the SNR of chemotactic signals. This may provide a molecular basis for the polarity observed in the chemotactic response of *Dictyostelium* cells [30]. When the second messenger is produced by a reaction with cooperativity, the gain, g_X, may become larger, causing noise reduction by decreasing the intrinsic noise [28].

It is important to emphasize that signal transduction systems can carry out signal processing under the strong influence of molecular noise. In order to fulfill their functions, cells must have some mechanism which makes them resistant to such strong noise. In particular, chemotactic cells must overcome the noise in order to gain high sensitivity for shallow gradients within a wide dynamic range. In addition, cells could also be taking advantage of the molecular noise to undertake their functions, which could not be achieved without noise. We show here that the *gain-fluctuation relationship* can be applied successfully to a stochastic signaling system. The signaling processes such as association/dissociation, enzymatic catalysis and chemical modification of signaling molecules have been described by Michaelis–Menten methodology and its extended equations. However these only explain the relationship between signal inputs and outputs on average. The *gain-fluctuation relationship* can be used to describe not only the signal but also the noise propagation along signaling cascades. According to the relationship, it is important to determine the time constants and the gains of the signaling reactions experimentally in order to reveal the signal and noise propagation along a stochastic signaling system. Single-molecule imaging techniques provide a unique tool for elucidating the stochastic nature of signaling molecules at work in living cells.

5.10
Conclusions

In RTK signaling across the plasma membrane, single-molecule imaging has revealed that the formation of signaling dimers of EGFR was facilitated by a high rate of association and positive cooperative binding of EGF to predimers of EGFR. The EGF signal was amplified by the secondary activation of EGFR using a dynamic interaction between occupied and unoccupied receptor molecules. Signaling complexes of NGFR were formed repeatedly during the immobile phase of NGF/NGFR dynamics suggesting that signals transduce as packets. These results indicate that cell signaling is improved at the plasma membrane using dynamic molecular systems. Also, we provide a theoretical framework to describe how signals and noise are propagated along stochastic signaling systems in living cells. It has long been suggested that the plasma membrane is not merely a simple entrance for external information to pass into cells but is the site where sophisticated signal processing takes place. Single-molecule imaging is now uncovering the mechanism of signal processing at the plasma membrane.

References

1 Sako, Y. and Yanagida, T. (2003) *Nat. Rev. Mol. Cell Biol.* **4** Suppl, SS1–SS5.

2 Wazawa, T. and Ueda, M. (2005) *Adv. Biochem. Eng. Biotechnol.* **95**, 77–106.

3 Sako, Y. (2006) *Mol. Syst. Biol.* doi:10.1038/msb4100100.

4 Schlessinger, J. (2000) *Cell* **103**, 211–225.

5 Burgess, A.W., Cho, H.-S., Eigenbrot, C. *et al.* (2003) *Molecular Cell* **12**, 541–552.

6 Chung, J.C., Sciaky, N. and Gross, D.J. (1997) Biophys. J. **73**, 1089–1102.

7 Klein, P., Mattoon, D., Lemmon, M.A. and Schlessinger, J. (2004) *Proc. Natl. Acad. Sci. USA* **101**, 929–934.

8 Teramura, Y., Ichinose, J., Takagi, H. *et al.* (2006) *EMBO J.* **25**, 4215–4222.

9 Sako, Y., Minoguchi, S. and Yanagida, T. (2000) *Nature Cell Biol.* **2**, 168–172.

10 Ichinose, J., Murata, M., Yanagida, T. and Sako, Y. (2004) *Biochem. Biophys. Res. Comm.* **324**, 1143–1149.

11 Kano, F., Sako, Y. and Tagaya, M. (2000) *et al. Mol. Biol. Cell* **11**, 3073–3087.

12 Huff, K., End, D. and Gordon, G. (1981) *J. Cell Biol.* **88**, 189–198.

13 Shibata, S.C., Hibino, K., Mashimo, T. *et al.* (2006) *Biochem. Biophys. Res. Comm.* **342**, 316–322.

14 Simson, R., Sheets, E.D. and Jacobson, K. (1995) *Biophys. J.* **69**, 989–993.

15 Zigmond, S.H. (1977) *J. Cell Biol.* **75**, 606–616.

16 Fisher, P.R., Merkl, R. and Gerisch, G. (1989) *J. Cell Biol.* **108**, 973–984.

17 Song, L., Nadkarni, S.M., Bödeker, H.U., Beta, C., Bae, A., Franck, C., Rappel, W.-J.,

Loomis, W.F. and Bodenschatz, E. (2006) *Eur. J. Cell Biol.* **85**, 981–989.

18 Berg, H.C. and Purcell, E.M. (1977) *Biophys. J.* **20**, 193–219.

19 Tranquillo, R.T. (1990) In *Biology of the Chemotactic Response* (eds J.P. Armitage and J.M. Lackie), Cambridge University Press, pp. 35–75.

20 Van Haastert, P.J.M. (1997) in *Dictyostelium* (eds Y. Maeda, K. Inouye and I. Takeuchi), Universal Academy Press, Tokyo, pp. 173–191.

21 Bialek, W. and Setayeshgar, S. (2005) *Proc. Natl. Acad. Sci. USA*, **102**, 10040–10045.

22 Ueda, M., Sako, Y., Tanaka, T., *et al.* (2001) *Science* **294**, 864–867.

23 Matsuoka, S., Iijima, M., Watanabe, T.M., *et al.* (2006) *J. Cell Sci.* **119**, 1071–1079.

24 Vazquez, F., Matsuoka, S., Sellers, W.R., *et al.*, (2006) *Proc. Natl. Acad. Sci. USA* **103** 3633–3638.

25 Ueda, M. and Shibata, T. (2007) *Biophys. J.* **93**, 11–20.

26 Van Haastert, P.J.M. and Devreotes, P.N. (2004) *Nat. Rev. Mol. Cell Biol.* **5**, 626–634.

27 Franca-Koh, J., Kamimura, Y. and Devreotes, P. (2006) *Curr. Opin. Gene. Dev.* **16**, 1–6.

28 Shibata, T. and Fujimoto, K. (2005) *Proc. Natl. Acad. Sci. USA* **102**, 331–336.

29 Rosoff, W.J., Urbach, J.S., Esrick, M.A., *et al.* (2004) *Nat. Neurosci.* **7**, 678–682.

30 Swanson, J.A. and Taylor, D.L. (1982) *Cell* **28**, 225–232.

6
Dynamics of Membrane Receptors: Single-molecule Tracking of Quantum Dot Liganded Epidermal Growth Factor

Guy M. Hagen, Keith A. Lidke, Bernd Rieger, Diane S. Lidke, Wouter Caarls, Donna J. Arndt-Jovin, and Thomas M. Jovin

6.1
Introduction

The erbB family of receptor tyrosine kinases (RTKs) includes erbB1 (the classical epidermal growth factor (EGF) receptor, hereafter referred to as EGFR), erbB2, erbB3 and erbB4. Activation of these transmembrane proteins initiates signaling cascades controlling numerous cellular processes such as DNA replication and division. Binding of specific peptide ligands to the ectodomains of the RTKs leads to auto- and transactivation of the cytoplasmic protein kinase domains. The activated receptors bind adaptor proteins, initiating several signal transduction cascades, such as those mediated by MAP kinases. The fate of the activated receptors is complex: endocytosis via coated pits, covalent modification (deactivation by enzymatic dephosphorylation and ubiquitinylation), and endosomal trafficking leading to proteosomal and/or lysosomal degradation or recycling to the plasma membrane. The overexpression and unrestrained activation of the erbB family are implicated in many types of cancer [1].

We have shown in previous publications [2, 3] that quantum dots (QDs) bearing natural ligands function as effector molecules and provide the means for prolonged real-time visualizations of erbB molecules on living cells. The multiple steps of the signaling pathways can be followed, and detailed movies of image sequences of the underlying mole-cular processes can be generated (available as supplementary information in [2, 3]).

For the studies reported here, biotinylated EGF was bound to commercial streptavidin-conjugated QDs. QDs have unique features providing many advantages for cellular imaging: (i) high absorption cross-sections and quantum yields, permitting detection down to the single nanoparticle level and reliable quantitation of binding and transport phenomena; (ii) extreme photostability, allowing imaging over prolonged periods; (iii) a broad excitation spectrum rising toward the UV, allowing the simultaneous excitation of visible fluorescent proteins (VFP) and QDs; and (iv) narrow emission bands across the visible spectrum. QDs can be regarded as single molecule probes [4].

In this chapter we feature (i) the retrograde transport of activated EGFR; (ii) the diffusional behavior of non-activated, kinase-inhibited single receptors on the cell body and filopodia using a high speed and high sensitivity electron multiplying CCD (emCCD) camera; and (iii) the rapid, light efficient three-dimensional (3D) imaging of the early binding events of individual QD-EGF ligands with a commercial prototype of a new generation, optically sectioning programmable array microscope (PAM).

6.2
Single QD Imaging

The excitation and emission spectra of commercially available QDs from Invitrogen are shown in Figure 6.1. Due to their high absorption cross-section in the low visible and UV range and emission in the red to far red, QDs are ideal emitters for signal acquisition by CCD cameras. However, the relatively long lifetimes (10–20 ns) implies that their maximal rate of fluorescence emission (in terms of photon flux) is somewhat limited [5]. Imaging single QDs in raster scanning systems, particularly with PMT detectors, requires long pixel dwell times, and high laser powers. Such systems are not ideally suited for observing diffusion rates or rapid, live processes. On the other hand, CCD cameras have high sensitivity at long wavelengths, and emCCDs in particular have superior sensitivity for short acquisition times at low light levels [6]. These features were exploited in the experiments presented here. In addition, a versatile optical sectioning microscope, the PAM, was used to measure the initial steps of QD-EGF binding and EGFR activation, and the diffusion of individual (mono-liganded) QD-EGFR complexes in three dimensions.

In order to take full advantage of the potential afforded by the unique characteristics of QDs, wide-field imaging systems generating continuous emission spectra at every pixel position are also highly desirable. Fourier encoding and CCD detection combines very efficient detection and spectral reconstruction of low-light level images. To record single QD spectra, we used a commercial Fourier interferometric spectrograph, the SpectraCube manufactured by Applied Spectral Imaging (Migdal Haemek, Israel). With the very high magnification afforded by a unique Olympus 150×1.45 NA objective mounted in an IX71 microscope, we obtained very distinct signals attributable to individual QDs either attached to a surface transpore or in cells [7] (Figures 6.1b and c).

6.3
Retrograde Transport of Activated EGFR Dimers

The epidermal growth factor receptor (EGFR) is a single chain integral transmembrane protein with an ectodomain encompassing the EGF peptide binding site. The receptor is distributed on the cellular plasma membrane, including the filopodia (Figure 6.2). Filopodia are fine processes extending from the cell body with a core of actin bundles, the filaments of which have pointed ends oriented towards the cell

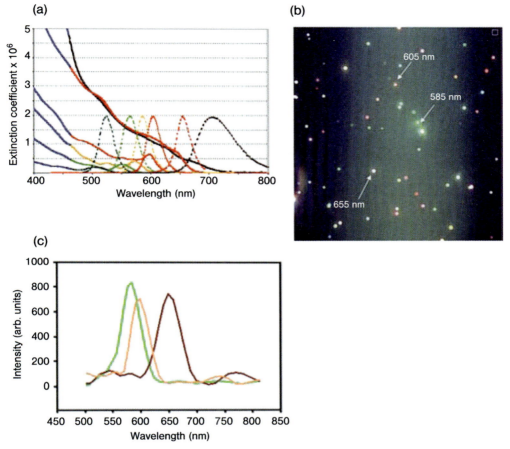

(a)

(b)

(c)

Figure 6.1 Absorption and fluorescence emission spectra of commercially available QDs in bulk solution, and hyperspectral image and emission spectra of single QDs. (a) Excitation spectra, solid curves; emission spectra, dotted curves. Figure from Invitrogen, used with permission. (b) Hyperspectral image of fluorescence emission from single QDs in a mixture as recorded with the Applied Spectral Imaging SpectraCube. Colors in the image are assigned by the spectral imaging software using an arbitrary table and are not necessarily representative of the true color of objects in the image. (c) Fluorescence emission spectra of the individual QDs indicated in panel b.

interior. Addition of complexes of QD–streptavidin conjugated with biotinylated EGF at sub-nM concentrations results in rapid binding of the QD–EGF to receptors on the cell body and along the filopodia (Figures 6.2 and 6.3). In a previous study [3] Lidke *et al.* showed that incubation of cells with 5 pM of QD–EGF for a few minutes caused individual QDs to attach to single EGFR molecules on filopodia and undergo diffusion. Addition of excess free EGF led to concerted activation of neighboring receptors and the QD–EGF–EGFR complex, which immediately underwent directed retrograde transport toward the cell body. An intact actin cytoskeleton as well as

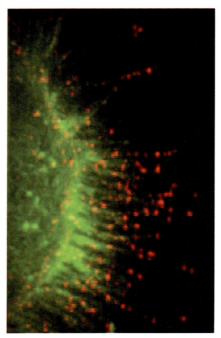

Figure 6.2 Reconstructed, sectioned image from the programmable array microscope (PAM) showing distribution of EGF-bound EGFR on A431 cells. The cells stably express EGFR–eGFP (green). QD655–EGF (red) bind to the EGFR on the filopodia and the cell membrane. The image is a maximum intensity projection of 19 focal planes acquired using an exposure of ~ 16 ms per slice with a spacing of 0.5 μm.

an activated EGF receptor dimer was necessary to maintain the transport process (Figure 6.3, taken from Figure 4 in [3]).

The actin bundles in filopodia undergo growth and exchange by addition of monomers to the plus ends and depolymerization from the minus ends, a process referred to as treadmilling. In combination with the active pulling of actin filaments by myosin, treadmilling results in a net flow of F-actin towards the interior of the cell. Association of macromolecules with an actin filament leads to a translocation of the cargo towards the cell body. Specific inhibitors of the EGFR RTK as well as cytochalasin D, a disruptor of the actin cytoskeleton, abolish transport but not free diffusion of the receptor–ligand complex. The coupling of the QD–EGF–EGFR complex to this retrograde flow was further corroborated by photobleaching of actin–EGFP bundles during transport and correlation of the movement of the QDs and the bleached actin segment (see Figure 3 in [3]). From data such as these we were able to calculate the velocity of retrograde transport to be 20 ± 8 nm/s.

In the same studies, it was demonstrated that retrograde transport precedes receptor internalization, which occurs at the base of the filopodia. The fact that initiation of transport requires the cooperative interaction of two or more activated

A

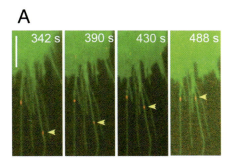

B

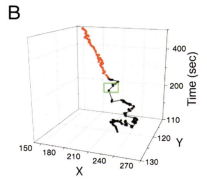

Figure 6.3 Time series and particle track showing movement of a QD–EGF–EGFR on an A431–eGFP cell. (A) Selected frames from a time series after binding 5 pM EGF–QD (red) followed by addition of free EGF (50 ng/ml) at 300 s. Bar, 5 μm. Images are contrast enhanced and were taken with a Zeiss LSM 510. (B) Trajectory of the indicated QD–EGF–EGFR complex (A, arrowhead) on a filopodium that exhibits random diffusional movement (black) until the addition of unlabeled EGF (green box), after which the complex begins active retrograde transport (red). (Figure from [3], reproduced by permission).

receptors suggests that filopodia function as sensory organelles for the cell, probing for the presence and concentration of effector molecules distant from the cell body, and coupling sensing to cellular response via a directed transport of receptors activated upon achieving a given threshold of ligand [3].

6.4
Single QD–EGF–EGFR Tracking

Measurements of QD–EGF–EGFR diffusion using single particle tracking methods in laser scanning confocal microscopes yielded diffusion constants an order of magnitude slower than those reported in studies of EGF–EGFR based on fluorescence recovery after photobleaching (FRAP) or fluorescence correlation spectroscopy (FCS) [3]. In this chapter, we present new data acquired with a fast emCCD camera of kinase-inhibited QD–EGF–EGFR under conditions permitting visualization of single molecules.

Although QD blinking can be advantageous, i.e. for identifying single particles and achieving spatial superresolution [8], the phenomenon complicates the analysis of tracking due to the transient interruption of the signal. We have developed special data processing routines to take into account the random blinking periods and to achieve automated identification and tracking of the QD–EGF–EGFR complexes. Definitions and procedures are as follows:

(a) An "on" QD signal is defined as an intensity at least 50% of the running average of all QD intensities. Intensities below this threshold are considered "off".

(b) Positional information is only stored when the QD is "on".

(c) To avoid losing the QD during the "off" period and switching the track to another "on" QD, we compare the newly found track position against the probability of a jump distance equal to or greater than the observed jump distance, r, by diffusion in the time Δt, $P(r, \Delta t) = e^{-r^2/4D\Delta t}$ where D is an assumed diffusion constant. If this probability is less than 5%, we search future time-frames for an "on" position with higher probability. If one is found, we mark the QD as "off" until tracking continues at the future time-fame. If no higher probability events are found, we accept the unlikely jump position.

Using this algorithm we imaged at video rates QD–EGF–EGFR on the cell body and filopodia of cells treated with the kinase inhibitor PD153035. As shown in Figure 6.4, analysis of the trajectories from single QDs yield mean square displacement (MSD) curves that could be fit to retrieve diffusion constants [9]. The MSD calculated from the trajectory in Figure 6.4 fits well to a "corralled diffusion" model $MSD(\Delta t) = \textit{offset} + \frac{L^2}{3} e^{-\Delta t/T}$ where L is the diameter of the confinement zone, T is the time required to explore the zone, $D = L^2/(12T)$ and the offset accounts for localization error. The parameters derived from this fit were: $D = 0.05\,\mu m^2/s$; $T = 0.39\,s$ and $L = 0.48\,\mu m$. A histogram of the values derived from these data is shown in Figure 6.4e. The mean diffusion constant of QD–EGF–EGFR was $0.021 \pm 0.022\,\mu m^2/s$ on the cell body and $0.015 \pm 0.013\,\mu m^2/s$ on filopodia; i.e. there were no discernable differences in the motion at short time scales for the two cases. The MSD plots indicate that in the presence of kinase inhibitor, the receptors underwent corralled diffusion, and not transport. A complete analysis of single molecule diffusion of the EGFR on the cell body as well as on filopodia has been conducted and compared with data obtained by FRAP of the unliganded receptor (K. A. Lidke *et al.*, unpublished data).

6.5
Programmable Array Microscopy

Dual pass programmable array microscopes (PAMs) are defined by the use of a spatial light modulator (SLM) in a primary image plane of a standard fluorescence microscope. The SLM provides for structured illumination as well as conjugate descanning of the image to achieve optical sectioning. The major advantages of the PAM are: (i) simple, inexpensive design with no moving parts; (ii) speed-up in optical sectioning due to an illumination duty cycle for each pixel of up to 50%; (iii) optimal detection sensitivity, e.g. using emCCD cameras; (iv) continuously programmable, arbitrary, and adaptive optical sectioning modes between or within images using libraries of dot, line, or pseudo-random (Sylvester) sequence patterns; (v) efficient and sensitive optical sectioning due to simultaneous detection and processing of both conjugate and non-conjugate light; (vi) compatibility with polarization, hyperspectral, lifetime-resolved, and other imaging modes; and (vii) minimal photobleaching.

Our initial implementation of the PAM [10–13] used a digital micromirror device (DMD) for optical sectioning or transmissive liquid crystal SLMs for imaging

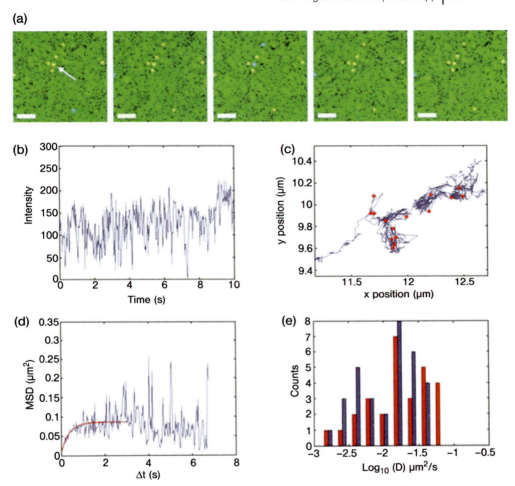

Figure 6.4 Analysis of single particle tracking data taken with an emCCD camera. (a) Sequential frames from a time series. Yellow spots indicate successful tracking. Blue spots indicate quantum dot "off" states. Only spots with a sufficiently high correlation to the microscope PSF were chosen for tracking. Quantum dots with 655-nm emission were imaged using an Andor iXon emCCD camera with illumination from a Hg lamp (435/20 bandpass filter) using 20-ms exposures at 33 frames/s.) (b) The intensity trace of the indicated QD over time demonstrates blinking. (c) Trajectory (blue) of the indicated (arrow) QD in A. Filled (red) circles designate "off" periods. (d) Diffusion constants were retrieved by fitting a parameterized model to the Mean Square Displacement (MSD) curve. (e) Distribution of diffusion constants found for receptors on the cell body (red) or filopodia (blue). Diffusion constants are derived by a linear fit to the first 10 points (0.3 s) of the MSD curve.

spectroscopy. Other reported PAMs are based on DMD [14–17] and liquid crystal-on-silicon (LCoS) [18] SLMs. Of these, only one [15] was applied in fluorescence microscopy, using a fringe projection technique similar to that used by Wilson *et al.* [19]. The DMD-based PAM suffered from several serious limitations leading to degradation of the acquired confocal images: lack of a suitable, easy to use video interface for scanning pattern definition, diffraction effects, limited VGA format, and a low micromirror tilt angle. In addition, the imaging of conjugate (in-focus light from "on" pixels) and non-conjugate (out-of-focus light from "off" pixels) on a single camera was not feasible.

We have developed a new PAM in collaboration with Cairn Research Ltd. (Faversham, UK) based on a ferroelectric LCoS microdisplay (Forth Dimension Displays, Dunfermline, Scotland). This microdisplay has several favorable characteristics for use in a PAM: SXGA resolution, high fill factor (93%), good contrast (400:1), and a fast liquid crystal switch time (40 µs). When used for video projection, this microdisplay operates in a 24-bit color-sequential mode; each video frame is broken into 24 adjustable-length bitplanes, eight each for red, green and blue. For PAM operation, the color-sequential mode is disabled and each of the 24 bitplanes is displayed for the same length of time, refreshing at 1.44 kHz. Scanning patterns are defined by simple display of a 24-color Windows bitmap file via a standard computer DVI video interface. Possible scanning patterns include those based on dot lattices, line arrays, or pseudorandom (Sylvester) sequences or arrays [10, 13, 20]. An "on" pixel is one in which plane polarized light is reflected with a 90-degree rotation in the plane of polarization. "Off" pixels reflect light with no change in polarization state. Light sources (LEDs, lasers, lamps) must thus be linearly polarized, either intrinsically, or by means of a polarizing beam splitter cube. Light of both linear polarization states (horizontal and vertical) is projected onto the microdisplay. Since both "on" and "off" pixels are reflected along the optical axis, an image splitter arrangement allows the conjugate (in-focus) and non-conjugate (out-of-focus) light to be imaged on a common detector, in our case an emCCD (Ixon DV885 or DV887, Andor Technology, Belfast, Northern Ireland). Spinning disk confocal microscopes also generate conjugate images; these are optically sectioned images with an offset due to cross-talk between the many pinholes. The PAM has the advantage that a scaled subtraction can be performed using both the conjugate and non-conjugate images to generate a confocal image with no DC offset [13, 21]. In addition, the sectioning capability is fully programmable and can achieve very high duty cycles. To generate the final image, the conjugate and non-conjugate images are registered and subtracted, with scaling factors applied to both images that depend on the pattern used [13]. Image registration, scaling, and if desired, background subtraction, filtering, and other image processing operations are performed in real time (processing and display require 10 ms per image) using a workstation-class graphics card and BrookGPU, a graphics processing unit (GPU) programming library freely available on the Internet [22].

To demonstrate the sensitivity and high speed optical sectioning ability of the PAM, we imaged single QDs bound to A431 cells expressing the EGFR. The QDs were added at a concentration of 200 pM, allowed to bind for 3 min at room temperature, then removed by washing with Tyrode's buffer. Imaging was started immediately at a rate of ∼6 Hz using an exposure time of 33 ms. Three selected frames of a 300-frame film are shown in Figure 6.5. Binding is similar to that shown in Figure 6.2.

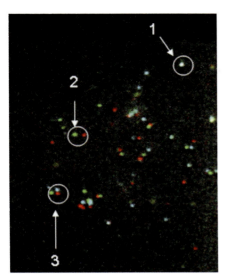

Figure 6.5 Movement of a QD–EGF–EGFR on an A431 cell taken with the programmable array microscope (PAM). Frames 1 (green, 0 s), 80 (red, 13.3 s), and 160 (blue, 26.6 s) of a 300-frame film (50 s total) shown as a three-color overlay. Diffusion and blinking of single QD–EGF–EGFR are visualized. Arrow 1 indicates a QD–EGF–EGFR that did not move; arrow 2, a QD–EGF–EGFR that has blinked "off" in frame 160; and arrow 3 a mobile QD–EGF–EGFR. Data were recorded with an exposure time of 33 ms at a rate of ~6 Hz.

6.6
Concluding Remarks

The field of cellular imaging has benefited enormously from parallel technological developments leading to dramatic increases in sensitivity, spatial and temporal resolution, and selectivity. Luminescent quantum dots, as well as silica-based nanoparticles and nanodot clusters of noble metals not featured in this chapter (see [23]), provide single-molecule sensitivity in imaging systems designed for studies of living cells. We favor wide-field microscope systems for this application because of their much higher acquisition speed, particularly in combination with electron multiplying CCD cameras that afford the ultimate performance at detected light levels of <100 photons/pixel. The programmable array microscope featured here offers optical sectioning in combination with resolution in time (fluorescence lifetimes), anisotropy (FRET, diffusion), space (hyperspectral imaging, diffusion), and "chemistry" (photoreactions) [24].

Note added in proof: This article was completed in November 2006.

Acknowledgments

The authors acknowledge support for the project from EU FP5 project QLG1-CT-2000-01260 and for development of the PAM through EU FP6 Project 037465

FLOUROMAG. D.S.L. was the recipient of a postdoctoral fellowship from the EU FP5 grant QLG2-CT-2000-02278 awarded to T.M.J. and B.R. was supported by a TALENT fellowship from the Netherlands OSR.

Appendix 6.A: Materials and Methods

6.A.1
Reagents

Biotin–EGF, EGF, and streptavidin-conjugated, -pegylated Qdots were obtained from Invitrogen. PD153035 was purchased from Calbiochem. Live cell labeling was carried out in Tyrode's buffer with 20 mM glucose and 0.1% BSA. PBS is phosphate buffered saline.

6.A.2
Cell Lines

HeLa cells, expressing 60 000 EGF receptors per cell, A431 cells, an epidermal carcinoma cell line expressing 2×10^6 EGF receptors per cell, and the same cell line stably transfected with EGFR–eGFP were maintained in DMEM with 10% fetal calf serum.

6.A.3
Cell Treatments

Cells were typically starved (0.1% FCS) overnight and were treated with $1 \mu M$ PD153035 kinase inhibitor for 2 h at 37 °C prior to and during diffusion measurements.

6.A.4
QD Conjugation to Epidermal Growth Factor

Monobiotinylated-EGF was coupled to the streptavidin QDs by incubation with 10 nM QD concentrations in PBS containing 1% BSA for at least 30 min at 4 °C. Ratios of EGF: QD were usually 3 : 1 or 1 : 1 unless otherwise specified.

6.A.5
Wide-field Microscopy

Wide-field detection of QDs was performed with an Andor iXon DV887 emCCD camera attached to an Olympus IX71 microscope with a 60×1.45 NA oil immersion or 60×1.2 NA water objective. QDs were excited by a mercury arc lamp at 436 nm with a bandpass filter and appropriate QD (20 nm) emission filters (Chroma

Technology). Fast tracking of QDs was performed with 20-ms exposure times at 30-ms intervals with the emCCD using 655QD–EGF conjugates.

6.A.6
PAM

The new generation programmable array microscope (PAM) is a prototype developed in collaboration with Cairn Research Ltd. (Faversham, Kent, UK). The stand-alone module, including light source(s) and detector(s), features an innovative catadioptric design and a ferroelectric liquid-crystal-on-silicon (LCoS, SXGA-R2D, Forth Dimension Displays, Dunfermline, Scotland) SLM instead of the original DMD used in the first PAM design. The LCoS-based PAM can be attached to a camera port of any unmodified fluorescence microscope. The prototype system currently operated at the Max Planck Institute for Biophysical Chemistry (Göttingen, Germany) incorporates a 6-position high-intensity LED illuminator as well as modulated laser light sources (both diode lasers and AOM-modulated argon-ion lasers), and an Andor iXon emCCD camera. The system is mounted on an Olympus IX71 inverted microscope with 60–150 × objectives, a high precision X,Y,Z stage (Nanoscan Z), and high speed filter wheels (Prior Scientific, Cambridge, UK).

6.A.7
Hyperspectral Imaging

Hyperspectral imaging was performed with an Applied Spectral Imaging (Migdal Haemek, Israel) SpectraCube imaging spectrograph equipped with a VDS cool-1300 camera and mounted on the IX71microscope. QD samples were imaged with a 150×1.45 NA objective. Excitation was with a Hg arc lamp and the epi-illumination filter set consisted of 435-nm narrow bandpass excitation, 505-nm dichroic, and 510-nm longpass emission filters. Streptavidin-coated QDs emitting at 585, 605, or 655 nm were diluted from the original stock concentration to final concentrations of $\sim 1\,pM$ (QD655, PEG coated, QD585, and QD605) in PBS. Approximately 25 µl of the diluted solution was pipetted onto an Ibidi (Munich, Germany) 18-well, poly-L lysine-coated slide and allowed to bind to the surface for 5 min. The samples were washed once with water and maintained in water. A mixture of all three QD types was also prepared. Hyperspectral images in the range 500–800 nm were acquired in 185 s, using 128 frames (1-s exposure) and 45 interferometer steps between each frame. After Hanning windowed-Fourier transformation, single QDs were marked and classified with the ASI SpectraView software. The edge of each full field image was selected as background, and the spectrum of this area was subtracted from the individual QD spectra. For cellular experiments not discussed in this chapter [7], a 1:1 mixture of 200 pM EGF-QD605 and 200 pM EGF–QD655 were incubated with F1-4 CHO cells expressing EGFR eGFP. These cells were not treated with kinase inhibitor as in the single particle tracking experiments, allowing the QDs to internalize into the cells after binding.

Appendix 6.B: Software and Image Processing

6.B.1
Single Particle Tracking

Data for single particle tracking (SPT) were acquired using wide-field detection as above. Video rate series consisted of 900 frames of 20-ms exposure at 33 frames/s. QD–EGF labeling was performed in the presence of a kinase inhibitor, PD153035 that inhibits the activation and dimerization of the EGFR receptor [25]. A focal plane with relatively low curvature in the area at the top of the cell was chosen for imaging. We devised an automated method for selecting QDs for tracking and analysis that fit high intensity regions to a Gaussian profile approximating the point spread function (PSF) of the microscope (see text). The quality of the fit was used to determine if a spot was chosen for analysis. This allowed batch processing of many acquired series. The low labeling and the selection of in-focus QDs yielded 1–10 tracked QDs per video series. The tracking routines were written in DIPimage (TU Delft, www.qi.tnw.tudelft.nl/DIPimage) a toolbox for Matlab (The Mathworks, Massachusetts, USA). The tracking was performed offline after acquisition.

6.B.2
Real Time Optically-sectioned Imaging with the PAM

In the LCoS-based PAM, the conjugate and the non-conjugate images are recorded simultaneously on the same CCD camera side-by-side. To yield a properly optically sectioned image, subtraction of these images is required. Before acquisition, the two images are registered by optimizing the two-dimensional correlation coefficient by a search procedure using a fixed step size. The registration parameters are translation (using a step size of 0.1 pixels), rotation (step size 0.001 radians), and magnification (step size 0.001). During acquisition, a background image is first subtracted from the image pair. Then, the non-conjugate image is transformed to overlap the conjugate image by applying the established registration parameters. For subpixel transformations, bilinear interpolation is used. The images are then subtracted using a weighting factor which is dependent on the duty cycle of the pattern used for acquisition [13]. The final image is scaled, and offset removed. If desired, a Gaussian filter with an adjustable sigma is applied. In order to perform the transformation and subtraction with sufficient speed, it is carried out on a NVidia Quadro FX 4400 GPU board using the BrookGPU GPU library [22] and DirectX 9 runtime. The GPU board was coupled to an Intel Xeon 3.2 GHz processor. By offloading the image processing to the GPU, the computation time is less than the fastest possible PAM exposure time required for a full scan (~ 16 ms). Interaction with the PAM is thus real-time, i.e. optically-sectioned images are displayed at video rate on the screen.

References

1 Yarden, Y. and Sliwkowski (2001) Untangling the ErbB signaling network. *Nat. Rev. Mol. Cell Biol.* **2**, 127–137.

2 Lidke, D.S., Nagy, P., Heintzmann, R., Arndt-Jovin, D.J., Post, J.N., Grecco, H.E., Jares-Erijman, E.A. and Jovin, T.M. (2004) Quantum dot ligands provide new insights into erbB/HER receptor-mediated signal transduction. *Nat. Biotech.* **22**, 198.

3 Lidke, D.S., Lidke, K.A., Rieger, B., Jovin, T.M. and Arndt-Jovin, D.J. (2005) Reaching out for signals: filopodia sense EGF and respond by directed retrograde transport of activated receptors. *J. Cell Biol.* **170**, 619–626.

4 Grecco, H.E., Lidke, K.A., Heintzmann, R., Lidke, D.S., Spagnuolo, C., Martinez, O.E., Jares-Erijman, E.A. and Jovin, T.M. (2004) Ensemble and single particle photophysical properties (two-photon excitation, anisotropy, FRET, lifetime, spectral conversion) of commercial quantum dots in solution and in live cells. *Microsc. Res. Tech.* **65**, 169–179.

5 Jares-Erijman, E.A. and Jovin, T.M. (2003) FRET imaging. *Nat. Biotech.* **21**, 1387.

6 Coates, C.G., Denvir, D.J., McHale, N.G., Thornbury, K.D. and Hollywood, M.A. (2004) Optimizing low-light microscopy with back-illuminated electron multiplying charge-coupled device: enhanced sensitivity, speed, and resolution. *J. Biomed. Opt.* **9**, 1244–1252.

7 Miskoski, S., Giordano, L., Etchehon, M.H., Menendez, G., Lidke, K.A., Hagen, G.M., Jovin, T.M. and Jares-Erijman, E.A. (2006) Spectroscopic modulation of multifunctionalized quantum dots for use as biological probes and effectors. *Proc. SPIE 6096, 60960X.*

8 Lidke, K.A., Rieger, B., Jovin, T.M. and Heintzmann, R. (2005) Superresolution by localization of quantum dots using blinking statistics. *Opt. Express* **12**, 7052–7062.

9 Suzuki, K., Ritchie, K., Kajikawa, E., Fujiwara, T. and Kusumi, A. (2005) Rapid hop diffusion of a G-protein-coupled receptor in the plasma membrane as revealed by single-molecule techniques. *Biophys. J.* **88**, 3659–3680.

10 Hanley, Q.S., Verveer, P., Gemkow, M., Arndt-Jovin, D.J. and Jovin, T.M. (1999) An optical sectioning programmable array microscope implemented with a digital micromirror device. *J. Microsc.* **196**, 317–331.

11 Hanley, Q.S. and Jovin, T.M. (2001) Highly multiplexed optically sectioned spectroscopic imaging in a programmable array microscope. *App. Spectrosc.* **55**, 1115.

12 Hanley, Q.S., Lidke, K.A., Heintzmann, R., Arndt-Jovin, D.J. and Jovin, T.M. (2005) Fluorescence lifetime imaging in an optically sectioning programmable array microscope (PAM). *Cytometry Part A* **67A**, 112–118.

13 Heintzmann, R., Hanley, Q.S., Arndt-Jovin, D. and Jovin, T.M. (2001) A dual path programmable array microscope (PAM): simultaneous acquisition of conjugate and non-conjugate images. *J. Microsc.* **204**, 119–135.

14 Liang, M., Stehr, R.L. and Krause, A.W., (1997) Confocal pattern period in multiple-aperture confocal imaging systems with coherent illumination. *Opt. Lett.* **22**, 751–753.

15 Fukano, T. and Miyawaki, A. (2003) Whole-field fluorescence microscope with digital micromirror device: imaging of biological samples. *Appl. Opt.* **42**, 4119–4124.

16 Cha, S., Lin, P.C., Zhu, L., Sun, P.-C. and Fainman, Y. (2000) Nontranslational three-dimensional profilometry by chromatic confocal microscopy with dynamically configurable micromirror scanning. *Appl. Opt.* **39**, 2605–2613.

17 Lane, P.M., Dlugan, A.L.P., Richards-Kortum, R. and MacAulay, C.E. (2000) Fiber-optic confocal microscopy using a spatial light modulator. *Opt. Lett* **25**, 1780–1782.

18 Smith, P.J., Taylor, C.M., Shaw, A.J. and McCabe, E.M. (2000) Programmable array microscopy with a ferroelectric liquid-crystal spatial light modulator. *Appl. Opt.* **39**, 2664–2669.

19 Neil, M.A.A., Juskaitis, R. and Wilson, T., (1997) Method of obtaining optical sectioning by using structured light in a conventional microscope. *Opt. Lett* **22**, 1905–1907.

20 Harwit, M. and Sloane, N.J.A., (1979) *Hadamard Transform Optics*, Academic Press, New York.

21 Heintzmann, R. and Benedetti, P.A. (2006) High-resolution image reconstruction in fluorescence microscopy with patterned excitation. *Appl. Opt.* **45**, 5037–5045.

22 Buck, I., Foley, T., Horn, D., Sugerman, J., Fatahalian, K., Houston, M. and Hanrahan, P. (2004) Brook for GPUs: Stream Computing on Graphics Hardware. *ACM Trans. Graph.* **23**, 777–786.

23 Arndt-Jovin, D.J., Lopez-Quintela, M.A., Lidke, D.S., Rodriguez, M.J., Santos, F.M., Lidke, K.A., Hagen, G.M. and Jovin, T.M. (2006) *In vivo* cell imaging with semi-conductor quantum dots and noble metal nanodots. *Proc. SPIE 6096 60960P*.

24 Fulwyler, M.J. Hanley, Q.S., Schnetter, C., Young, I.T., Jares-Erijman, E., Arndt-Jovin, D.J. and Jovin, T.M. (2005) Selective photoreactions in a Programmable Array Microscope (PAM): Photoinitiated polymerization, photodecaging and photochromic conversion. *Cytometry* **67A**, 68–75.

25 Fry, D.W., Kraker, A.J., McMichael, A., Ambroso, L.A. Nelson, J.M., Leopold, W.R., Connors, R.W. and Bridges, A.J., (1994) A specific inhibitor of the epidermal growth factor receptor tyrosine kinase. *Science* **265**, 1093–1095.

7
Studying the Dynamics of Ligand–Receptor Complexes by Single-Molecule Techniques

Christophe Danelon and Horst Vogel

7.1
Introduction

The specific recognition and binding between molecules is fundamental to biological function. The first step in essentially all cellular signaling activities is the interaction between a ligand and a protein to form a defined complex that triggers downstream intracellular reactions.

Here, we will focus on ligand–receptor interactions on cellular membranes. Signal transduction across cellular membranes relies on a complex biochemical network involving chemical or physical stimuli, membrane receptors and cytosolic proteins. The signal initiated by ligand binding is transduced across the receptor protein to modulate subsequent interactions on the opposite side of the membrane. Membrane receptors can be classified on the basis of the mechanism by which they transfer the information:

(i) G protein-coupled receptors (GPCRs) mediate the detection of photons or diverse chemical compounds such as hormones, neurotransmitters, odorants, and nucleosides [1] to the activation of G proteins located on the cytosolic face of the plasma membrane, which finally stimulate other membrane effectors. Rhodopsin, opioid receptors, olfactory receptors are prototypical representatives of GPCRs.

(ii) Ion channels and transporters allow the translocation of ions or molecules across cellular membranes. Ligand-activated ion channels expose one or several affinity sites for agonist molecules, the binding of which gates the channel to an open state for specific ions [2]. The ligand can be extracellular such as a neurotransmitter, or intracellular e.g. a cyclic nucleotide or inositol-3-phosphate. There is another type of receptor channels that comprise a binding site at the interior of the pore. These include specific bacterial channels which have been studied in detail with regard to the translocation mechanism.

Single Molecule Dynamics in Life Science. Edited by T. Yanagida and Y. Ishii
Copyright © 2009 WILEY-VCH Verlag GmbH & Co. KGaA, Weinheim
ISBN: 978-3-527-31288-7

(iii) Enzyme-linked receptors possess a catalytic site on their interior which is activated by the binding of signaling ligands in the external region. The response of the receptor to growth factors and hormones is to trigger a guanylate cyclase or a kinase activity. Members of the receptor tyrosine kinases include epidermal growth factor receptor, insulin receptor, and vascular endothelial growth factor receptor.

The pharmacology of ligand binding has recently been documented elsewhere and is not addressed here (for a recent review see [3]). This chapter is devoted to recent advances related to the application of single-molecule techniques to the study of ligand interactions with membrane-bound receptors. Current progresses in micro- and nanofabrication have led to the development of new single-molecule methods that have proved to be valuable tools for investigating molecular interactions in biological systems. A number of these emerging technologies will be presented.

7.2
Labeling Methods for Cell Surface Receptors

7.2.1
General Considerations

Novel optical microscopies and spectroscopies play an important role in the elucidation of biochemical interactions in live cells. In order to take full advantage, the selective labeling of cellular components is required. Current orthogonal labeling strategies suited for single-molecule imaging and spectroscopy of proteins and other components in live cells or *in vitro*, involve the use of (i) fluorescent analogs of ligands, (ii) fluorescent antibodies binding to selective epitopes on the target protein or fluorescent streptavidin binding to strep-tags or biotin on the target protein, (iii) fusion to autofluorescent proteins (AFPs), or direct labeling of proteins either (iv) posttranslationally by chemical reactions of activated chromophores (chemical labeling of OH, -COOH, -NH_2 or -SH comprising amino acid side chains; reversible labeling of polyhistidine sequences with nitrilotriacetate (NTA) comprising chromophores) or (v) introducing non-natural fluorescent amino acids during protein synthesis using the suppressor tRNA technology, and finally (vi) posttranslational labeling of carrier proteins fused to the protein under investigation. Each labeling technique has particular advantages and disadvantages:

(i) Receptor ligands can be coupled to diverse high-performance organic fluorophores but this is a cost- and labor-intensive process requiring a different synthesis strategy for each particular ligand. The conjugate often has substantially modified properties when compared to the unlabeled ligand and only allows for the study of the ligand-bound state of the receptor. Nevertheless, a number of groups have undertaken this approach with some effort and considerable success [4, 5].

(ii) Fluorescent antibodies are available for a broad range of epitopes on proteins, but they are often larger than the protein of interest, may crosslink the target proteins and therefore may interfere with its mobility and functionality. In addition to antibodies raised against epitopes specific for the particular protein [6], there are antibodies which have been raised against generally used sequence-tags, for instance those used for protein purification such as strep-, myc-, flag- or polyhistidine-tags [7, 8] or even anti-GFP antibodies which can be used to double-label GFP fusion proteins with additional or more suitable fluorophores. Once it has been established that these antibodies do not interfere with the function of the target, they may constitute handy labels in situations where signal strength is an issue, because they can carry multiple fluorophores (up to 10). In a comparable approach, target proteins have been selectively labeled with fluorescent streptavidin either at strep-tags or at biotinylated protein sites [7–9]. In this case, the size of the probe and its potential crosslinking capability may also change the functionality of the target proteins.

(iii) Green fluorescent protein (GFP) and its variants [10, 11] can be attached genetically at different positions to the target protein in a 1:1 stoichiometry, but it suffers from poor photostability, the tendency to form oligomers and spectral overlap with other cellular luminescent compounds leading to substantial autofluorescence. Also, for the study of proteins in the plasma membrane it is preferable to label only the properly translocated fraction, especially in cases where this fraction is small. Although AFPs are not the best choice for SPT in terms of their photophysical properties, their beneficial biochemical properties are responsible for their impact in intracellular FCS applications [12].

(iv) Direct covalent labeling of particular amino acid side chains bearing -COOH, -OH, -SH and -NH2 groups by activated chromophores are an ideal method for imaging purified single proteins but usually lack selectivity in the case of live cells, resulting in a large fluorescence background [13]. Nevertheless, single voltage-gated ion channels, labeled by this approach, have been investigated in living frog oocytes thereby detecting structural changes of the channel during voltage-gating [14–16]. For technical details we refer to the comprehensive collection of examples, mostly on ensemble measurements, in *The Handbook – A Guide to Fluorescence and Labeling Technologies* from Molecular Probes/Invitrogen (http://probes.invitrogen.com/handbook/).

A novel covalent labeling of specific tetracystein sequences by biarsenical-bearing chromophores was developed by Tsien *et al.* [17, 18]. Although yet only applied to ensemble imaging of proteins in living cells, they offer interesting potential for single-molecule microscopy: the reactive yet non-fluorescent chromophore permeates the cellular membrane and becomes fluorescent after reacting with the tetracystein sequence motive on a particular target protein inside of a living cell.

Several strategies have been developed in recent years to overcome some of the limitations within the methods mentioned above. A few recent reviews summarize these efforts [19–21]. In the following we report on our efforts to use the novel approaches for single-molecule microscopies.

7.2.2
Suppressor tRNA Technology

Several methods have been developed to incorporate unnatural amino acids site-specifically into proteins in mammalian cells. Chemically aminoacylated suppressor tRNAs have been microinjected or electroporated into CHO cells and neurons, and used to suppress nonsense amber mutations with a series of unnatural amino acids [22]. Thereby it was possible to incorporate unnatural amino acids with diverse physicochemical and biological properties into defined positions of the sequence of target proteins expressed in mammalian cells. This method has been widely used to probe channel proteins by electrophysiology [23] and was used for the first time by Turcatti *et al.* [24] to insert non-natural fluorescent amino acids at specific sites in a GPCR (NK2 receptor) in frog oocytes for exploiting by fluorescence resonance energy transfer (FRET) the receptor structure. Meanwhile, Schultz *et al.* extended the method in an elegant, general approach (for a recent review see [25]): a mutant *Escherichia coli* aminoacyl-tRNA synthetase (aaRS) is first evolved in yeast to selectively aminoacylate its tRNA with the unnatural amino acid of interest. This mutant aaRS together with an amber suppressor tRNA is then used to site-specifically incorporate the unnatural amino acid into a protein in mammalian cells in response to an amber nonsense codon. This and other approaches [26] overcome the originally low efficiency of expressed proteins.

7.2.3
O6-Alkylguanine–DNA Alkyltransferase (AGT)

The labeling is based on the irreversible and specific reaction of human O6-alkylguanine–DNA alkyltransferase (hAGT) with fluorescent derivatives of O^6-benzylguanine (BG), leading to the transfer of the synthetic probe to a reactive cysteine residue on hAGT which is fused to a protein of interest [27, 28]. Wild-type hAGT is a monomeric protein of 207 aa, the 30 C-terminal residues of which can be deleted without affecting the reactivity against BG, making it smaller than autofluorescent proteins. Importantly, the rate of the reaction of AGT fusion proteins with BG derivatives is independent of the nature of the label, opening up the possibility of labeling a single AGT fusion protein with a variety of different probes. Specific labeling of AGT fusion proteins in mammalian cells can be achieved by using AGT-deficient cell lines, and it has been shown that nuclear-localized AGT can be labeled specifically with fluorescein in such cells [27]. Synthetic BG derivatives can be used for the sequential labeling of AGT fusion proteins with a variety of different fluorophores in live mammalian cells and the approach can be used for multicolor analysis of cellular processes and FRET measurements [27, 28]. A mutant form of AGT was recently created that can be specifically labeled with substrates that are not accepted by wild-type AGT [29]. This permits the labeling of two different AGT fusion proteins with various fluorophores in the same cell or *in vitro*.

7.2.4
Acyl-carrier Protein (ACP)

This labeling comprises the enzymatic transfer (using specific phosphopanteinyl transferases) of the phosphopantetheine of a fluorescent Coenzyme A (CoA) conjugate to an ACP fused to the target protein [30–32]. This technique enabled us to track single G protein-coupled receptors (GPCRs) during signaling because several obstacles had been overcome: (i) proteins not properly inserted in the plasma membrane were not labeled and hence did not contribute to out-of-focus background fluorescence. (ii) The fraction of labeled protein could be precisely controlled in time (pulse labeling) with the possibility of labeling repetitively at subsequent times, ensuring a low and well defined dye concentration required for single-molecule detection. (iii) Free choice of probes allows protein labeling with long-wavelength dyes (for example Cy5, Atto dyes) of high absorption cross section, high quantum yield and high photo-stability to be used as label for an improved signal-to-background ratio and observation time. (iv) Multicolor labeling of defined ratio(s) between different probes is easily possible by controlling the composition of substrates in the bulk medium. (v) Non-specific binding of the label to the plasma membrane, the most crucial obstacle for the successful application of single-molecule microscopies, was reduced by the presence of the hydrophilic CoA-moiety. (vi) Recent developments reduced the size of the ACP tag from 76 to 11 residues without losing too much specificity and efficiency of the labeling reactions, making this approach even more attractive [33].

7.2.5
Nitrilotriacetate (NTA)

This is a generic method for selectively labeling proteins *in vivo* and *in vitro*, rapidly (within seconds) and reversibly, with small molecular probes that can have a wide variety of properties [34]. The probes comprise a chromophore and a metal ion-chelating NTA moiety, which binds reversibly and specifically to engineered oligo-histidine sequences in proteins of interest. The feasibility of the approach was demonstrated by binding NTA–chromophore conjugates to a representative ligand-gated ion channel and a GPCR, each containing a polyhistidine sequence. In contrast to the transient binding of such a conventional mono-NTA-bearing fluorophore, multivalent-NTA-bearing fluorophores form complexes with oligohistidine sequences in proteins, which show increased lifetimes of more than 1 h [35, 36].

Hence, the residence time of the NTA label can be tuned according to the specific needs from seconds to hours.

7.2.6
Reversible Sequential Labeling (ReSeq)

A novel method for performing reversible sequential (ReSeq)-binding assays on particular neuroreceptors has recently been developed [37]. With this assay, a series of investigations can be performed on the same cell by repetitively applying specific, fluorescently labeled ligands that have fast association–dissociation kinetics. Complete saturation ligand-binding and competition ligand-binding assays have been obtained on a single cell with excellent accuracy and reproducibility. This new approach offers several advantages as it (i) substantially reduces the number of cells needed, (ii) allows the investigation of cell-to-cell variations because extensive data can be collected from individual cells, and (iii) circumvents problems related to low expression levels of receptors and photobleaching of fluorescent ligands, since measurements can be repetitively performed on the same cell to enhance accuracy. Moreover, ReSeq-binding assays can be easily automated and implemented in on-chip analysis, which offers a substantial improvement in reliability, efficiency, and reduction of sample consumption. In a single-molecule imaging experiment, typically tens of frames are recorded before fluorophore photobleaching occurs. Because a reversibly binding fluorescently-labeled ligand can be washed off completely and added fresh once again, single-molecule experiments can be repeated more than 30 times with one cell, and data with excellent statistics can be collected.

7.3
Functional Mobility of Receptors in Cell Membranes

7.3.1
Organization and Dynamics of Cell Membranes

The organization of biological membranes and the lateral diffusion of membrane proteins are involved in a large number of signaling processes. For example, the distribution and translational diffusion of membrane receptors regulate the plasticity and function of neuronal (review: [38]) and immunological synapses [39], growth and guidance of axons [40] and the gradient sensing of attractant molecules by chemotactic cells [41]. Signal transduction at cellular membranes is often initiated by ligand binding and followed by the formation of dynamic molecular networks involving the active receptors. Interactions with structuring elements such as lipid shells and membrane-associated or cytoskeletal components [42–44], or with other membrane proteins [45, 46], microtubule-associated proteins [47–49] or molecules in the membrane of adjacent cells, inevitably restrict receptor movements, as compared to Brownian diffusion, and participate in their localization to specialized micro- and nanometer-sized environments.

7.3.2
Techniques

Classical techniques for investigating the lateral diffusion of fluorescently-labeled lipids and proteins are fluorescence recovery after photobleaching (FRAP or FRP) and post-electrophoresis relaxation (PER) that measure the local translational diffusion coefficient and the mobile fraction of diffusing species on surfaces of living cells [50]. More recently, the trajectories of individual molecules on cellular membranes have been monitored using a variety of reporting labels such as colloidal particles and latex beads by nanovideomicroscopy [51–53], fluorescent organic probes [54, 55] and quantum dots [44, 56, 57] by fluorescence imaging, and gold nanoparticles by photothermal interference contrast [58, 59]. Tracking of various membrane receptors on different cell types or organelles revealed the heterogeneity of molecular motions on cell membranes as reflected by the presence of various diffusion modes and the broad distribution of (restricted) mobilities in length and time scales [60, 61].

Another complementary approach to the investigation of the rapid single-molecule dynamics in living cells is fluorescence correlation spectroscopy (FCS). Although the principles of FCS were formulated in the 1970s, only recent technological advances providing high sensitivity and specific protein labeling allowed non-invasive applications to live cells (reviews: [62, 63]). Diffusion of molecules across a diffraction-limited focal volume introduces fluctuations in the fluorescence intensity characteristic of their velocities (Figure 7.1A, B). The autocorrelation function G (τ) that describes the two-dimensional Brownian diffusion of multiple membrane components is given by:

$$G(\tau) = 1 + \frac{1}{N} \sum_{i=1}^{n} \frac{\gamma_i}{1 + \frac{\tau}{\tau_i}}, \qquad (7.1)$$

where γ_i and τ_i are the fraction and the diffusion time constant of the species i, respectively, and N is are the average number of molecules in the detection volume. The diffusion coefficient D_i of the species i is related to the diffusion time by:

$$D_i = \frac{w_{xy}^2}{4\tau_i}, \qquad (7.2)$$

where w_{xy} is the radius of the detection volume in the focal plane. Complex diffusional behaviors that originate from non-planar membrane topology or interaction networks can also be investigated by FCS.

Therefore, the effects of ligand binding to its membrane receptor can be observed as changes in the diffusional behavior of the active receptor along the different molecular events of the signaling pathway. In the Table 7.1, we present some examples that stress the utility of single particle/molecule tracking and FCS to better understand the mechanism of signal transduction mediated by membrane receptors on cultured cells and to address the question of the functional implication of the receptor mobility.

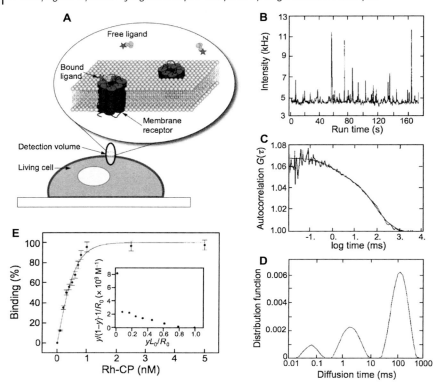

Figure 7.1 Application of FCS to membrane receptor mobility and ligand binding on living cells (Adapted from [137]). (A) The detection volume is positioned on the cell surface for monitoring the binding of fluorescent ligands to membrane receptors. (B) Fluorescence time course for tetramethylrhodamine-labeled C-peptide bound to membrane receptors on cultured human renal tubular cells. (C) Autocorrelation function of the fluorescence intensity fluctuations shown in B. The observed autocorrelation was fitted by the diffusion model described by Equation 7.3. Two components of receptor-bound C-peptide were observed. Diffusion times and corresponding fractions were $\tau_{b1} = 80$ ms, $y_1 = 0.75$; $\tau_{b2} = 1$ ms, $y_2 = 0.15$. A small fraction of unbound C-peptide that diffuses into the volume element extending into the extracellular medium was characterized by $\tau_f = 0.15$ ms and $y_3 = 0.10$. (D) Distribution of diffusion times calculated from the auto-correlation function in C. (E) Percent of bound ligand versus the total Rh-CP concentration. Inset: Scatchard plot according to Equation 7.4.

7.4
Investigating Kinetics and Thermodynamics of Ligand–Receptor Interactions by FCS

7.4.1
Principles

The most common methods to evaluate receptor–ligand characteristics are the radioligand binding assays, which require the use of radioactively-labeled

Table 7.1 Single-molecule experiments investigating ligand binding and receptor diffusion in living cells.

Receptor	Ligand	Technique	Receptor mobility and ligand binding properties	Reference
GPCR				
A1-AR in CHO cells	Antagonist **xanthine amine congener** labeled with Bodipy 630	FCS	*Diffusion:* One population of slow-diffusing receptor-bound ligands was observed. *Binding:* The fluorescent competitive antagonist binds to the intramembrane region of A1-AR. Binding isotherms yield $K_d = 33$ nM and $B_{max} = 75$ nM.	[139]
A1-AR in CHO cells	N^6-**aminobutyl adenosine** labeled with Bodipy 630	FCS	*Diffusion:* Two populations of ligand–receptor complexes with different mobility were observed.	[140]
β$_2$-**AR** in hippocampal neurons and alveolar epithelial type II cell line A549	**Arterenol** labeled with Alexa 532	FCS	*Diffusion:* Two populations of ligand–receptor complexes with different mobility were observed. The fast-diffusion component appears immediately after ligand addition, whereas the slow-diffusing component is delayed, which is likely due to the internalization of ligand–receptor complexes. *Binding:* Equilibrium and kinetic experiments revealed $K_d = 1.3$ nM, $B_{max} = 6.0$ nM, and $k_{ass} = 4.9 \times 10^8$ M^{-1} s^{-1} in neurons, and $K_d = 6.0$ nM, $B_{max} = 15.1$ nM, and $k_{ass} = 1.2 \times 10^8$ M^{-1} s^{-1} in A549 cells.	[141]

(Continued)

Table 7.1 (Continued)

Receptor	Ligand	Technique	Receptor mobility and ligand binding properties	Reference
cAMPR in Dictyostelium discoideum cells	cAMP labeled with Cy3	SMI (TIRFM)	Diffusion: Regulation of chemotactic signaling by receptor mobility. Binding: Studies of the time course of single fluorescent spots and release curves revealed different dissociation kinetics at the anterior and posterior regions, with the main fraction of bound ligands having $k_{diss} = 1.1\,s^{-1}$ and $0.4\,s^{-1}$, respectively. The difference derived from intrinsic cell polarity was caused by altered interactions between the receptor and functional G proteins rather than by receptor phosphorylation. The GTP-dependent decrease in the lifetime of bound cAMP molecules was related to functional G proteins. It was postulated that tight association of the receptor and G protein stabilizes agonist binding.	[41]
C-peptide receptor in several human cell types	Human C-peptide and insulin labeled with tetramethylrhodamine	FCS	Diffusion: Two populations of ligand–receptor complexes with different mobility. Binding: C-peptide binding isotherm yielded two binding processes: a low affinity with stoichiometry 1:1 ($n = 1$), and possibly a high affinity ($K_d < 0.3\,nM$) with $n < 1$. Time-dependent displacement showed a single dissociation rate constant of $k_{diss} = 4.5 \times 10^{-4}\,s^{-1}$. The results suggest the existence of at least two different complexes: one of low affinity and high mobility and one of high affinity and low mobility. Pertussis toxin abolished binding indicating that C-peptide binds to a GPCR.	[137]

C-peptide receptor in human skin fibroblasts	C-peptide labeled with tetramethylrhodamine	FCS	*Diffusion:* Two populations of ligand–receptor complexes with different mobility. *Binding:* C-peptide binding isotherm yielded $K_d = 0.3$ nM. Functional C-peptide receptors could be solubilized with Chaps detergent.	[142]
C-peptide receptor in tubular cells	C-peptide labeled with tetramethylrhodamine	FCS	*Binding:* Displacement of bound tetramethylrhodamine-labeled C-peptide by synthetic C-peptide fragments. Residue Glu27 at the C-terminal part is particularly important for specific binding.	[143]
ET$_A$R in virus-like particles	Endothelin-1 labeled with tetramethylrhodamine	FCS and FIDA	*Binding:* Equilibrium binding experiments yielded $K_d \sim 0.3$ nM similar to values measured in living cells.	[144]
Galanin receptor in cultured insulinoma cells	Galanin labeled with tetramethylrhodamine	FCS	*Diffusion:* Two populations of ligand–receptor complexes with different mobility that might belong to different subpopulations or receptor subtypes. *Binding:* Binding isotherm yielded one binding site with $K_d = 1.2$ nM ($n = 1$). Displacement experiments yielded a single dissociation rate constant $k_{diss} = 3.7 \times 10^{-4}$ s^{-1}. Pertussis toxin abolished binding indicating that galanin binds to a GPCR.	[145, 146]
mGluR5 in hippocampal neurons and Ptk2 cells tagged with 0.5-μm latex beads coated with anti-myc antibodies	DHPG	SPT	*Diffusion:* Regulation of mGluR5 diffusion by the sub-membranous scaffolding protein Homer and by receptor activity. In presence of DHPG the diffusion coefficient was increased by a factor of three but the relative distribution of diffusive and confined states was not modified.	[147]
NK1R labeled with EGFP in HEK 293 cells	Tetramethylrhodamine-labeled substance P	SMI	*Binding:* The mobility of the ligand–receptor complexes was slowed down during the first 1 s after ligand binding and became more confined.	[148]

(Continued)

Table 7.1 (Continued)

Receptor	Ligand	Technique	Receptor mobility and ligand binding properties	Reference
OR17-40 fused to the acyl carrier protein and fluorescently labeled with CoA-Cy5 in HEK 293 cells	Agonist **helional** Antagonist α-methyl-**cinnamaldehyde**	SMI	*Diffusion*: Regulation of the internalization dynamics of OR17-40. Both agonist and antagonist induced confinement in ~190-nm-sized domains which are probably precursors of clathrin-coated pits.	[149]
SSTR1 and **SSTR5** in CHO-K1 cells	**Somatostatin** labeled with FITC or Texas red	FRET and FCS	*Diffusion*: SSTR5 homooligomerization and SSTR5/SSTR1 heterooligomerization was regulated by ligand binding.	[150]
Ion channels				
5-HT$_3$R in HEK 293 cells	Antagonist **GR** labeled with Cy5	FCS	*Diffusion*: The majority of the clustering receptors are immobile, a minority shows diffusion.	[151]
5-HT$_3$R tagged with a polyhistidine sequence and labeled with Ni^{2+}-NTA-Atto 647 in HEK 293 cells	Antagonist **GR** or GR labeled with Cy5 Agonist **serotonin**	SMI	*Diffusion*: Serotonin slightly decreased the fraction of mobile receptors.	[152]
AMPAR containing the glutamate receptor subunit GluR2 in cultured hippocampal neurons and tagged with 0.5-μm latex beads coated with anti-GluR2 antibodies	Glutamate	SPT	*Diffusion*: Regulation of diffusion by neuronal maturation, synaptic sites, basal neuronal activity (action potentials, intracellular calcium) and activation with glutamate. The mobility of GluR2 was slowed down by glutamate and calcium.	[153]

AMPAR containing the glutamate receptor subunit GluR2 in cultured hippocampal neurons and tagged with anti-GluR2 antibody labeled with Cy5 or Alexa 647	Glutamate	SMI	*Diffusion*: Glutamate application increased both receptor mobility inside synapses and the proportion of mobile receptors in the juxta-synaptic region.	[154]
Benzodiazepine receptor in hippocampal neurons	Benzodiazepine Ro7-1986/602 (N-des-diethyl-fluor-azepam) labeled with Alexa 532	FCS	*Diffusion*: Two populations of ligand–receptor complexes with different mobility. A clustering preceding internalization of the GABA$_A$R–ligand complexes could explain the slower diffusion time. *Binding*: Saturation experiments yielded $B_{max} = 38$ nM and $K_d = 9.9$ nM. Time course displacement with an excess of non-labeled agonist, midazolam, showed a mono-exponential decrease with $k_{diss} = 1.3 \times 10^{-3}$ s^{-1}.	[155][a]
GABA$_A$R in hippocampal neurons	Muscimol labeled with Alexa 532	FCS	*Diffusion*: Two populations of ligand–receptor complexes with different mobility. *Binding*: Binding isotherm yielded one binding site per receptor with $K_d = 3.4$ nM and $B_{max} = 18.4$ nM. Time-dependent displacement experiments showed a mono-exponential decrease with $k_{diss} = 5.4 \times 10^{-2}$ s^{-1}. A positive cooperativity of ligand binding yielding to a selective increase in the fraction of ligand–receptor complexes with hindered mobility was observed by co-incubation with midazolam [156], protoberberine type 2 alkaloids [157], or xanthohumol [158].	[156–158]
GABA$_A$R in human cortical neurons	Kavain labeled with tetramethylrhodamine	FCS	*Diffusion*: One fraction of bound ligand.	[159][b]

(Continued)

Table 7.1 (*Continued*)

Receptor	Ligand	Technique	Receptor mobility and ligand binding properties	Reference
			Binding: Non-linear Scatchard plot and sigmoid Hill plot with $n > 1$ indicate more than one binding site with different affinities. The mean $K_d = 1.6$ nM.	
Enzyme-linked				
$\alpha_5\beta_1$**Integrin** in fibroblasts	**Fibronectin** (FN7-10 recombinant fragment) conjugated to 1-µm beads	SPT and optical trapping	*Diffusion*: Low density antibody-coated beads displayed free diffusion (monomeric unliganded integrins). Fibronectin- or high density antibody-coated beads induced rearward motion on lamellipodia. Cross-linking of integrins promoted attachment to the cytoskeleton. Interaction of Integrins with rigid sites in the extracellular matrix was modeled by optical trapping of the beads. Restriction of integrin movement caused localized reinforcement of the links with the cytoskeleton.	[160]
$\alpha_L\beta_2$**integrin (LFA-1)** tagged with 1-µm beads conjugated with antibody directed against conformation-dependent epitopes in T-cells	**ICAM-1**	SPT	*Diffusion*: Regulation of LFA-1 receptor mobility by ligand-induced conformational changes. LFA-1 exists in at least two distinct populations on both resting and PMA-activated cells. Cell activation and disruption of cytoskeletal interactions increased the fraction of mobile receptors indicating that on resting cells a pool of receptors was constrained by cytoskeletal interactions. Receptors in the active, open conformation became largely immobile and experienced directed motion upon cell activation. ICAM-1-ligated LFA-1 receptors were largely immobile on both resting and activated states, due to cytoskeletal attachment.	[161]

System	Ligand	Method	Description	Reference
β1 integrin tagged with 0.2-μm beads coated with antibody in sympathetic neurons	GRGDS peptide	SPT	*Diffusion:* The effect of NGF on the movement of β1 integrin, a subunit of the receptor for laminin and fibronectin, was studied. Binding of GRGDS peptide to β1 integrin caused rearward movement. NGF stimulated rapid displacement of unliganded (low-crosslinked) receptors and forward excursions toward the tips of filopodia. Forward transport was dependent on actin filaments and myosin ATPase activity. NGF stimulated the coupling of crosslinked β1 integrins to the retrograde flow of actin.	[162]
β1 integrin tagged with 40-nm gold particles coated with antibody or fibronectin in fibroblasts	RGD peptide Fibronectin	SPT	*Diffusion:* The effect of RGD peptide or fibronectin binding to integrins on the receptor lateral diffusion was studied. Ligand binding regulates integrin association with dynamic components of the cytoskeleton, which is crucial for directed receptor movement. Directed motion was mediated by crosslinking between the cytoplasmic domain of the β1 subunit and the rearward-moving cytoskeleton.	[163]
Class II MHC receptor in CHO cells	Moth Cytochrome C peptide labeled with Cy5	SMI	*Diffusion:* Both GPI-linked and native class II MHC receptors exhibit predominant Brownian motion with no influence of actin or tubulin cytoskeletal networks.	[164]
EGFR in human diploid fibroblasts	EGF labeled with rhodamine	FCS	*Diffusion:* Two populations of ligand–receptor complexes with different mobility, representative of different forms (dimers, oligomers) or subtypes of EGFR. *Binding:* Equilibrium saturation binding experiment yielded $K_d = 0.7\,\text{nM}$. The dissociation kinetics follow a single exponential function with $k_{diss} = 2.9 \times 10^{-4}\,\text{s}^{-1}$.	[165]

(Continued)

Table 7.1 (*Continued*)

Receptor	Ligand	Technique	Receptor mobility and ligand binding properties	Reference
EGFR in A431 carcinoma cells	**EGF** labeled with Cy3 or Cy5	SMI (TIRFM) and single molecule FRET	*Diffusion:* Single-molecule tracking revealed that in most cases, EGFR dimerization preceded the binding of the second EGF molecule. *Binding:* EGF binding was measured on both basal and apical surfaces. Formation of EGF–EGFR dimeric complexes was deduced by measuring the intensity distribution of the fluorescent spots at various times after the addition of EGF-Cy3. Dimerization initiated intracellular calcium response. Single-molecule FRET from Cy3-EGF to Cy5-EGF proved dimer formation. After dimerization, EGFR activation was demonstrated by the use of an antibody specific to the phosphorylated cytoplasmic domain.	[166]
EGFR (erbB1) in A431 epidermal cells	**EGF** conjugated to QD	SMI	*Diffusion:* Binding on filopodia and retrograde transport of activated receptors toward the cell body were observed. Directed transport of surface receptors proceeds via actin flow in filopodia. Accessibility experiments with exposure to an acidic medium or quenching by FRET demonstrated that transport precedes receptor internalization. *Binding:* Tracking with two-color EGF-QD ligands showed that retrograde transport requires dimerization of erbB1–ligand complexes (2:2 dimer complex).	[167]
IgE receptor FcεRI in RBL-2H3 cells	**IgE** labeled with Alexa 488 or 546	Two-photon FCS	*Diffusion:* Upon stimulation (IgE receptor cross-linking by multivalent antigen) both FcεRI and the Src family tyrosine kinase Lyn had a lower lateral mobility.	[168]

			Binding: Interaction between Lyn and activated FcεRI was revealed by FCS cross-correlation.	[169]
Insulin receptor in renal tubular cells	**Insulin** labeled with tetramethylrhodamine	FCS	*Diffusion:* Two populations of ligand–receptor complexes with different mobility. *Binding:* Insulin binding isotherm yielded two binding processes: a low affinity with $K_d = 1$ nM and stoichiometry 1:1 ($n=1$), and a high affinity with $K_d = 5 \times 10^{-2}$ nM and $n<1$.	
N-cadherin receptor in myogenic cells	**Ncad-Fc** ligand conjugated to 1-μm latex beads	SPT	*Diffusion:* The kinetics of the anchoring of N-cadherin to the actin cytoskeleton on lamellipodia was studied. The movement of ligand-coated beads showed an initial freely diffusing phase. The anchoring latency is lower at high ligand density at the bead surface.	[170]
NGFR (TrkA or/and p75 receptors) in dorsal root ganglion growth cones	**NGF** labeled with Cy3	SMI	*Diffusion:* Two diffusion modes for NGFR were observed: Brownian motion and one-directional movement toward the central region of the growth cone. Ligand–receptor complexes were transported toward the central region by the rearward flow of the actin network. The endocytosis of individual complexes (no need for cluster formation) was evidenced by exposure to an acidic medium. *Binding:* Counting the number of fluorescent dots revealed that lamellipoidal expansion was initiated by 40 bound Cy3-NGF. On high-affinity receptors, $K_d = 2.7 \times 10^{-2}$ nM and $B_{max} = 175$ bound ligands on a single growth cone.	[171]

(Continued)

Table 7.1 (*Continued*)

Receptor	Ligand	Technique	Receptor mobility and ligand binding properties	Reference
TCR and **CD8** tagged with fluorescently labeled antibody in T-cells	**Peptide MHC tetramer**	FCS and time-resolved fluorescence	*Diffusion:* Heterogeneous diffusion of both CD8 and TCR in the plasma membrane. Two populations of TCR and CD8 receptors with different mobility were observed. Some of the CD8 and TCR molecules are colocalized in membrane domains. *Binding:* The time-course of peptide MHC binding to live T cells was biphasic. A fast association to CD8 preceded association of the peptide MHC-CD8 complex with TCR.	[172]
TrkA receptor labeled with QD coated with anti-TrkA antibody in neural PC12 cells	RTA anti-TrkA that acts as a partial agonist **NGF**	SMI	*Diffusion:* Tracking of activated TrkA receptors revealed endocytosis and active transport via microtubules to different subcellular locations.	[173]

[a]The nature of the benzodiazepine receptor was not described in this reference.

[b]No evidence for direct interaction between GABA$_A$R and kavain was reported in this study, only the specific binding of kavain to cortical neurons was demonstrated. Abbreviations for receptors: 5-HT$_3$R, serotonin type 3 receptor; A1-AR, A1-adenosine receptor; AMPAR, α-amino-3-hydroxy-5-methyl-4-isoxazole propionic acid receptor; β$_2$-AR, β$_2$-adrenergic receptor; cAMPR, cyclic adenosine 3′,5′-monophosphate receptor; EGFR, epidermal growth factor receptor; ET$_A$R, endothelial A receptor; GABA$_A$R, γ-aminobutyric acid class A receptor; IgE receptor, immunoglobulin E receptor; LFA-1, leukocyte function-associated antigen-1; mGluR 5, metabotropic glutamate receptor 5; MHC receptor, major histocompatibility complex receptor; NGFR, nerve growth factor receptor; NK1R, neurokinin 1 receptor; OR17-40, human odorant receptor 17-40; SSTR1, somatostarin type 1 receptor; SSTR5, somatostarin type 5 receptor; TCR, T-cell receptor.

Other abbreviations: CHO, Chinese hamster ovary; FCS, fluorescence correlation spectroscopy; FIDA, fluorescence intensity distribution analysis; FRET, fluorescence resonance energy transfer; GPI, glycosyl phosphatidylinositol; HEK, human embryonic kidney; ICAM-1, intercellular adhesion molecule-1; PMA, phorbol-12-myristate-13-acetate; Ncad-Fc, N-cadherin ectodomain fused to the IgG Fc fragment; QD, quantum dots; SMI, single molecule imaging; SPT, single particle tracking; TIRFM, total internal reflection fluorescence microscopy.

compounds, usually tritium or iodine-125 isotopes. After incubating receptor-comprising membrane preparations or whole cells with radioligands, the unbound fraction of the labeled ligand is separated from the receptor-bound fraction. Major drawbacks of these binding assays are their low sensitivity, which excludes the investigation of cell surface receptors at low expression, and requires the requirement of long-living receptor–ligand complexes that should be stable during the separation and the washing steps.

FCS is not only a powerful technique for measuring the lateral mobility of membrane proteins, it is also a sensitive and non-invasive tool to determine the thermodynamic and kinetic parameters of ligand–receptor interactions in living cells with no need for the physical separation of the receptor-bound from the free ligand fraction (Table 7.1). Moreover, the small detection volume allows measurements in sub-cellular compartments such as the nucleus [64], the endoplasmic reticulum or the Golgi [65].

Quantitative analysis of the diffusion time and fraction of the different species is given by the following simplified autocorrelation function $G(\tau)$, which describes the three-dimensional diffusion of free ligands in solution and the two-dimensional diffusion of bound ligands in different populations of cell membrane receptors having different mobilities:

$$G(\tau) = 1 + \frac{1}{N}\left[\left(1 - \sum \gamma_i\right)\left(\frac{1}{1 + \frac{\tau}{\tau_f}}\right)\left(\frac{1}{1 + \left(\frac{w_{xy}}{w_z}\right)^2 \frac{\tau}{\tau_f}}\right)^{1/2} + \sum \gamma_i\left(\frac{1}{1 + \frac{\tau}{\tau_{bi}}}\right)\right],$$

(7.3)

where w_z is the vertical half axis of the detection volume, τ_f and τ_{bi} are the diffusion time constants of the free ligands and ligands bound to the population i of membrane receptors, respectively. It is assumed in the multicomponent diffusion model of Equation 7.3 that the quantum yield of the fluorescently-labeled ligand is the same in the free and bound states. Since the diffusion time scales with the cubic root of the molecular weight (Stokes–Einstein relation), the diffusion time of free ligands is substantially lower than the diffusion time of ligands bound to comparatively large receptors. This has been observed in binding experiments performed on detergent-solubilized membrane receptors [66]. The two components are more easily discernable for receptors diffusing in cell membranes due to the higher viscosity of the lipid bilayer (Figure 7.1C). Typically, the receptor-bound ligand diffuses more slowly, by about two orders of magnitude, than the free ligand in solution. The fraction of different species can be further examined by evaluating the distribution of diffusion times (Figure 7.1D and [137]).

The ligand-binding isotherm is obtained by determining at equilibrium the fraction of bound ligand from the autocorrelation amplitude for different ligand concentrations (Figure 7.1E). At saturation, the maximum concentration of bound ligand B_{max} equals the total number of binding sites. Assuming one binding site for one receptor molecule, the receptor density in the cell membrane is calculated from the average observed membrane area. It is useful to linearize the equation for the binding curve to display the Scatchard representation of the mass action law (Figure 7.1E, inset) according to:

$$\frac{\sum \gamma_i}{(1 - \sum \gamma_i) R_0} = \sum K_i \left(n_i - \frac{\gamma_i L_0}{R_0} \right),$$

(7.4)

where R_0 is the total receptor concentration at the cell surface, K_i is the association constant of the ligand–receptor complex, and n_i is the number of binding sites per receptor. Assuming $n_i = 1$, R_0 is equivalent to B_{max}.

The specificity and kinetics of ligand–receptor binding can be analyzed by competitive displacement of the bound ligand by an excess of non-labeled ligand. A simple monoexponential dissociation model is given by:

$$\sum \gamma_i = \gamma_{unspecific} + \gamma_0 \exp(-k_{diss} t),$$

(7.5)

where $\gamma_{unspecific}$ is the fraction of non-specific binding, γ_0 is the fraction of specific binding, and k_{diss} is the dissociation rate constant. The association rate constant of the complex, k_{ass}, is calculated by multiplying K_i by k_{diss}. More complex binding reactions, such as those involving several binding sites of different affinities as observed in ligand-binding isotherms and Scatchard plots, or multi-exponential displacement curves, can also be measured by FCS (Table 7.1).

In the particular case of large ligand molecules (i.e. with a similar mass as the receptor), binding to membrane receptors may not produce a sufficient effect on the diffusion time monitored by FCS. In this case fluorescence cross-correlation spectroscopy (FCCS) is the method of choice. In dual-color FCCS, the two interacting partners are labeled with spectrally separated fluorophores. Cross-correlation analysis of the two signals allows us to assess binding properties independently of the relative diffusion times [63].

7.4.2
FCS at High Fluorophore Concentrations

In a classical FCS experiment, the diffraction-limited observation volume restricts fluorophore concentrations to the pico and nanomolar ranges. Two different strategies have recently been presented to reduce the effective observation volume, thus enabling single-molecule detection at micromolar concentrations:

(i) The excitation light was confined at the bottom of subwavelength apertures that act as zero-mode waveguides made of a metal film on a fused silica coverslide [67]. The technique has been successfully applied to probe the dynamics of lipids and proteins on artificial and cell membranes [68–70].

(ii) The effective detection volume was reduced by allowing the fluorescent molecules to flow in submicrometer-sized fluidic channels [71].

With these methods, ligand–receptor interactions with micromolar equilibrium binding constants can now be analyzed with high spatial and temporal resolutions.

7.5
Modulation of Ion Channel Current by Ligand Binding

7.5.1
Ligand-activated Ion Channels: Decoupling Ligand Binding and Channel Gating with Single-molecule Patch-clamp

Ligand-gated ion channels switch between a closed (non-conducting) conformation and an open (conducting) conformation. In most cases, ligand binding increases the open probability due to higher affinity to the open state, which mainly arises from a smaller dissociation rate constant from the open channel. A simplified allosteric model that separates the ligand-binding step from the channel gating step is usually represented by the mechanism described by del Castillo and Katz [72]:

$$
\underbrace{R \underset{k_{diss}}{\overset{k_{ass}}{\rightleftharpoons}} LR}_{\text{binding}} \underbrace{\overset{\beta}{\underset{\alpha}{\rightleftharpoons}} LR^*}_{\text{gating}} \tag{7.6}
$$

where R represents a closed channel, R^* an open channel, L denotes a ligand (agonist molecule), k_{ass} and k_{diss} are the association and dissociation rate constants, α and β the closing and opening rate constants of the channel, respectively. This model assumes fast equilibrium binding versus the subsequent conformational changes occurring during channel gating.

Single-channel patch-clamp recordings revealed the fine structure of the effective openings, referred to as bursts, composed of transient closing events. This allowed the evaluation of the binding and gating rate constants separately [73]. The rate constants k_{diss}, α and β, referring to the liganded receptor, can be accurately estimated by using different methods. Because the association reaction is not usually diffusion limited and the number of channels present in the patch of membrane cannot be unequivocally defined [73], single-channel current measurement is not however the most reliable approach to determine k_{ass}.

7.5.2
The Nicotinic Acetylcholine Receptor as a Prototypical Example

The nicotinic acetylcholine receptor (nAChR) is a heteropentameric, cation-selective channel that is activated by binding of the natural neurotransmitter acetylcholine. Two ligand binding sites are located at subunit interfaces in the large extracellular domain. The muscle type nAChR is concentrated at neuromuscular junctions, where it mediates fast synaptic transmission by depolarizing the post-synaptic membrane. Single-channel current recordings from patches of end-plate membranes activated by acetylcholine revealed a two-state reaction scheme, which reflects transient openings of nAChR channels. Burst analysis of elementary currents allowed the derivation of an activation mechanism with microscopic binding and gating rate constants. Figure 7.2A shows single-channel currents simulated with a two-binding site mechanism in which only the diliganded receptor initiates channel opening as

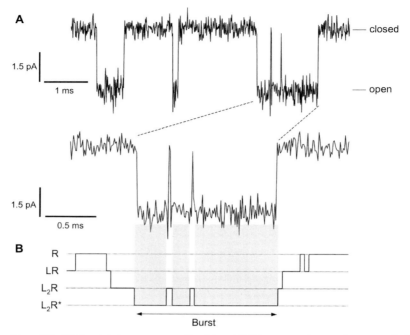

Figure 7.2 (A) Simulated current trace of a wild-type nAChR channel. The sampling frequency used was 100 kHz, the Gaussian filter was 40 kHz, and the elementary current was 3 pA. The simulation was carried out using QUB software (www.qub.buffalo.edu). (B) Transitions of the nAChR between microscopic states, which illustrate the molecular events underlying the burst shown on the expanded time scale in A. In the standard kinetic model, receptor activation follows the sequential binding of two ligand molecules.

illustrated in the occupancy state diagram in Figure 7.2B. The values of the rate constants used in the kinetic model have been experimentally estimated for wild-type mouse AChRs (adult-type) elicited by 1 μM acetylcholine at −100 mV applied voltage [74]: the closing rate constant α is 1700 s^{-1}, the opening rate constant β is 48000 s^{-1}, the dissociation rate constant from diliganded to monoliganded receptor, k_{diss}, is 46 000 s^{-1}, and the association rate constant from monoliganded to diliganded receptor, k_{ass}, is 2×10^8 M^{-1}s^{-1}. With these parameters, the mean time of openings per burst, $1/\alpha$, is 0.59 ms, the mean time of closings per burst, $1/(\beta + k_{diss})$, is 11 μs, and the mean time of bursts, $(1 + \beta/k_{diss})/\alpha + (\beta/k_{diss})/(\beta/k_{diss})$, is 1.2 ms.

It has been demonstrated that the two nAChR binding sites have different affinities for ligands. On the other hand, the nAChR can be activated by various agonists, whose specific interactions not only influence the binding rate constants but also the gating kinetics [75, 76]. Current recordings of single nAChR channels activated by ligands of different size, charge distribution, or hydrophobicity afforded a clearer understanding of the distinct contribution of agonist-specific chemical groups on docking or channel gating [76].

The central question of how ligand binding triggers channel opening has been investigated during recent years at the single amino acid level by combining protein engineering and single-channel electrophysiology. Key residue interactions engaged in the molecular pathway coupling ligand binding to gating have been identified by producing mutations at the interfacial region between the extracellular ligand-binding domain and the transmembrane channel domain [77, 78]. The propagation of conformational changes between intermediate pre-opening states was studied on a series of nAChR mutants affecting the binding site, the interfacial region, and transmembrane segments lining the pore [79, 80]. A picture of the sequential molecular events was suggested from the position of the gating transition state along the reaction coordinate estimated at the different sites in the protein [81, 82].

7.5.3
Chemical Gating by Specific Ligand Binding inside Ion Channels

Diffusion of hydrophilic solutes across membranes of cells and organelles occurs through protein channels comprising an aqueous pore. A variety of biological processes rely on channel-based transport systems, such as the transfer of nucleic acids and proteins across the nucleus membrane, the export of nucleotides across the mitochondria membrane, and the uptake of nutrients across the outer membrane of bacteria, to mention a few.

Selectivity of ions or metabolites is achieved through a network of attractive interactions between the diffusing particles and protein residues lining the pore. As a consequence of the presence of an internal binding site, both the occupation probability and the mean residence time of a solute inside the channel increase. Diffusion models have demonstrated that the net flux, or translocation probability via passive diffusion, of particles can be larger in the presence of a potential well compared with that in its absence [83–87]. Moreover, two important facts have been described. First, there is an optimal well depth that enhances channel efficiency depending on the solute concentration [83, 85]. Second, the asymmetric position of the binding site inside the channel can lead to asymmetric transport [86].

Although the binding of solutes in the pore does not initiate subsequent molecular events mediated by the channel protein itself, the above considerations show that facilitated diffusion through membrane channels can be viewed as a biological process resulting from specific interactions between a ligand (the permeating molecule) and a membrane receptor (the channel).

In some cases, the interactions are so strong that the transient occupancy of a pore by a diffusing molecule can be resolved in time at the single-molecule level by means of electrophysiology techniques. A simple kinetic scheme of channel-facilitated membrane transport assumes a two-state model, in which the channel is either open or (partially) closed to ionic current by a diffusing molecule. In the standard one-site two-barrier model, the occupied channel corresponds to the bound state [88]; in the diffusion model, the ligand is not necessarily bound to obstruct the pore (e.g. [84]).

The reaction may be thought of as a stochastic process, which generates a random sequence of microscopic pulses of ionic current. The pattern of current fluctuations induced by the transient occupancy of the pore by a diffusing molecule is an electrical signature of the molecular interactions that occur inside the protein channel [89]. The magnitude of the current decrease reflects the extent to which the channel is blocked by the translocating molecule, while the duration of the current pulse refers to the strength of interactions between the molecule and the amino acids lining the pore. This approach, known as the molecular Coulter counter, provides rapid and accurate characterization in real-time of molecules interacting with single ion channels [90, 91]. The passage of carbohydrates [92], nucleic acids [93], and antibiotics [94], could be resolved on a single molecular level.

Statistical analysis of the ligand-binding events can be undertaken by computing the autocorrelation of the current fluctuations. For a simple two-state Markovian process, the autocorrelation takes the form of a single exponential, and the inverse of the relaxation time is a linear function of the association (closing) and dissociation (opening) rate constants, k_{ass} and k_{diss}, respectively. In practice, a similar but more convenient approach consists of converting time-dependent fluctuating current into a frequency domain by using power spectrum analysis [94]. Spectra of the ligand-induced current fluctuations can be described by a single Lorentzian function (the Fourier transform of an exponential is a Lorentzian function) of the form:

$$S(w) = \frac{4\sigma^2 \tau}{1 + w^2 \tau^2}, \tag{7.7}$$

where τ is the relaxation time of the reaction, and σ^2 is the variance. Expressions of τ and σ^2 as a function of the binding rate constants are represented as follows:

$$\tau = (k_{ass} C + k_{diss})^{-1}, \tag{7.8}$$

$$\sigma^2 = \frac{N(\Delta i)^2 C k_{ass} k_{diss}}{(k_{ass} C + k_{diss})^2}, \tag{7.9}$$

where C is the ligand concentration, N is the total number of channels and Δi is the current through one channel.

7.5.4
Facilitated Translocation of Sugars through Bacterial Porins

The outer cell membrane of bacteria is permeable to smaller solutes below a molecular weight of about 400 Da. Such substances can freely permeate under a concentration gradient through general diffusion porins. However, under conditions of low nutrients the simple diffusion process is too slow and the bacterial cells need to enhance the efficiency of translocation. For this purpose, channels specific for a certain compound are present in the outer cell wall. The most extensively studied examples of specific porins are the maltooligosaccharide-specific channel LamB, also known as maltoporin, from *Escherichia coli* [95] and the sucrose-specific channel ScrY of *Salmonella typhimurium* [96].

High-resolution crystal structures of several of these specific porins are available. Although Lamb and Scry display only moderate sequence homology, their structures can be superimposed [97, 98]. They are composed of trimeric proteins when each monomer consists of 18 membrane spanning β-strands forming a barrel with a central water filled poor, short turns on the periplasmic side and large irregular loops on the outside of the cell. Based on structures of LamB complexed with a sugar, three adjacent subsites of binding were observed at the constriction zone located in the middle of the pore (Figure 7.3B), and a specific translocation pathway has been postulated [99]. The hydrophobic face of the sugar rings is in van der Waals contact with a slide of aromatic residues, while hydrogen bonds are formed between the sugar's hydroxyl groups and polar residues.

Maltoporin has been extensively studied using the black lipid membrane technique [100]. Single maltoporin trimers were reconstituted in lipid bilayers and the temporary binding events of maltodextrin molecules inside the channel were observed as transient closures of the individual monomers to ion current [101–104]. Analysis of the occupancy probability for the four particular states (fully open, single blocked, double blocked, fully blocked) shows that all three channels of the maltoporin protein transport sugars independently [101]. Figure 7.3A shows current recordings from single channels in solutions of maltodextrins of various lengths, containing from three (maltotriose) to seven (maltoheptaose) glucose moieties [102]. It can be seen that shorter carbohydrates block the current over a shorter period. However, they all totally obstruct one channel as indicated by the amplitude of the current transitions of one-third of the total current.

Power spectra of the sugar-induced current fluctuations can be described by a Lorentzian function (Figure 7.3C), which is characteristic of a two-state Markov process. The association and dissociation rate constants governing the open–close sequence experienced by one channel can be determined using Equation 7.8 by means of a linear plot of the reciprocal relaxation times versus sugar concentrations (see e.g. [105]). For maltohexose, values of k_{ass} and k_{diss} are about $10^7\,\mathrm{M}^{-1}\,\mathrm{s}^{-1}$ and $1000\,\mathrm{s}^{-1}$, respectively, at a 150-mV applied voltage in the symmetrical addition of the sugar. The asymmetry of binding of sugar molecules entering the channel from either the extracellular or the periplasmic side has been investigated in detail [88, 102, 103].

According to basic principles of thermodynamics, an electric field strength E, is a variable of state just like temperature and external pressure. Therefore, equilibrium constants depend on E according to the appropriate van't Hoff equation that involves the molar dipole moment of the reaction (an analogous relationship can be applied to elementary rate constants, which involves the reaction dipole moment in the transition state). Actually the effect is negligible in ordinary bulk reactions since the values of E are usually much too small. However, this field readily becomes very large during reactions of oriented molecules in biological membranes [106],

A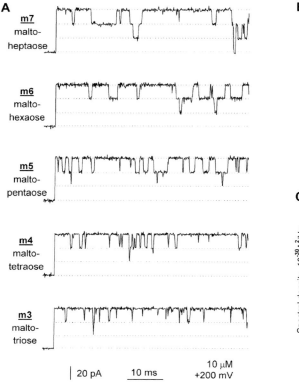

B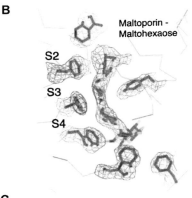

C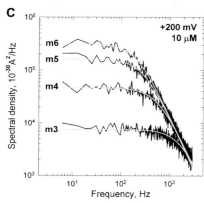

Figure 7.3 (A) Current recordings of single maltoporin proteins obtained from experiments where maltodextrins of various lengths were used. Downward current steps with the amplitude of one-third of the total current correspond to time-resolved binding events. The sugar concentration was 10 μM, the applied transmembrane voltage was 200 mV, and the data were low-pass filtered at 15 kHz (taken from [102]). (B) Crystal structure of maltoporin complexed with maltohexose. The helical arrangement defined by the aromatic lateral chains lining the pore allows the translocation of the left-handed sugar helix in a screw-like manner. The three binding subsites S2, S3 and S4 have been shown (taken from [99]).
(C) Power spectral densities of the current fluctuations induced by reversible binding of different sugar molecules. The Lorentzian type spectra indicate a two-state Markovian process for sugar binding. Spectra were obtained after subtracting the open channel noise in the absence of the sugar (taken from [102]).

such as those occurring in receptor channels. Indeed, a pronounced asymmetry of channel properties induced by the polarity of the applied voltage has been demonstrated [88, 92].

Maltoporin serves as the receptor for bacteriophage lambda [95, 107]. Phage docking to the bacterial surface is followed by injection of the viral DNA. Real-time formation of the phage–receptor complex was recently monitored in planar lipid bilayers and new hypotheses regarding the binding regions have been postulated [108, 109].

7.5.5
Combined Electrical and Fluorescence Measurements

Pioneering work on combined optical and electrical recordings originated from Yanagida's laboratories [110, 111, 186]. The activity of single ryanodine receptor channels reconstituted in planar lipid bilayers [112, 187] and of single nAChRs expressed in *Xenopus*oocytes has been imaged using a fluorescent indicator of calcium flux [113]. The formation of single gramicidin dimers was monitored by combined electrical and FRET measurements [114]. Planar patch-clamp instruments based on microstructured devices are used in the pharmaceutical industry for high-throughput screening of compounds targeted to ion channels. Great benefits can be derived from the planar design of the chips used to perform single-molecule fluorescence measurements [115]. With further improvements, simultaneous electrical and fluorescence recordings will enable the correlation between local structural fluctuations or ligand binding and channel activity at the single-molecule level to be elucidated. Good candidates for this procedure would be ligand-gated ion channels [116].

7.6
Forces of Ligand–Receptor Interactions in Living Cells

7.6.1
Principles of Single-molecule Dynamic Force Spectroscopy and Applications to Cell Surface Receptors

Various techniques have been developed to measure molecular forces. Most prominent force transducers are optical and magnetic tweezers, atomic force microscopy (AFM) cantilevers, and biomembrane force probes. In recent years, atomic force microscopy has become a powerful technique to visualize and manipulate single biomolecules in native-like environments, making it possible to study the structure of complex biological systems at sub-nanometer resolution [117, 118], the folding/unfolding trajectories of single proteins [119–121], and the dynamics of chemical bond rupture/formation between a ligand and its receptor [122, 123]. In the following section, particular emphasis will be given to the analysis of ligand–receptor interactions in living cells by dynamic force spectroscopy. For recent reviews on the principles of single molecule force measurements, preparation of tips, and recognition imaging, the reader can refer to Kienberger *et al.* [124], and Hinterdorfer and Dufrêne [125]. (For reviews on the use of AFM in living cells, see [126–130]).

Different configurations for the immobilization of living cells have been used in single-molecule force measurements (Figure 7.4A): (i) the AFM tip is functionalized with ligand molecules and living cells expressing specific membrane receptors are immobilized on the supporting surface, (ii) a cell is mounted on a modified (tipless) AFM cantilever (Figure 7.4B) and ligands are attached on the supporting

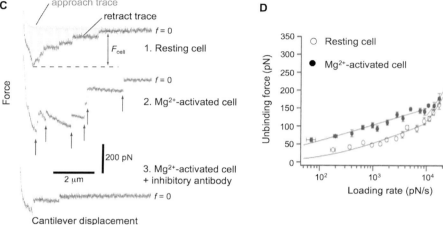

Figure 7.4 Single-molecule force spectroscopy on cell surface receptors. (A) Different strategies of functionalization of AFM probes and solid supports for investigating ligand–receptor interactions in living cells (see text for details). (B) Optical micrograph of a T-cell attached to the tip of an AFM cantilever (taken from [138]). (C) Force-displacement measurements between LFA-1 expressed on the surface of T-cells and immobilized ICAM-1. LFA-1/ICAM-1 interactions are important for leukocyte adhesion. Traces were acquired using a retraction speed of $2\,\mu m\,s^{-1}$. In trace 1, the shaded area represents de-adhesion between the cell and the substrate, and F_{cell} is the detachment force. In trace 2, T-cells were stimulated with Mg^{2+}/EGTA, which induced tighter adherence to ICAM-1 than that in resting cells. Arrows indicate single rupture events. In trace 3, addition of antibodies against either LFA-1 or ICAM-1 inhibited the adhesion (adapted from [138]). (D) Dynamic force spectrum of LFA-1/ICAM-1 interactions. Two loading regimes in the average unbinding forces were observed. For details on the extracted Bell model parameters (see Table 2; adapted from [138]).

surface, (iii) cells are attached to both the cantilever and the supporting surface for cell–cell contact experiments. Since bacterial cells cannot spread on solid surfaces, firm attachments must be achieved in order to carry out imaging and force experiments.

The general procedure for force spectroscopy consists of measuring the unbinding forces of single ligand–receptor complexes in force–distance cycles recorded at different loading rates. Figure 7.4C shows single molecular recognition events in the force curve. Since mechanical breaking is a stochastic process, several hundred force curves should be recorded to extract a statistically relevant value of the unbinding force. The distribution of unbinding forces measured at various loading rates, reffered as the dynamic force spectrum, yields the energy landscape for the interaction (Figure 7.4D).

Contrary to the techniques described above, force spectroscopy gives information on *non-equilibrium* reaction kinetics. A force F applied to a complex increases the dissociation rate constant as represented below [131]:

$$k_{\text{diss}}(F) = k_{\text{diss}}(0)\exp\left(\frac{x_\beta F}{k_B T}\right), \tag{7.10}$$

where $k_{\text{diss}}(0)$ denotes the dissociation rate constant in the absence of an external force, x_β is the distance between the bound state and the transition state along the reaction coordinate (direction of the applied force), k_B is the Boltzmann constant, and T is the absolute temperature. The product $x_\beta F$ corresponds to the decrease in the height of the energy barrier when a mechanical force is applied. Values of $k_{\text{diss}}(0)$ for different ligand–receptor complexes are shown in Table 7.2. Importantly, $k_{\text{diss}}(0)$ is similar to the thermal (equilibrium) dissociation rate constant under the unique conditions that ligand–receptor interactions are characterized by a one-barrier energy landscape. Examples of ligand–receptor dissociation processes with multiple energy barrier profiles are also presented in Table 7.2 and in Figure 7.4 D.

The association kinetics can be characterized using single-molecule force measurements by the adhesion probability P, i.e. the fraction of the force curves which show a specific binding event. The association rate constant k_{ass} can be determined by integrating the differential equation with the appropriate boundary conditions:

$$\frac{dP}{dt} = -k_{\text{diss}}P + k_{\text{ass}}c_{\text{eff}}(1-P), \tag{7.11}$$

where c_{eff} denotes the effective concentration of binding partners in the accessible volume between the AFM tip and the cell surface [132].

Interestingly, single-molecule force spectroscopy has essentially been applied to the study of adhesion receptors. To our knowledge, the interaction between a GPCR and its ligand has not been investigated in living cells. The mechanical unfolding of single membrane proteins has recently been investigated by combining AFM imaging and force spectroscopy [120]. Because the spectra showing unfolding events contain information on the strength and location of molecular interactions within the protein [133], the binding between a GPCR and its ligand could be studied using this technique.

7.6.2
Novel AFM-based Techniques

Recently, we emphasized the potential of planar cell membranes deposited on submicrometer-sized apertures in the study of molecular interactions with both sides of integral membrane proteins [134]. In a similar approach, a promising two-chamber AFM set-up has been designed, which allowed combined fluorescence microscopy and high spatial resolution imaging of proteins spanning suspended

Table 7.2 Single-molecule force spectroscopy experiments investigating ligand–receptor interactions in living cells.

Receptor	Ligand	Immobilization strategy	Properties of the binding/ unbinding process	Reference
Enzyme-linked				
$\alpha_5\beta_1$integrin in endothelial cells	Fibronectin	Fibronectin was cross-linked onto the AFM probe with PEG. Cells were grown on gelatin-coated dishes.	The mean unbinding force of a single fibronectin–integrin bound was about 34 pN at a retraction speed of 0.8 μm s^{-1}. After histamine treatment, the probability of adhesion increased and the mean adhesion force increased to about 39 pN.	[174]
$\alpha_5\beta_1$integrin in K562 cells	Fibronectin (fragment FN7-10)	A K562 cell was attached to the end of an AFM cantilever FN7-10 was adsorbed on Petri dishes.	The mean unbinding force was 69 pN at loading rates of about 2000 pN s^{-1}. Force spectra were obtained at loading rates between 10 and 5×10^4 pN s^{-1}, which revealed a two-regime unbinding dynamics. Bell model parameters were $k_d(0) = 0.13$ s^{-1} and $x_\beta = 0.41$ nm at slow loading rates (outer energy barrier), and $k_d(0) = 33.5$ s^{-1} and $x_\beta = 0.09$ nm at fast loading rates (inner energy barrier). Mutations to the synergy site of fibronectin yielded $k_d(0) = 0.85$ s^{-1} and $x_\beta = 0.37$ nm at slow loading rates, and $k_d(0) = 25.0$ s^{-1} and $x_\beta = 0.10$ nm at fast loading rates. Deletion of the RGD loop of fibronectin resulted in a single regime with $k_d(0) = 0.13$ s^{-1} and $x_\beta = 0.46$ nm. Activation of integrin by TS2/16 antibody involved interactions with both the RGD and synergy sites of fibronectin.	[175]

αvβ3 integrin receptor in osteoclast and osteoblast cells	RGD peptides and proteins	The cantilever tips were coated with PEG prior to passive chemisorption of ligands. Cells were fixed on glass cover slides	Rupture forces ranged between 40 and 100 pN for the different ligands. Separation rates were reduced from 50 to 1 μm s^{-1}.	[176]
β1 integrin in endothelial cells (HUVECs) and leukemia HL-60 cells	**P-selectin** **E-selectin** **ICAM-1** **VCAM-1**	Individual HL-60 cells were attached to the tip of a cantilever through Con A-mediated linkage Soluble forms of VCAM-1 and P-selectin were adsorbed onto Petri dishes Endothelial cells were grown on fibronectin-coated Petri dishes.	The detachment of leukocyte–HUVEC complexes involved a series of rupture events with force transitions between 40 and 100 pN at a retraction speed of 3 μm s^{-1}. Similar rupture magnitudes were observed on immobilized P-selectin and VCAM-1 proteins. Adhesion was inhibited by cyclic RGD peptides.	[177]
Fibronectin adhesin in *Staphylococcus epidermis* bacteria	**Fibronectin**	Fibronectin molecules were bound to gold-coated AFM tips by chemisorption of thiol groups. Bacteria were grown on fibronectin-coated surfaces.	A rupture force of 85 pN at a separation rate of 1 μm s^{-1} was measured. Specific interactions are located at the heparin-binding site of fibronectin. A single linear regime in the force spectrum was observed. Bell model parameters were $k_d(0) = 4.8$ s^{-1} and $x_\beta = 0.33$ nm. The value of $k_d(0)$ was six orders of magnitude larger than that estimated for the thermally activated dissociation, which demonstrated that the potential energy curve of the interaction was controlled by more than one activation barrier. Low adhesion probability at short contact time, and $k_a c_{eff} = 11.8$ s^{-1}.	[178]

(Continued)

Table 7.2 (Continued)

Receptor	Ligand	Immobilization strategy	Properties of the binding/ unbinding process	Reference
LFA-1 in T cell hybridoma 3A9	ICAM-1	A LFA-1 expressing 3A9 cell was attached to the end of a cantilever coated with Con A (successive incubations with biotin-BSA, streptavidin and biotin-Con A). A soluble form of ICAM-1 was adsorbed onto a culture dish	Adhesion between 3A9 cells and ICAM-1 is mainly due to the formation of LFA-1/ICAM-1 complexes. Mg^{2+} and EGTA strengthened the interactions. The average unbinding forces of individual LFA-1/ICAM-1 complexes calculated at loading rates ranging from 60 to $>10^4$ pN s^{-1} led to a two-regime force spectrum. Mg^{2+}-stabilized interactions resulted in higher unbinding forces in the slow loading regime. The energy landscape for the dissociation process consists of two activation barriers. The outer energy barrier was characterized at slow loading rates by $k_d(0) = 4$ s^{-1} (in agreement with the value reported from equilibrium experiments) and $x_\beta = 0.15$ nm for resting cells, and $k_d(0) = 0.17$ s^{-1} and $x_\beta = 0.21$ nm for Mg^{2+}-activated cells. The inner activation barrier was characterized at fast loading rates by $k_d(0) = 57$ s^{-1} and $x_\beta = 0.018$ nm for resting cells, and $k_d(0) = 40$ s^{-1} and $x_\beta = 0.024$ nm for Mg^{2+}-activated cells.	[138]

LFA-1 in T cell hybridoma 3A9	**ICAM-1**	See [138] for protein immobilization and AFM probe functionalization with cells	Individual LFA-1/ICAM-1 unbinding forces were measured at various loading rates. Cell adhesion was enhanced by stimulating the cells with PMA as evidenced by the larger maximum force required to detach a cell. The mean unbinding force of single complexes was about 30 pN for both resting and PMA-stimulated cells at a rate of 100 pN s^{-1} and their dynamic response was similar. Instead, the observed enhanced cell adhesion induced by PMA arises from a softening of the cell, which increased the contact area between the interacting surfaces.	[179]
LFA-1 in Jurkat T cells	**ICAM-1** **ICAM-2**	See [138] for protein immobilization and AFM probe functionalization with cells	Two-regime dynamic force spectrum for both LFA-1/ICAM-1 and LFA-1/ICAM-2 interactions were observed for loading rates ranging from 50 to 6×10^4 pN s^{-1}. For LFA-1/ICAM-1, Bell model parameters for resting cells were $k_d(0) = 0.55$ s^{-1} and $x_\beta = 0.26$ nm at slow loading rates (outer energy barrier), and $k_d(0) = 19$ s^{-1} and $x_\beta = 0.05$ nm at fast loading rates (inner energy barrier). For LFA-1/ICAM-2, Bell model parameters for resting cells were $k_d(0) = 0.31$ s^{-1} and $x_\beta = 0.45$ nm at slow loading rates, and $k_d(0) = 10$ s^{-1} and $x_\beta = 0.16$ nm at fast loading rates. For both complexes, Mg^{2+} had a stabilizing effect only at slow loading rates, whereas PMA had little if no influence on the dissociation reaction.	[180]

(Continued)

Table 7.2 (*Continued*)

Receptor	Ligand	Immobilization strategy	Properties of the binding/unbinding process	Reference
PSGL-1 in human PMN cells and LS174T human colon adenocarcinoma cells	**P-selectin**	PMN and LS174T cells were immobilized on culture dishes. Silanized cantilevers were coated with anti-human IgG Fc monoclonal antibody and subsequently with P-selectin-IgG Fc chimera proteins	Single peak distributions on rupture force histograms were observed at loading rates from 0.5 to 30 μm s^{-1}. Force spectra deviated from linearity in the lower limit of the loading rate. Values for $k_d(0)$ and x_β were 0.20 s^{-1} and 0.14 nm respectively for P-selectin/PSGL-1 interactions, and 2.78 s^{-1} and 0.13 nm respectively for P-selectin/LS174T ligand interactions. The lower binding strength of P-selectin to its ligand on LS174T cells explains the larger rolling velocity measured in flow-chamber experiments.	[181]
Transporter				
SGLT in intact brush border membrane vesicles	Inhibitor **phlorizin**	Phlorizin was coupled to the tips via a flexible cross-linker. Vesicles were adsorbed onto a gold surface	The unbinding force for phlorizin-sodium/glucose co-transporter was 120 pN and x_β was 0.5 nm. Analysis of the wash-out recovery time of phlorizin binding probability yielded $k_d(0) = 10^{-5}$ s^{-1}. The on rate k_a was $\sim 5 \times 10^3$ M^{-1} s^{-1} leading to an equilibrium dissociation constant of ~ 0.2 μM. Recognition was sodium dependent and inhibited by D-glucose.	[182]
SGLT1 in stably transfected CHO cells	**D-glucose**	Ligands were conjugated to AFM tips with a flexible PEG-crosslinker. Cells were seeded on poly-L-lysine-coated glass slides	The mean unbinding force of D-glucose to SGLT1 in the presence of sodium was about 50 pN at a loading rate of 1 μm s^{-1}. Binding was inhibited in the presence of free substrate D-galactose, free inhibitor phlorizin but not in the presence of L-glucose.	[183]

Non-integral membrane receptor

Con A in *Saccharomyces* strains	**Glucose Mannose**	Gold-coated AFM probes were functionalized with oligoglucose or Con A using thiol chemistry. Cells were immobilized by mechanical trapping in a polycarbonate membrane	Individual lectin–carbohydrate interactions were recorded in living yeast cells in aggregating conditions. Unbinding forces of about 120 pN have been measured at a rate of 0.5 μm s^{-1}.	[184]
csA in aggregating cells of *Dictyostelium discoideum*	**csA** in aggregating cells of *Dictyostelium discoideum*	Tipless cantilevers were functionalized with a lectin for firm attachment of a cell. The targeted cells were grown in Petri dishes	The EDTA-stable cell adhesion was mediated by the csA glycoprotein. A rupture force of 23 pN at a separation rate of 2.5 μm s^{-1} was measured.	[185]

Abbreviations for proteins: Con A, concanavalin A; csA, contact site A; LFA-1, leukocyte function-associated antigen-1; PSGL-1, P-selectin glycoprotein ligand-1; SGLT, sodium/glucose co-transporter.

Other abbreviations: BSA, bovin serum albumin; CHO, Chinese hamster ovary; EGTA, ethylene glycol-bis(2-aminoethylether)-N,N,N′,N′-tetraacetic acid; HUVECs, human umbilical vein endothelial cells; ICAM-1, intercellular adhesion molecule-1; ICAM-2, intercellular adhesion molecule-2; IgG, immunoglobulin G; PEG, poly (ethylene glycol); PMA, phorbol myristate acetate; PMN, polymorphonuclear leukocyte; VCAM-1, vascular cell adhesion molecule-1.

native membranes [135]. In a second method, also based on microstructured devices, AFM imaging was combined with electrical recording of ion channels in lipid bilayers using a conducting tip [136].

Acknowledgments

This work was supported by the European Commission via contract LSHG-CT-2004-504601 (E-MeP).

References

1 Ji, T.H. *et al.* (1998) *J. Biol. Chem.*, **273**, 17299–17302.

2 Colquhoun, D. (2006) *Br. J. Pharmacol.*, **147**, S17–S26.

3 Rang, H.P. (2006) *Br. J. Pharmacol.*, **147**, S9–S16.

4 Freudenthaler, G. *et al.* (2002) *Histochem. Cell. Biol.*, **117**, 197–202.

5 Grandl, J. *et al.* (2007) *Angew. Chem.*, **46**, 3505–3508.

6 Bouzigues, C. and Dahan, M. (2007) *Biophys. J.*, **92**, 654–660.

7 Hassaine, G. *et al.* (2006) *Protein Expr. Purif.*, **45**, 343–351.

8 Ehrensperger, M.-V. *et al.* (2007) *Biophys. J.*, **92**, 3706–3718.

9 Charbonniere, L.J. *et al.* (2006) *J. Am. Chem. Soc.*, **128**, 12800–12809.

10 Shaner, N.C. *et al.* (2005) *Nat. Methods*, **2**, 905–909.

11 Giepmans, B.N. *et al.* (2006) *Science*, **312**, 217–224.

12 Bacia, K. and Schwille, P. (2003) *Methods*, **29**, 74–85.

13 Asamoah, O.K. *et al.* (2003) *Neuron*, **37**, 85–97.

14 Sonnleitner, A. *et al.* (2002) *Proc. Natl Acad. Sci. USA*, **99**, 12759–12764.

15 Sonnleitner, A. and Isacoff, E. (2003) *Methods Enzymol.*, **361**, 304–319.

16 Blunck, R. *et al.* (2007) *Biophys. J.*, 526A–526A. Suppl. S JAN 2007.

17 Griffin, B.A. *et al.* (1998) *Science*, **281**, 269–272.

18 Martin, B.R. *et al.* (2005) *Nat. Biotechnol.*, **23**, 1308–1314.

19 Miller, L.W. and Cornish, V.W. (2005) *Curr. Opin. Chem. Biol.*, **9**, 56–61.

20 Foley, T.L. and Burkart, M.D. (2007) *Curr. Opin. Chem. Biol.*, **11**, 12–19.

21 Johnsson, N. and Johnsson, K. (2007) *ACS Chem. Biol.*, **2**, 31–38.

22 Wang, L. *et al.* (2006) *Annu. Rev. Biophys. Biomol. Struct.*, **35**, 225–249.

23 Monahan, S.L. *et al.* (2003) *Chem. Biol.*, **10**, 573–580.

24 Turcatti, G. *et al.* (1996) *J. Biol. Chem.*, **271**, 19991–19998.

25 Liu, W.S. *et al.* (2007) *Nat. Methods*, **4**, 239–244.

26 Ilegems, E. *et al.* (2002) *Nucleic Acids Res.*, **30**, e128.

27 Keppler, A. *et al.* (2003) *Nat. Biotechnol.*, **21**, 86–89.

28 Keppler, A. *et al.* (2004) *Proc. Natl Acad. Sci. USA*, **101**, 9955–9959.

29 Heinis, C. *et al.* (2006) *ACS Chem. Biol.*, **1**, 575–584.

30 Yin, J. *et al.* (2004) *J. Am. Chem. Soc.*, **126**, 7754–7755.

31 George, N. *et al.* (2004) *J. Am. Chem. Soc.*, **126**, 8896–8897.

32 Meyer, B.H. *et al.* (2006) *Proc. Natl Acad. Sci. USA*, **103**, 2138–2143.

33 Yin, J. *et al.* (2005) *Proc. Natl Acad. Sci. USA*, **102**, 15815–15820.

34 Guignet, E.G. *et al.* (2004) *Nat. Biotechnol.*, **22**, 440–444.

35 Lata, S. *et al.* (2006) *J. Am. Chem. Soc.*, **128**, 2365–2372.

36 Hauser, C.T. and Tsien, R.Y. (2007) *Proc. Natl Acad. Sci. USA*, **104**, 3693–3697.

37 Schreiter, C. *et al.* (2005) *Chembiochem*, **6**, 2187–2194.

38 Choquet, D. and Triller, A. (2003) *Nat. Rev. Neurosci.*, **4**, 251–265.

39 Grakoui, A. *et al.* (1999) *Science*, **285**, 221–227.

40 Huber, A.B. *et al.* (2003) *Annu. Rev. Neurosci.*, **26**, 509–563.

41 Ueda, M. *et al.* (2001) *Science*, **294**, 864–867.

42 Meier, J. *et al.* (2001) *Nat. Neurosci.*, **4**, 253–260.

43 Ritchie, K. *et al.* (2003) *Mol. Membr. Biol.*, **20**, 13–18.

44 Bates, I.R. *et al.* (2006) *Biophys. J.*, **91**, 1046–1058.

45 Rocheville, M. *et al.* (2000) *Science*, **288**, 154–157.

46 Daumas, F. *et al.* (2003) *Biophys. J.*, **84**, 356–366.

47 Zhou, X. *et al.* (2001) *J. Biol. Chem.*, **276**, 44762–44769.

48 Jin, T. and Li, J. (2002) *J. Biol. Chem.*, **277**, 32963–32969.

49 Sergé, A. *et al.* (2003) *J. Cell Sci.*, **116**, 5015–5022.

50 Axelrod, D. *et al.* (1976) *Biophys. J.*, **16**, 1055–1069.

51 Geerts, H. *et al.* (1987) *Biophys. J.*, **52**, 775–782.

52 Sheetz, M.P. *et al.* (1989) *Nature*, **340**, 284–288.

53 Kucik, D.F. *et al.* (1989) *Nature*, **340**, 315–317.

54 Barak, L.S. and Webb, W.W. (1981) *Cell Biol.*, **90**, 595–604.

55 Schmidt, T. *et al.* (1996) *Proc. Natl Acad. Sci. USA*, **93**, 2926–2929.

56 Dahan, M. *et al.* (2003) *Science*, **302**, 442–445.

57 Charrier, C. *et al.* (2006) *J. Neurosci.*, **26**, 8502–8511.

58 Boyer, D. *et al.* (2002) *Science*, **297**, 1160–1163.

59 Cognet, L. *et al.* (2003) *Proc. Natl Acad. Sci. USA*, **100**, 11350–11355.

60 Feder, T.J. *et al.* (1996) *Biophys. J.*, **70**, 2767–2773.

61 Cherry, R.J. *et al.* (1998) *FEBS Lett.*, **430**, 88–91.

62 Vukojevic, V. *et al.* (2005) *Cell. Mol. Life Sci.*, **62**, 535–550.

63 Bacia, K. *et al.* (2006) *Nat. Methods*, **3**, 83–89.

64 Jankevics, H. *et al.* (2005) *Biochemistry*, **44**, 11676–11683.

65 Weiss, M. *et al.* (2003) *Biophys. J.*, **84**, 4043–4052.

66 Wohland, T. *et al.* (1999) *Biochemistry*, **38**, 8671–8681.

67 Levene, M.J. *et al.* (2003) *Science*, **299**, 682–686.

68 Edel, J.B. *et al.* (2005) *Biophys. J.*, **88**, L43–L45.

69 Samiee, K.T. *et al.* (2006) *Biophys. J.*, **90**, 3288–3299.

70 Wenger, J. *et al.* (2007) *Biophys. J.*, **92**, 913–919.

71 Foquet, M. *et al.* (2004) *Anal. Chem.*, **76**, 1618–1626.

72 Del Castillo, J. and Katz, B. (1957) *Proc. R. Soc. Lond. B*, **146**, 369–381.

73 Colquhoun, D. and Hawkes, A.G. (1995) In *Single-channel Recording*, 2nd edn (eds B. Sakmann and E. Neher), Plenum Publishing Cooperation, pp. 397–482.

74 Hille, B. (2001) *Ionic Channels of Excitable Membranes*, 3rd edn Sinauer, Sunderland, MA.

75 Marshall, C.G. *et al.* (1991) *J. Physiol.*, **433**, 73–93.

76 Zhang, Y. *et al.* (1995) *J. Physiol.*, **486**, 189–206.

77 Lee, W.Y. and Sine, S.M. (2005) *Nature*, **438**, 243–247.

78 Lummis, S.C.R. *et al.* (2005) *Nature*, **438**, 248–252.

79 Grosman, C. *et al.* (2000) *Nature*, **403**, 773–776.

80 Mitra, A. *et al.* (2005) *Proc. Natl Acad. Sci. USA*, **102**, 15069–15074.

81 Auerbach, A. (2003) *Sci. STKE*, **188**, re11.

82 Auerbach, A. (2005) *Proc. Natl Acad. Sci. USA*, **102**, 1408–1412.

83 Berezhkovskii, A.M. *et al.* (2002) *J. Chem. Phys.*, **116**, 9952–9956.

84 Berezhkovskii, A.M. *et al.* (2003) *J. Chem. Phys.*, **119**, 3943–3951.

85 Berezhkovskii, A.M. and Bezrukov, S.M. (2005) *Biophys. J.*, **104**, L17–L19.

86 Bauer, W.R. and Nadler, W. (2006) *Proc. Natl Acad. Sci. USA*, **103**, 11446–11451.

87 Kasianowicz, J.J. *et al.* (2006) *Proc. Natl Acad. Sci. USA*, **103**, 11431–11432.

88 Schwarz, G. *et al.* (2003) *Biophys. J.*, **84**, 2990–2998.

89 Bezrukov, S.M. *et al.* (1994) *Nature*, **370**, 279–281.

90 Bezrukov, S.M. (2000) *J. Membrane Biol.*, **174**, 1–13.

91 Bayley, H. and Martin, C.R. (2000) *Chem. Rev.*, **100**, 2575–2594.

92 Kasianowicz, J.J. *et al.* (1996) *Proc. Natl Acad. Sci. USA*, **93**, 13770–13773.

93 Nestorovich, E.M. *et al.* (2002) *Proc. Natl Acad. Sci. USA*, **99**, 9789–9794.

94 DeFelice, L.J. (1981) *Introduction to Membrane Noise*, Plenum Press, New York.

95 Szmelcman, S. and Hofnung, M. (1975) *J. Bacteriol.*, **124**, 112–118.

96 Schmid, K. *et al.* (1988) *Mol. Microbiol.*, **2**, 1–8.

97 Schirmer, T. *et al.* (1995) *Science*, **267**, 512–514.

98 Forst, D. *et al.* (1998) *Nat. Struct. Biol.*, **5**, 37–46.

99 Dutzler, R. *et al.* (1996) *Structure*, **4**, 127–134.

100 Hanke, W. and Schlue, W.-R. (1993) *Planar Lipid Bilayers: Methods and*, 152.

101 Bezrukov, S.M. *et al.* (2000) *FEBS Lett.*, **476**, 224–228.

102 Kullman, L. *et al.* (2002) *Biophys. J.*, **82**, 803–812.

103 Danelon, C. *et al.* (2003) *J. Biol. Chem.*, **278**, 35542–35551.

104 Kullman, L. *et al.* (2006) *Phys. Rev. Lett.*, **96**, 038101.

105 Nekolla, S. *et al.* (1994) *Biophys. J.*, **66**, 1388–1397.

106 Schwarz, G. (1978) *J. Membr. Biol.*, **43**, 127–148.

107 Schwarz, M. (1976) *J. Mol. Biol.*, **103**, 521–536.

108 Berkane, E. *et al.* (2006) *Biochemistry*, **45**, 2708–2720.

109 Gurnev, P.A. *et al.* (2006) *J. Mol. Biol.*, **359**, 1447–1455.

110 Ide, T. and Yanagida, T. (1999) *Biochem. Biophys. Res. Commun.*, **265**, 595–599.

111 Ide, T. *et al.* (2002) *Japan. J. Physiol.*, **52**, 429–434.

112 Peng, S. *et al.* (2004) *Biophys. J.*, **86**, 134–144.

113 Demuro, A. and Parker, I. (2005) *J. Gen. Physiol.*, **126**, 179–192.

114 Borisenko, V. *et al.* (2003) *Biophys. J.*, **84**, 612–622.

115 Danelon, C. *et al.* (2006) *Chimia*, **60**, A754–A760.

116 Biskup, C. *et al.* (2007) *Nature*, **446**, 440–443.

117 Muller, D.J. *et al.* (1995) *Biophys. J.*, **68**, 1681–1686.

118 Schabert, F.A. *et al.* (1995) *Science*, **268**, 92–94.

119 Marszalek, P.E. *et al.* (1999) *Nature*, **402**, 100–103.

120 Oesterhelt, F. *et al.* (2000) *Science*, **288**, 143–146.

121 Fernandez, J.M. and Li, H. (2004) *Science*, **303**, 1674–1678.

122 Florin, E.-L. *et al.* (1994) *Science*, **264**, 415–417.

123 Merkel, R. *et al.* (1999) *Science*, **397**, 50–53.

124 Kienberger, F. *et al.* (2006) *Acc. Chem. Res.*, **39**, 29–36.

125 Hinterdorfer, P. and Dufrêne, Y.F. (2006) *Nature Methods*, **3**, 347–355.

126 Lehenkari, P.P., *et al.* (2000) *Expert Rev. Mol. Med.*, **2**, 1–19.

127 Horton, M. *et al.* (2002) *J. Recept. Signal Transduct. Res.*, **22**, 169–190.

128 Dufrêne, Y.F. (2003) *Curr. Opin. Microbiol.*, **6**, 317–323.

129 Wojcikiewicz, E.P. *et al.* (2004) *Biol. Proced. Online*, **6**, 1–9.

130 Evans, E.A. and Calderwood, D.A. (2007) *Science*, **316**, 1148–1153.

131 Bell, G.I. (1978) *Science*, **200**, 618–627.

132 Fritz, J. *et al.* (1998) *Proc. Natl Acad. Sci. USA*, **95**, 12283–12288.

133 Kedrov, A. *et al.* (2005) *EMBO Rep.*, **6**, 668–674.

134 Danelon, C. *et al.* (2006) *Langmuir*, **22**, 22–25.

135 Gonçalves, R.P. *et al.* (2006) *Nat. Methods*, **3**, 1007–1011.

136 Quist, A.P. *et al.* (2007) *Langmuir*, **23**, 1375–1380.

137 Rigler, R. *et al.* (1999) *Proc. Natl Acad. Sci. USA*, **96**, 13318–13323.

138 Zhang, X. *et al.* (2002) *Biophys. J.*, **83**, 2270–2279.

139 Briddon, S.J. *et al.* (2004) *Proc. Natl Acad. Sci. USA*, **101**, 4673–4678.

140 Briddon, S.J. *et al.* (2004) *Faraday Discuss.*, **126**, 197–207.

141 Hegener, O. *et al.* (2004) *Biochemistry*, **43**, 6190–6199.

142 Henriksson, M. *et al.* (2001) *Biochem. Biophys. Res. Commun.*, **280**, 423–427.

143 Pramanik, A. *et al.* (2001) *Biochem. Biophys. Res. Commun.*, **284**, 94–98.

144 Zemanová, L. *et al.* (2004) *Biochemistry*, **43**, 9021–9028.

145 Pramanik, A. *et al.* (1999) *Biomed. Chromatogr.*, **13**, 119–121.

146 Pramanik, A. *et al.* (2001) *Biochemistry*, **40**, 10839–10845.

147 Sergé, A. *et al.* (2002) *J. Neurosci.*, **22**, 3910–3920.

148 Lill, Y. *et al.* (2005) *Chem. Phys. Chem.*, **6**, 1633–1640.

149 Jacquier, V. *et al.* (2006) *Proc. Natl Acad. Sci. USA*, **103**, 14325–14330.

150 Patel, R.C. *et al.* (2002) *Proc. Natl Acad. Sci. USA*, **99**, 3294–3299.

151 Pick, H. *et al.* (2003) *Biochemistry*, **42**, 877–884.

152 Guignet, E.G. *et al.* (2007) *ChemPhysChem.*, **8**, 1221–1227.

153 Borgdorff, A.J. and Choquet, D. (2002) *Nature*, **417**, 649–653.

154 Tardin, C. *et al.* (2003) *EMBO J.*, **22**, 4656–4665.

155 Hegener, O. *et al.* (2002) *Biol. Chem.*, **383**, 1801–1807.

156 Meissner, O. and Häberlein, H. (2003) *Biochemistry*, **42**, 1667–1672.

157 Halbsguth, C. *et al.* (2003) *Planta Med.*, **69**, 305–309. *Applications*, Academic Press, San Diego.

158 Meissner, O. and Häberlein, H. (2006) *Planta Med.*, **72**, 656–658.

159 Boonen, G. *et al.* (2000) *Planta Med.*, **66**, 7–10.

160 Choquet, D. *et al.* (1997) *Cell*, **88**, 39–48.

161 Cairo, C.W. *et al.* (2006) *Immunity*, **25**, 297–308.

162 Grabham, P.W. *et al.* (2000) *J. Cell Sci.*, **113**, 3003–3012.

163 Felsenfeld, D.P. *et al.* (1996) *Nature*, **383**, 438–440.

164 Vrljic, M. *et al.* (2002) *Biophys. J.*, **83**, 2681–2692.

165 Pramanik, A. and Rigler, R. (2001) *Biol. Chem.*, **382**, 371–378.

166 Sako, Y. *et al.* (2000) *Nat. Cell. Biol.*, **2**, 168–172.

167 Lidke, D.S. *et al.* (2005) *J. Cell. Biol.*, **170**, 619–626.

168 Larson, D.R. *et al.* (2005) *J. Cell Biol.*, **171**, 527–536.

169 Zhong, Z.H. *et al.* (2001) *Diabetologia*, **44**, 1184–1188.

170 Lambert, M. *et al.* (2002) *J. Cell Biol.*, **157**, 469–479.

171 Tani, T. *et al.* (2005) *J. Neurosci. B*, 2181–2191.

172 Gakamsky, D.M. *et al.* (2005) *Biophys. J.*, **89**, 2121–2133.

173 Rajan, S.S. and Vu, T.Q. (2006) *Nano Lett.*, **6**, 2049–2059.

174 Trache, A. *et al.* (2005) *Biophys. J.*, **89**, 2888–2898.

175 Li, F. *et al.* (2003) *Biophys. J.*, **84**, 1252–1262.

176 Lehenkari, P.P. and Horton, M.A. (1999) *Biochem. Biophys. Res. Commun.*, **259**, 645–650.

177 Zhang, X. *et al.* (2003) *Am. J. Physiol.*, **286**, H359–H367.

178 Bustanji, Y. *et al.* (2003) *Proc. Natl Acad. Sci. USA*, **100**, 13292–13297.

179 Wojcikiewicz, E.P. *et al.* (2003) *J. Cell Sci.*, **116**, 2531–2539.

180 Wojcikiewicz, E.P. *et al.* (2006) *Biomacromolecules*, **7**, 3188–3195.

181 Hanley, W. *et al.* (2003) *J. Biol. Chem.*, **278**, 10556–10561.

182 Wielert-Badt, S. *et al.* (2002) *Biophys. J.*, **82**, 2767–2774.

183 Puntheeranurak, T. *et al.* (2006) *J. Cell Sci.*, **119**, 2960–2967.

184 Touhami, A. *et al.* (2003) *Microbiology*, **149**, 2873–2878.

185 Benoit, M. *et al.* (2000) *Nat. Cell. Biol.*, **2**, 313–317.

186 Ide, T., *et al.* (2002) *Single Mol.*, **3**, 33–42.

187 Peng, S. *et al.* (2004) *Biophys. J.*, **86**, 145–151.

8
RNA in cells

Valeria de Turris and Robert H. Singer

8.1
Why Study RNA?

In terms of movements and interactions, RNA is one of the most dynamic and flexible molecules in the cell. These characteristics are the keys to understanding its biological roles, in particular in terms of regulation of gene expression. In the cell, the expression of a gene is controlled at different levels. Checkpoints are spread out all along the way, from the triggering of a signal to "open" the chromatin until post-translational modifications and degradation of the protein. In this complex network, RNA occupies a central position since it is the "messenger" (mRNA) sent from the site of information storage (DNA in the nucleus) to everywhere in the cell. Some RNA molecules are coding RNAs, "read" by the ribosome in the cytoplasm to produce functional proteins by ordered loading of the correct amino acids. Other functional RNAs are never translated into proteins (non-coding RNAs, ncRNAs). They can be involved in ribosome assembly and function, such as ribosomal RNA (rRNA), transfer RNA (tRNA) and the small nucleolar RNA (snoRNA); others are implicated in splicing pieces of mRNA together, as are small nuclear RNAs (snRNAs) important components of the spliceosome, the macro-complex that accomplishes this splicing. Moreover, in the last few years new classes of regulatory non-coding RNAs have been discovered: small interfering RNAs (siRNAs) and microRNAs (miRNAs). These ncRNAs are characterized by their tiny dimensions (varying between 21 and 24 nucleotides) and new members are continuously being found. They are involved in multiple functions, from the protection against parasitic nucleic acids, such as viruses and transposons [1], to the control of the expression of specific mRNAs in development and cancer [2]. The discovery of these new RNAs that has strongly improved our understanding of cell defense and regulation also provides tools to manipulate and study gene expression. The relevance of this finding is evident from the 2006 Nobel Prize where Fire and Mello shared the Medicine Prize for the discovery of RNA interference by these small RNAs.

Single Molecule Dynamics in Life Science. Edited by T. Yanagida and Y. Ishii
Copyright © 2009 WILEY-VCH Verlag GmbH & Co. KGaA, Weinheim
ISBN: 978-3-527-31288-7

Another important characteristic of RNA molecules is that they are never "naked" in the cell. They are always associated with proteins, forming ribonucleoprotein particles (RNPs). As a consequence, the RNP is much higher in molecular weight than the RNA alone. Moreover, the composition of the complex is modified over time due to exchange of binding partners, therefore increasing the complexity and the spectrum of possible interactions and functions. These characteristics make the RNA molecule, in all its forms and functions, an exciting and important object of study.

Many questions have been raised about RNP dynamics in the nucleus and in the cytoplasm. Some of them concern how RNAs move, whether they follow rules of free diffusion or energy-dependent movement, and to what extent the environment, such as chromatin in the nucleus or filament networks in the cytoplasm, constrains RNA movements. RNA dynamics range from the sites of nuclear transcription, where maturation occurs, to the specific localization of particular RNAs in the cytoplasm, which creates defined gradients by enrichment in, or exclusion from, particular areas.

8.2
RNA Visualization inside Cells

Because "seeing is believing", during the last decade efforts have been focused on observing the actual dynamics of RNA movement inside a single living cell.

In the next section we outline the most important components to be considered when imaging mRNAs during their movements in living cells: the development of suitable methods to label the RNA, generating a sufficient signal to detect specific individual transcripts, and improvements in imaging technologies. In the subsequent sections we will describe the travel of RNA molecules from transcription sites until their final destination in their respective translational compartments.

8.2.1
Techniques to Label RNA

The intrinsic complexity of the cellular system gives rise to many issues. To address them, different methods have been developed to visualize RNAs. Before choosing a particular approach, pros and cons have to be considered taking into account the cellular system, the target and the aim of the project.

One among the first techniques utilized to study the RNA dynamics in living cells was Fluorescent *In Vivo* Hybridization (FIVH, [3]). This method was developed on the basis of the Fluorescent *In Situ* Hybridization (FISH, [4–6]) and relies on intrinsic abilities of oligonucleotides to recognize and hybridize to a complementary target sequence. The main difference between the two techniques is that FISH applies to fixed cells, while FIVH allows the study of transcripts in living cells. The first methodological improvement of the FIVH technique was the optimization of protocols for oligonucleotide uptake and hybridization *in vivo*. Fluorescent [3] or caged-fluorescent oligo-dT [7] were used to probe the poly(A) tails of all mRNAs and study their movements. The ability to obscure the fluorochrome on the probe by a protecting, "caging", group allows the movement of mRNA to be studied at a higher resolution. The photolytic unmasking of the fluorochrome, "uncaging", activates the fluorescent probes only at the illuminated region of the nucleus. This distinguishes hybridized probes from the free oligo-dT, because the latter diffuse faster and therefore disappear by dilution in the cellular volume.

An additional tool to label endogenous RNAs based on oligonucleotide hybridization is provided by molecular beacons [8]. These molecules are characterized by a particular stem–loop structure that maintains the fluorophore and its quencher bound close together at each end of the probe. The aim of this system is to overcome the background signal derived from unbound probes, since the fluorescent signal will be visible only when the annealing of the molecular beacon to its target separates the quencher from the fluorophore. However, the stem–loop structure could be destroyed *in vivo* by nuclease activity or protein binding, enabling fluorescence without hybridization. Therefore, an improvement of this tool has been developed exploiting the advantages of Fluorescence Resonance Energy Transfer (FRET, [9–11]). Briefly FRET occurs when two spectrally-matched fluorescent pairs are sufficiently close (<10 nm) and in the correct orientation. The fluorophore excited (donor) by an external source (lamp) does not disperse all the energy in the emission instead it transfers activation energy to the second longer-wavelength fluorophore (acceptor) that in turn will emit. With this technique, opportunely designed pairs of molecular beacons anneal to adjacent sequences on the same RNA target, thus recruiting the donor and the acceptor of the FRET pair close enough to generate the FRET signal [12].

Another method to resolve the background due to unbound probes is the direct *in vitro* labeling of the target RNA before introduction into the cell [13, 14]. In this case, unlike previous techniques, the target RNA is not endogenously produced possibly eliminating some steps in the normal maturation pathway. Since mRNA injected into the cytoplasm will not have contact with the nuclear environment it could assemble a different mRNP complex. For example, mRNAs injected in the cytoplasm may lack all the nuclear factors usually recruited during their travel in this compartment. Furthermore, even if injected into the nucleus, they may be deficient in all the factors deposited during transcription and maturation; processes like splicing and polyadenylation. Nonetheless, these features are not always a con and they have been exploited, for instance, to determine the involvement of nurse cell factors in *bicoid* RNA dynamics in *Drosophila* embryos [15].

A completely different approach relied on the power of fluorescently tagged proteins [16–18]. Green Fluorescent Protein (GFP) and other fluorescent proteins derived from the jellyfish *Aequorea victoria* are extensively used to tag RNA binding proteins. In this case, the binding of the chimeric protein will indirectly label the transcript. GFP-poly(A) binding protein 2 (GFP-PABP2) and the GFP-TBP export factor [19] were used to study the movement of the bulk of endogenous mRNA by Fluorescent Recovery after photobleaching (FRAP) experiments [20]. General RNA binding proteins, like FIVH with oligo-dT, can be used to address endogenous mRNAs but they do not discriminate one transcript from another, showing the dynamics of a population and not of a specific transcript. Furthermore, there is the additional complexity of the off- and on-rates of the protein and its recycling to another transcript. In any case, the approach was not appropriate for studying the dynamics of a single RNA molecule. The key for this advance came from the bacteriophage MS2 coat protein (MCP) coupled with FP tagging [21]. This phage protein is an RNA-binding protein recognizing specifically a distinctive binding site on a stem–loop folded RNA (Figure 8.1). The high affinity interaction (<1 nM) between the stem–loop and the phage protein make this method highly specific. Since two MCPs bind each stem–loop as dimers, the insertion of several MS2 binding sites (MBS) into the target gene will recruit multiple florescent tagged MCPs on a single molecule (Figure 8.1), providing a powerful system for the detection of single mRNPs distinguishable from the GFP-MCP background.

Developing this system in yeast [21] and in human cells [22, 23] made it possible to probe RNA expression of a specific sequence. A recent study focused on the synthesis of a specific gene array of MBS containing-transcripts by pol II and demonstrated the advantages of the MS2 system in mammalian cells [24]. In this work, integration of results obtained by FRAP, photoactivation, mathematical modeling and computation analysis allowed the quantification of the *in vivo* dynamics and kinetics of pol II transcription.

Coupling this method with other emerging tools like Fluorescence Correlation Spectroscopy (FCS), a means of resolving molecular events within rapid time frames, for example the dynamics of the specific steps of transcription or for splicing, will be likely to yield valuable kinetic data.

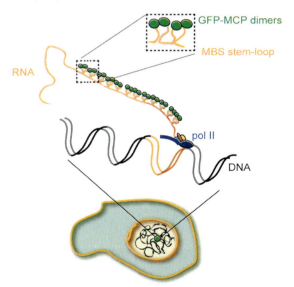

Figure 8.1 Schematic representation of a specific RNA labeled with the MS2 system. Upon stable insertion of the modified gene in the cell, its active transcription site will be visualized by fluorescence due to the binding of the GFP-MCP dimers at the MCB stem–loop regions. For simplicity only a single transcribing polymerase is depicted. Multiple active polymerases at the site will increase the RNA production therefore the intensity, allowing the site to be detected over the signal background.

The concept behind the MCP system led to the creation of an alternative system based on the U1A splicing factor and its recognition sequences [25–27]. This approach works well in yeast but is not exploitable in mammalian cells, which contain the endogenous U1A protein.

8.2.2
Advancements in Imaging Technologies

A critical aspect in single-molecule imaging is the development of the best optical system to match the requirements for imaging single moving particles. This includes acquiring images at rates as fast as, or faster than, the particle movements projected optically onto the capture chip; applying a minimal amount of light to avoid phototoxicity and bleaching of the sample, and all of this while maximizing the signal-to-noise ratio and tracking the particles in three dimensions (3D) and in time ("4D" imaging). In the past, the classical confocal microscope was preferred for imaging fluorescently-labeled cells, but now alternative options are available. The confocal platform itself has been recently modified to increase speed and enable fast imaging in the z-plane, in time and with different wavelengths simultaneously. More advances have also been made in the wide-field epifluorescent microscope. The improvements in the Charge-Coupled Device (CCD) camera, that converts incident photons at the detector into electronic information, and the technology of electron-multiplying CCDs (EMCCDs) which detect very weak signals, are providing higher levels of sensitivity than the photomultiplying tubes (PMT) used in the confocal microscope. In this latter instrument, photons are first converted into electrons which then converge in an electron multiplier where a system of electrodes amplifies the signal by a secondary emission process. However, the major difference between confocal and wide-field imaging is that the confocal microscope discards photons that are not within the image plane and this loss of information reduces the sensitivity required to detect single molecules. Instead, with the wide-field microscope all the information from "out of focus" photons collected during the acquisition process becomes important and useful. In fact applying the images deconvolution algorithms, a "point spread function" will allow "reconstructing" them in three dimensions. In this way the light is reassigned to its point of origin to recreate an image with a signal-to-noise ratio much higher than that in a confocal microscope [28]. In conclusion, the developments in biophotonics, imaging technologies, bioinformatics and computational analysis are continuously increasing their relevance and indispensable roles in the discovery of new principles of cellular and molecular biology in living cells.

8.3
RNA Dynamics in the Nucleus

The birth site of RNAs, namely the transcription site or "RNA factory", is located in the nucleus. Transcription by RNA polymerase I (pol I) occurs inside the nucleoli, while pol II and pol III are active in the nucleoplasm. During transcription by those enzymes, RNAs are matured and released from the sites. They then move towards the nuclear envelope to translocate in the cytoplasm through the nuclear pore structure. Beyond this elementary information, what are the actual RNA dynamics in all these processes?

8.3.1
Dynamics in Transcription

Most of our knowledge about transcription comes from ensemble measurements using methods such as the Northern blot, RT-PCR and microarray. These analyses only provide results that are averages for specific molecules in a population and obscure the differences among all cells. Therefore, only a single cell approach can provide insight into the dynamic behavior and responses to specific stimuli of an individual cell. Exploiting the MS2 system in bacteria, *E. coli* transcripts were tracked and details of prokaryotic transcription revealed [29, 30]. It was demonstrated that transcription in prokaryotes occurred in "bursts" with an average of 6 min of activation for approximately 37 min of inactivity and that RNA partitioning during cell division was random, decreasing the correlation between RNA and protein at the beginning of the cell cycle. Recently, the same approach was used in the eukaryote *Dictyostelium discoideum* for the characterization of the transcription of an endogenous developmental gene [31]. Discrete "pulses" of gene activity were found with an estimated mean time of 5–6 min on or off. The length and intensity of the pulse were consistent during development. This was surprising, considering the strong changes in transcriptional stimuli occurring throughout differentiation of this organism. The important conclusion for this developmental system was that the number of pulses during development did not increase, but rather there was an increase in the number of cells that became committed to transcribing the gene. Initiation of synchronous transcription in neighboring cells was observed to be more frequent than predicted by random events. Furthermore, a "transcriptional memory" existed in cells that had already transcribed that gene; they were more prone to restart transcription than cells that had never expressed it.

The study of transcriptional dynamics is at its beginning and yet very promising. Developing of sensitive systems to observe specific transcription in real-time in mammalian cells, will open new frontiers and most likely reveal new insights into gene expression. A first step in this direction has been made in the study of pol II transcription *in vivo* [24]. Transcription of a specific locus was monitored by FRAP of YFP-pol II recruited to the active site as well as both FRAP and photo-activation of GFP-MCP labeling the transcribed RNA. A systems-modeling approach combined with quantification and testing of the model using transcription inhibitors provided sufficient resolution to demonstrate a faster transcription rate coupled with pausing steps during elongation. Variations in the period length and percentage of pol II pausing could possibly correlate with the appearance of transcriptional "bursting" after relief of the block caused by the upstream pausing polymerase.

8.3.2
A Journey from the Transcription Site to the Nuclear Envelope

The mechanism of RNA movement in the nucleus has been addressed by various approaches [32–35]. One of the hypotheses was that the mRNPs moved from transcription sites to the nuclear envelope guided by some internal structures similar to a railroad, and driven by receptors or a transporting complex. Since the elements of the cytoskeleton, such as actin, nuclear myosin and other related proteins are found in the nucleus and have even been shown to be involved in the transcriptional process [36–38], it has been proposed that nuclear transport machinery that relied on these skeletal structures, including nuclear motor proteins, might exist. Although FRAP experiments on GFP-actin protein show that actin polymerization occurs in the nucleus and that those structures are highly dynamic [39], their involvement in nuclear transport has never been observed to date. Instead many studies inferred that the movement of mRNPs is a combination of Brownian motion and ATP-dependent movements [19, 35]. Most of the questions focused on the relevance and the actual meaning of the ATP requirement in RNA movements inside the nuclear environment. Since directed movements are never observed in the nucleus, the energetic demand may not supply molecular motors but more likely could be used to release RNPs from stalling during random interactions with nuclear structures on their way, such as dense chromatin domains, chromatin scaffolds or the cytoskeleton. Rather than imagine RNP moving on tracks [40], we can envisage the particles moving by diffusion inside a system of interconnected sinusoidal "channels" of fluid phase bounded by dense chromatin domains [41]. The RNPs will travel in this network of interchromatin space and occasionally interact with other complexes and/or domains becoming trapped within areas of high-density chromatin. Reversion from stationary to mobile depends on the consumption of ATP [19, 35]. Single particle tracking shows that RNP motion is energy-independent and not directed [23]. The observation of corralled, and in rare cases, constrained movements highlights the existence of dense and inaccessible structures hindering the free diffusion of large molecular complexes such as mRNPs (Figure 8.2). ATP has an essential role in chromatin remodeling; decondensation of chromatin after energy depletion could be responsible for affecting motility by trapping mRNPs within high-viscosity regions of DNA strands. The caveat in the ATP-depletion experiments is that drug treatments have many pleiotropic effects that impair a clear discrimination between direct or indirect causes.

The RNP could dynamically interact with the environment and change its protein partners during the journey from transcription site to the nuclear envelope, eventually arriving at the proper composition to interact with the export machinery [42–44]. The correct processing of the mRNA will deposit specific proteins on the transcript, like flags indicating that the particle is ready to be exported, or whether it still needs processing or has to be retained and degraded. Some proteins involved in mRNA transport are (respectively yeast/mammalian homolog) Yra1p/Aly of the REF (RNA and Export Factor binding) family of hnRNP-like proteins and Mex64p/TAP [43–45]. The first pair is an RNA-binding protein and the second the

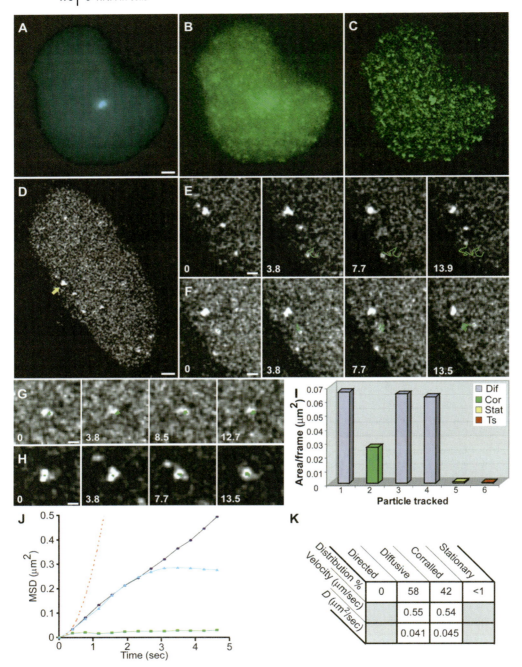

adaptor bridging the RNA/REF complex with the NPC component, namely nucleoporins with FG repeats (see below).

8.3.3
Transport through the Nuclear Pore Complex

The dynamics of particle translocation through the nuclear pore complex (NPC) are still unclear and the mechanism for transport is under investigation [46]. Three types of transport are associated with the pore in the nuclear envelope (NE): restricted diffusion, facilitated diffusion and unidirectional receptor-dependent transport. If the molecule does not interact with the nucleoporins, protein constituents of the NPC, it is defined as "inert" and permeates through the internal channel by restricted diffusion with a rate inversely proportional to its molecular mass, with a limiting size of 50 kDa. Particles interacting directly with the nucleoporin FG repeats, usually transport receptors like NTF2 and transportin 1, are subjected to facilitated translocation. Both these mechanisms are passive bidirectional processes while transport mediated by the receptor is an active unidirectional transport that proceeds against the concentration gradient of the cargo proteins. A cargo is an inert molecule that cannot diffuse freely through the pore. Instead they harbor specific signals (Nuclear Localization Signal, NLS and Nuclear Export Signal, NES) bound by the adaptor to be translocated to the right compartment [47]. Also in this case the translocation process *per se* is not an energy-consuming task, since is not directly coupled with ATP hydrolysis. The real energy driving the transport mediated by importin and exportin proteins is the chemical potential of the RanGTP gradient maintained by NTF2 and RanGEF. This latter protein recharges the RanGDP imported into the nucleus by NTF2 with GTP. The RanGTP gradient, higher in the nucleus than in the cytoplasm, is important for the correct directionality of the cargo transport, since association of the receptor with the cargo is influenced by its level. Importins load the cargo at low levels of RanGTP in the cytoplasm while in the nucleus high RanGTP levels trigger the replacement of the cargo with RanGTP. The exportins work in the opposite direction and with an opposite mechanism: they load the cargo only in combination

---

Figure 8.2 Live-cell imaging and single-particle tracking of individual mRNPs in the nucleus of a mammalian cell. Images from time-lapse films acquired from a cell co-transfected with (A) CFP-Lac repressor that marks the insertion locus and (B) YFP-MCP. (C) Reduction of noise for tracking of mRNPs was obtained by deconvolution. Bar, 2 μm. (D) Tracking of mRNP (arrow, transcription site) (bar, 2 μm) showed (E) diffusing particles, (F) corralled particles, (G) stationary particles, and (H) the transcription site. Tracks are marked in green, and time in seconds from the beginning of tracking for each particle that appears in each frame. Bars, 1 μm. (I) Plot of the area per frame traveled throughout the tracking period. Diffusive particles are shown in blue, corralled in green, stationary in yellow, and transcription site in red. (J) Mean-square displacement (MSD) of tracked nucleoplasmic particles versus time indicated the presence of three types of characterized movements: diffusive (black circles), corralled (blue triangles), and stationary (green squares). Directed movement was never detected (red dotted line). (K) Table summarizing the mean velocities and diffusion coefficients of tracked particles at 37 °C. (Adapted from [23]).

with RanGTP in the nucleus and release both when they reach the cytoplasm. In both cases, once in the cytoplasm, RanGTP is hydrolyzed to RanGDP to disassemble the complex.

The messenger RNAs rely on the adaptor protein TAP for their transport in the cytoplasm. The TAP-mediated export of the mRNAs appears to be unlinked from the concomitant binding of RanGTP [48], and its marginal role in the process is due to its involvement in nuclear import of TAP and other proteins rather than in the mRNA transport itself. In that case the unidirectional movement seems to be maintained by a highly conserved DEAD-box ATPase/RNA helicase essential for mRNA export, Dbp5p [49]. This is a shuttling protein that associates with the RNA early in transcription and translocates into the cytoplasm with the complex. On the cytoplasmic side of the NPC multiple binding sites for Dbp5p anchor the helicase in this region where Dbp5p is activated by the concomitant presence of Gle1 and Inositol-P_6 [50, 51]. Remodeling of the mRNP causes the release from the NPC and the recycling in the nucleus of the proteins involved in the transport, thus avoiding a possible backward movement of the complex. Although we have considerable information about different kinds of interactions and the mechanism of receptor-mediated cargo transport, how molecules actually translocate through the pore is still unclear.

A key role is suggested for the FG repeats in the nucleoporins [47, 52, 53]. Since these phenylalanine-rich domains are able to interact with each other and with the transport receptors, several models have been developed to describe the possible movements inside the NPC. A Brownian affinity-gating model proposes the formation of an internal channel with binding sites at the tunnel entrance that facilitate the access of the bound molecules but completely exclude those that are unbound [52]. Inside the channel, the particles move by Brownian motion. Macara [47] proposed instead that the channel walls are actually covered with the FG repeats allowing the molecules to jump from one repeat to another while inert molecules can diffuse in the channel. Another possibility is the formation of a meshwork by interaction among the FG domains creating a permeability barrier that restricts the passage for inert molecules [53]. This selective phase model proposes that the nucleoporins form this sieve-like structure within the pores, and transient interactions with the FG repeats would allow the bound particle to "dissolve" into this structure. The carrier would help the cargo to translocate by masking domains that enable them to interact positively with the meshwork.

Fundamental insights into the translocation process require further investigation and higher resolution structures of the intact NPC, a goal that can be achieved only by single molecule approaches. These methods can provide unique information on topographic properties and kinetic processes with excellent spatial and time resolution. To this end, a single-molecule far-field fluorescence microscopy approach was applied to the NPC of permeabilized human cells [54], allowing the measurement of dwell times of NTF2 and transportin with and without their specific cargo molecules bound. These data highlight that binding at the NPC is not the rate-limiting step and that particles can translocate simultaneously via multiple parallel pathways.

8.4
RNA Dynamics in the Cytoplasm

Once the transcripts reach the cytoplasm, they move from the pore to disperse in the environment. We can divide them into two different classes of RNAs based on their final distribution: non-localizing and localizing RNAs. The first will uniformly distribute in the cytoplasm while the latter will be confined or enriched in specific areas. Nevertheless both have the ability to move. Studies with inert tracers suggest that the cytosol is heterogeneous with viscoelastic behavior, allowing limited diffusion for particles of sizes similar to RNPs [55–58]. Therefore, unlike the situation in the nucleus, the particles may require active transport if diffusion is impaired or inefficient. However their dynamics should differ to allow the observed specific compartmentalization. In particular since distances are much larger in the cytoplasm, for instance the distal region of a neuron, a mechanism is required to facilitate transport.

8.4.1
Non-localizing RNA

Most mRNAs, such as housekeeping mRNAs, appear to belong to the non-localizing class although their distribution may in fact be non-homogeneous (for instance mRNAs for mitochondrial proteins appear to be near mitochondria [59]). Their role is to spread out in the cytosol to ensure that their protein products will be generally and uniformly available. The dynamics of single and specific RNAs in living cells has been observed and measured by exploiting the MCP system in COS cells [60].

Three reporter genes with the MBS inserted and different 3′UTR (3′ UnTranslated Region) sequences either from human growth hormone (hGH mRNA) gene, SV40 (SV mRNA) or β-actin (as a control for known localizing RNA) were used. The first two reporters exhibited four possible movements (Figure 8.3): static (33–40%), corralled (∼ 40%), diffusional (15–25%) and directed (2–5%). Interestingly, individual particles were able to switch between these movements and no correlation was observed between a specific behavior and a particular area in the cells. Since active transport is usually associated with cytoskeleton components, this hypothesis was investigated by treating cells with specific drugs against microtubules and microfilaments. The results confirmed the crucial role played by the cytoskeleton in anchoring static particles, supplying tracks for directed motion and creating restricted areas not accessible to the particles, possibly transforming their diffusion into corralled motion. The new finding that "non-localizing" RNPs are also subjected to directed movements suggests the involvement of active transport by molecular motors on microtubules similar to localized RNPs. Actually, both RNP classes moved with the same average speed (1–1.5 μm/s) but the localized RNP classes used active motion more frequently and for longer distances. Therefore, molecules switch stochastically between various movements, but each RNA will have a specific probability of displaying each of the four movements dependent on its sequence. If a sequence enables the recruitment of factors interacting directly with the motors

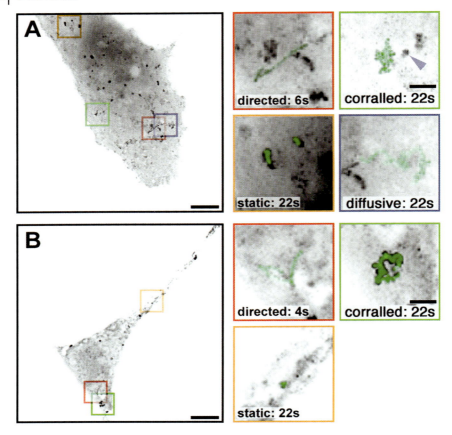

Figure 8.3 Dynamics of single mRNA molecules in the cytoplasm of mammalian cells. Direct movements are also observed in the cytoplasm. (A) Cos cells transiently expressing the reporter hGH mRNAs and the MCP-GFP were imaged live. Left: a maximum intensity image projection of 200 frames on one image. The scale bar represents 10 μm. Right: panel magnifications: the scale bar represents 2 μm. mRNA track superimposed (green) from each of the indicated boxed regions. The blue arrow points to a static particle in the vicinity of a corralled mRNA. (B) COS cells transiently expressing the reporter SV mRNA and MCP-GFP were imaged as in A. The scale bar represents 10 μm. Right: panel magnifications, track of mRNA movement superimposed (green) on an enlargement from each of the indicated boxed regions. The scale bar represents 2 μm. (Adapted from [57]).

or with adaptors, a specific outcome will be determined as the RNA distribution will be a result of the motor direction and its persistence.

8.4.2
RNA Localization

RNA localization is involved in the regulation of many processes inside the cell and often acts in concert with translation and RNA degradation. Most of its effects give rise to an asymmetric distribution of factors, creating in turn polarized cells [61]. This polarity has important consequences in many processes, such as development, differentiation, cell motility and neuron functionality. Different means are used to reach this goal but the most effective seems to be RNA localization associated with local translation to generate proteins only in the targeted area. The elements required for the localization are sequences *in cis* on the RNA, called "zipcodes" or Localization Elements (LEs), and the *trans*-acting factors recognizing and binding them [62–64]. Examples are: the β-actin localization zipcode [65–67] and its trans-acting factor, ZBP1 [68, 69] in fibroblasts; Vg1 LE with hnRNP I, Vera and 40 LoVe in *Xenopus* oocytes [70–73]; the A2RE signal in the Myelin Basic Protein (MBP) with hnRNP A2 in neurons [13, 74] and the *ASH1* zipcode with She2p in yeast *S. cerevisiae* [21, 75, 76]. In general, localizing mRNAs are shuttled to specific areas of the cell or the oocyte along cytoskeletal elements such as microtubules or actin filaments. They seem to be actively translocated by motor proteins of the myosin, kinesin and dynein families. A corollary of localization is that the mRNA must be translationally repressed during its movement. A number of *trans*-acting factors mediate translational repression by binding the RNA directly (ZBP1, [77]; Puf6p, [78]; Khd1p, [79, 80]).

8.4.2.1 Some Examples of Localization in Mammalian Cells and Drosophila

Localization is particularly important during development. The most characterized cellular systems to study RNA localization in mammalian cells are migrating fibroblasts, oligodendrocytes and neurons. In fibroblasts β-actin mRNAs are localized at the leading edge of the cell, a fact that correlates with the requirement of high protein levels for actin polymerization during cell movement. The complex of mRNA, ZBP1 and ZBP2 assembled in the nucleus [81] moves in the cytoplasm along actin filaments probably carried by a myosin motor [82, 83] to be anchored at the leading edge possibly by EF1α [84] where it is finally translated.

Neurons and oligodendrocytes are also a class of highly "polarized" cells since many mRNAs typically travel from the cell body to the extremities in dendrites and axons. RNAs travel in granules that may contain many copies of an mRNA or several types of mRNA. All this trafficking moves on cytoskeleton elements by motors: MBP mRNA is probably associated with microtubules through a kinesin [85]. The same motor is also responsible for the CamKIIα targeting in hippocampal dendrites [86] and tau mRNA in axons [87]. Moreover in neurons, β-actin is localized in the growth cone by ZBP1 along microtubules. Since the same motor can drive the movement of different RNAs, the recruitment of specific adaptors and RNA binding proteins in the locasome will allow the selection of the final "address" for the specific cargo in the complex.

During development, localization mechanisms are also used by *Drosophila* cells to create mRNA gradients, and consequently protein gradients, indispensable for generating specific patterns of expression essential for development of the oocyte and the embryo. One of the first determinants breaking the initial symmetry of the oocyte is *gurken* mRNA. It is involved in the specification of both the anterior–posterior and the dorsal–ventral axis by two rounds of signals at different times during oogenesis [88]. At the beginning *gurken* is localized at the future posterior pole of the oocyte, sending a signal back to the oocyte to initiate the formation of the anteroposterior axis. The signal leads to the repolarization of the oocyte microtubules and the migration of the oocyte nucleus to the dorsoanterior corner of the oocyte [89]. When *gurken* is localized in an anterodorsal cap near the oocyte nucleus the second round of signaling initiates the formation of the dorsoventral axis. The overlying follicle cells acquire dorsal fates, leading later to secretion of correct eggshell structures [88, 90]. The *gurken* mRNA first moves across the internal oocyte to the anterior and then turns towards the nucleus in the anterodorsal position. Both steps require dynein and microtubules, but they rely on different microtubule networks [91]. RNA binding proteins such as Squid and Hrp48 are involved in *gurken* dorsal movements [92] and its localized translation is restricted to the dorsal anterior region [91, 93]. After the initial signal from *gurken,* the further development of the anterior–posterior symmetry of the oocyte involves several other localized transcripts in addition to *gurken*: *bicoid* for anterior specification and *oskar* and *nanos* mRNAs, both localized in the posterior. This axial polarity is established by opposite gradients of these proteins maintained in the oocyte by the maternal determinants (from ovarian nurse cells), transported on cytoskeletal networks to their destination and then anchored and translated. At the anterior pole *bicoid* is recruited in two phases: an earlier phase in mid-oogenesis when microtubules are polarized towards the anterior pole, and a later phase, after nurse cell dumping, when ooplasmic streaming is thought to facilitate the mixing of the incoming material [94]. In the first phase, microtubules and the binding of the *trans*-acting factor Exuperentia are essential for the localization at the anterior pole [15]. Also in the late phase *bicoid* is localized by active transport [94] instead of diffusion and trapping as is the case with *nanos* (see below). This involves the binding of Staufen protein to *bicoid*, before nurse cell dumping, and transport of the complex on a subset of microtubules that originates at the anterior pole. Microtubules and actin filaments are responsible for the enrichment of *bicoid* at this pole not through anchoring, but instead by a continuous active dynein-driven transport [94].

At the posterior pole *oskar* is one of the first mRNAs recruited, probably by kinesin I. Interestingly, proteins in the exon-junction-complex (EJC) and the splicing reaction *per se* seem to be involved in its localization [95, 96]. Oskar protein in turn is required for *nanos* mRNA localization. The peculiarity of *nanos* localization is how the specific expression in the posterior pole is achieved. Indeed, *nanos* enters the oocyte during nurse cell dumping and is dispersed in the ooplasm by streaming movements and by diffusion in the whole oocyte [97]. Once at the posterior, it is anchored to the actin cytoskeleton and translated [97]. In contrast, outside this region *nanos* is translationally repressed by Glorund in the oocyte [98] and in the embryo by Smaug [99, 100] or also degraded [101]. Another mRNA which becomes localized at the posterior pole by degradation outside its target region is *hsp83* [102]. Both these posterior enrichments require two distinct *cis*-acting elements in the 3′ untranslated region (UTR) of the RNA: a degradation element that targets the mRNA for destruction in all regions of the egg or embryo, and a protection element that stabilizes the mRNA at the posterior [101].

In the early stages of *Drosophila* embryonic development, initial nuclear cleavages are not accompanied by cell division, creating a large multinucleate syncytium in a broad band of cortical cytoplasm (periplasm) where zygotic transcription begins at the blastoderm stage. The nuclei form a layer subdividing the periplasm into two compartments: the apical, above the nuclei, and the inner, basal periplasm below the nuclei. The pair-rule mRNAs, essential for the further segmentation of the embryo, are restricted in the apical compartment. Their localization mechanism requires specific sequences in the 3'UTR of the transcripts [103], Squid protein to promote apical transport [104] and dynein-mediated transport on microtubules [105, 106].

8.5
Conclusion

We now know much about how RNA travels from its birth place to its functional sites. However much more needs to be known and new tools need to be developed to fully understand the process. The study of single molecules in live cells is becoming essential to discover the connections between different pathways and the actual mechanisms for regulating gene expression in the cell.

References

1 Sontheimer, E.J. and Carthew, R.W. (2005) Silence from within: endogenous siRNAs and miRNAs. *Cell*, **122**, 9–12.

2 Bartel, D.P. (2004) MicroRNAs: genomics, biogenesis, mechanism, and function. *Cell*, **116**, 281–297.

3 Politz, J.C., Browne, E.S., Wolf, D.E. and Pederson, T. (1998) Intranuclear diffusion and hybridization state of oligonucleotides measured by fluorescence correlation spectroscopy in living cells. *Proc. Natl Acad. Sci. USA*, **95**, 6043–6048.

4 Lawrence, J.B. and Singer, R.H. (1985) Quantitative analysis of *in situ* hybridization methods for the detection of actin gene expression. *Nucleic Acids Res.*, **13**, 1777–1799.

5 Levsky, J.M. and Singer, R.H. (2003) Fluorescence *in situ* hybridization: past, present and future. *J. Cell Sci.*, **116**, 2833–2838.

6 Shav-Tal, Y., Shenoy, S.M. and Singer, R.H. (2004) Visualization and quantification of single RNA molecules in living cells, in *Live Cell Imaging: A Laboratory Manual*, (eds R.D. Goldman and D.L. Spector), Cold Spring Harbor Laboratory Press.

7 Politz, J.C., Tuft, R.A., Pederson, T. and Singer, R.H. (1999) Movement of nuclear poly(A) RNA throughout the interchromatin space in living cells. *Curr. Biol.*, **9**, 285–291.

8 Santangelo, P., Nitin, N. and Bao, G. (2006) Nanostructured probes for RNA detection in living cells. *Ann. Biomed. Eng.*, **34**, 39–50.

9 Wieb Van Der Meer, B., Coker, G. III and Chen, S.Y. (1994) *Resonance Energy Transfer: Theory and Data*, VCH, New York.

10 Pollok, B.A. and Heim, R. (1999) Using GFP in FRET-based applications. *Trends Cell Biol.*, **9**, 57–60.

11 Jares-Erijman, E.A. and Jovin, T.M. (2003) FRET imaging. *Nat. Biotechnol.*, **21**, 1387–1395.

12 Santangelo, P.J., Nix, B., Tsourkas, A. and Bao, G. (2004) Dual FRET molecular beacons for mRNA detection in living cells. *Nucleic Acids Res.*, **32**, e57.

13 Ainger, K., Avossa, D., Morgan, F., Hill, S.J., Barry, C., Barbarese, E. and Carson, J.H. (1993) Transport and localization of exogenous myelin basic protein mRNA microinjected into oligodendrocytes. *J. Cell Biol.*, **123**, 431–441.

14 Glotzer, J.B., Saffrich, R., Glotzer, M. and Ephrussi, A. (1997) Cytoplasmic flows localize injected *oskar* RNA in Drosophila oocytes. *Curr. Biol.*, **7**, 326–337.

15 Cha, B.J., Koppetsch, B.S. and Theurkauf, W.E. (2001) *In vivo* analysis of *Drosophila bicoid* mRNA localization reveals a novel microtubule-dependent axis specification pathway. *Cell*, **106**, 35–46.

16 Zhang, J., Campbell, R.E., Ting, A.Y. and Tsien, R.Y. (2002) Creating new fluorescent probes for cell biology. *Nat. Rev. Mol. Cell Biol.*, **3**, 906–918.

17 Lippincott-Schwartz, J. and Patterson, G.H. (2003) Development and use of fluorescent protein markers in living cells. *Science*, **300**, 87–91.

18 Shaner, N.C., Steinbach, P.A. and Tsien, R.Y. (2005) A guide to choosing fluorescent proteins. *Nat. Methods*, **2**, 905–909.

19 Calapez, A., Pereira, H.M., Calado, A., Braga, J., Rino, J., Carvalho, C., Tavanez, J.P., Wahle, E., Rosa, A.C. and Carmo-Fonseca, M. (2002) The intranuclear mobility of messenger RNA binding proteins is ATP dependent and temperature sensitive. *J. Cell Biol.*, **159**, 795–805.

20 Reits, E.A. and Neefjes, J.J. (2001) From fixed to FRAP: measuring protein mobility and activity in living cells. *Nat. Cell Biol.*, **3**, E145–E147.

21 Bertrand, E., Chartrand, P., Schaefer, M., Shenoy, S.M., Singer, R.H. and Long, R.M. (1998) Localization of ASH1 mRNA particles in living yeast. *Mol. Cell*, **2**, 437–445.

22 Janicki, S.M., Tsukamoto, T., Salghetti, S.E., Tansey, W.P., Sachidanandam, R., Prasanth, K.V., Ried, T., Shav-Tal, Y., Bertrand, E., Singer, R.H. and Spector, D.L. (2004) From silencing to gene expression: real-time analysis in single cells. *Cell*, **116**, 683–698.

23 Shav-Tal, Y., Darzacq, X., Shenoy, S.M., Fusco, D., Janicki, S.M., Spector, D.L. and Singer, R.H. (2004) Dynamics of single mRNPs in nuclei of living cells. *Science*, **304**, 1797–1800.

24 Darzacq, X., Shav-Tal, Y., de Turris, V., Brody, Y., Shenoy, S.M., Phair, R.D. and Singer, R.H. (2007) *In vivo* dynamics of polymerase II transcription. *Nat. Struct. Mol. Biol.*, **14**, 796–806.

25 Takizawa, P.A. and Vale, R.D. (2000) The myosin motor, Myo4p, binds Ash1 mRNA via the adapter protein. She3p. *Proc. Natl Acad. Sci. USA*, **97**, 5273–5278.

26 Brodsky, A.S. and Silver, P.A. (2000) Pre-mRNA processing factors are required for nuclear export. *RNA*, **6**, 1737–1749.

27 Brengues, M., Teixeira, D. and Parker, R. (2005) Movement of eukaryotic mRNAs between polysomes and cytoplasmic processing bodies. *Science*, **310**, 486–489.

28 Carrington, A. (1996) Microwave spectroscopy at the dissociation limit *Science*, **274**, 1327–1331.

29 Golding, I. and Cox, E.C. (2004) RNA dynamics in live Escherichia coli cells. *Proc. Natl Acad. Sci. USA*, **101**, 11310–11315.

30 Golding, I., Paulsson, J., Zawilski, S.M. and Cox, E.C. (2005) Real-time kinetics of gene activity in individual bacteria. *Cell*, **12**, 1025–1036.

31 Chubb, J.R., Trcek, T., Shenoy, S.M. and Singer, R.H. (2006) Transcriptional pulsing of a developmental gene. *Curr. Biol.*, **16**, 1018–1025.

32 Politz, J.C. and Pederson, T. (2000) Review: movement of mRNA from transcription site to nuclear pores. *J. Struct. Biol.*, **129**, 252–257.

33 Carmo-Fonseca, M., Platani, M. and Swedlow, J.R. (2002) Macromolecular mobility inside the cell nucleus. *Trends Cell Biol.*, **12**, 491–495.

34 Vinciguerra, P. and Stutz, F. (2004) mRNA export: an assembly line from genes to nuclear pores. *Curr. Opin. Cell Biol.*, **16**, 285–292.

35 Vargas, D.Y., Raj, A., Marras, S.A., Kramer, F.R. and Tyagi, S. (2005) Mechanism of mRNA transport in the nucleus. *Proc. Natl Acad. Sci. USA*, **102**, 17008–17013.

36 Pederson, T. and Aebi, U. (2005) Nuclear actin extends, with no contraction in sight. *Mol. Biol. Cell*, **1**, 5055–5060.

37 Visa, N. (2005) Actin in transcription. Actin is required for transcription by all three RNA polymerases in the, eukaryotic cell nucleus. *EMBO Rep.*, **6**, 218–219.

38 Percipalle, P. and Visa, N. (2006) Molecular functions of nuclear actin in transcription. *J. Cell Biol.*, **172**, 967–971.

39 McDonald, D., Carrero, G., Andrin, C., de Vries, G. and Hendzel, M.J. (2006) Nucleoplasmic beta-actin exists in a dynamic equilibrium between low-mobility polymeric species and rapidly diffusing populations. *J. Cell Biol.*, **172**, 541–552.

40 Lawrence, J.B., Singer, R.H. and Marselle, L.M. (1989) Highly localized tracks of specific transcripts within interphase nuclei visualized by in situ hybridization. *Cell*, **57**, 493–502.

41 Zachar, Z., Kramer, J., Mims, I.P. and Bingham, P.M. (1993) Evidence for channeled diffusion of pre-mRNAs during nuclear RNA transport in metazoans. *J. Cell Biol.*, **121**, 729–742.

42 Siomi, H. and Dreyfuss, G. (1997) RNA-binding proteins as regulators of gene expression. *Curr. Opin. Genet. Dev.*, **7**, 345–353.

43 Kim, V.N. and Dreyfuss, G. (2001) Nuclear mRNA binding proteins couple pre-mRNA splicing and post-splicing events. *Mol. Cells*, **2**, 1–10.

44 Dreyfuss, G., Kim, V.N. and Kataoka, N. (2002) Messenger-RNA-binding proteins and the messages they carry. *Nat. Rev. Mol. Cell. Biol.*, **3**, 195–205.

45 Zenklusen, D. and Stutz, F. (2001) Nuclear export of mRNA. *FEBS Lett.*, **498**, 150–156.

46 Rodriguez, M.S., Dargemont, C. and Stutz, F. (2004) Nuclear export of RNA. *Biol. Cell*, **96**, 639–655.

47 Macara, I.G. (2001) Transport into and out of the nucleus. *Microbiol. Mol. Biol. Rev.*, **65**, 570–594.

48 Clouse, K.N., Luo, M.J., Zhou, Z. and Reed, R. (2001) A Ran-independent pathway for export of spliced mRNA. *Nat. Cell Biol.*, **3**, 97–99.

49 Schmitt, C., von Kobbe, C., Bachi, A., Pante, N., Rodrigues, J.P., Boscheron, C., Rigaut, G., Wilm, M., Seraphin, B., Carmo-Fonseca, M. and Izaurralde, E. (1999) Dbp5, a DEAD-box protein required for mRNA export, is recruited to the cytoplasmic fibrils of nuclear pore complex via a conserved interaction with CAN/Nup159p. *EMBO J.*, **18**, 4332–4347.

50 Weirich, C.S., Erzberger, J.P., Flick, J.S., Berger, J.M., Thorner, J. and Weis, K. (2006) Activation of the DExD/H-box protein Dbp5 by the nuclear-pore protein Gle1 and its coactivator InsP6 is required for mRNA export. *Nat. Cell Biol.*, **8**, 668–676.

51 Alcazar-Roman, A.R., Tran, E.J., Guo, S. and Wente, S.R. (2006) Inositol hexakisphosphate and Gle1 activate the DEAD-box protein Dbp5 for nuclear mRNA export. *Nat. Cell Biol.*, **8**, 711–716.

52 Rout, M.P., Aitchison, J.D., Suprapto, A., Hjertaas, K., Zhao, Y. and Chait, B.T. (2000) The yeast nuclear pore complex: composition, architecture, and transport mechanism. *J. Cell Biol.*, **148**, 635–651.

53 Ribbeck, K. and Gorlich, D. (2001) Kinetic analysis of translocation through nuclear pore complexes. *EMBO J.*, **20**, 1320–1330.

54 Kubitscheck, U., Grunwald, D., Hoekstra, A., Rohleder, D., Kues, T., Siebrasse, J.P. and Peters, R. (2005) Nuclear transport of single molecules: dwell times at the nuclear pore complex. *J. Cell Biol.*, **168**, 233–243.

55 Luby-Phelps, K., Castle, P.E., Taylor, D.L. and Lanni, F. (1987) Hindered diffusion of inert tracer particles in the cytoplasm of mouse 3T3 cells. *Proc. Natl Acad. Sci. USA*, **84**, 4910–4913.

56 Luby-Phelps, K. (1993) Effect of cytoarchitecture on the transport and localization of protein synthetic machinery. *J. Cell Biochem.*, **52**, 140–147.

57 Luby-Phelps, K. (1994) Physical properties of cytoplasm. *Curr. Opin. Cell Biol.*, **6**, 3–9.

58 Janson, L.W., Ragsdale, K. and Luby-Phelps, K. (1996) Mechanism and size cutoff for steric exclusion from actin-rich cytoplasmic domains. *Biophys. J.*, **71**, 1228–1234.

59 Sylvestre, J., Vialette, S., Corral Debrinski, M. and Jacq, C. (2003) Long mRNAs coding for yeast mitochondrial proteins of prokaryotic origin preferentially localize to the vicinity of mitochondria. *Genome Biol.*, **4**, R44.

60 Fusco, D., Accornero, N., Lavoie, B., Shenoy, S.M., Blanchard, J.M., Singer, R.H. and Bertrand, E. (2003) Single mRNA molecules demonstrate probabilistic movement in living mammalian cells. *Curr. Biol.*, **13**, 161–167.

61 Mohr, E. and Richter, D. (2001) Messenger RNA on the move: implications for cell polarity. *Int. J. Biochem. Cell Biol.*, **33**, 669–679.

62 Kloc, M., Zearfoss, N.R. and Etkin, L.D. (2002) Mechanisms of subcellular mRNA localization. *Cell*, **10**, 533–544.

63 Chabanon, H., Mickleburgh, I. and Hesketh, J. (2004) Zipcodes and postage stamps: mRNA localisation signals and their trans-acting binding proteins. *Brief. Funct. Genomic Proteomic*, **3**, 240–256.

64 St Johnston, D. (2005) Moving messages: the intracellular localization of mRNAs. *Nat. Rev. Mol. Cell Biol.*, **6**, 363–375.

65 Lawrence, J.B. and Singer, R.H. (1986) Intracellular localization of messenger RNAs for cytoskeletal proteins. *Cell*, **45**, 407–415.

66 Kislauskis, E.H. and Singer, R.H. (1992) Determinants of mRNA localization. *Curr. Opin. Cell Biol.*, **4**, 975–978.

67 Kislauskis, E.H., Li, Z., Singer, R.H. and Taneja, K.L. (1993) Isoform-specific 3′-untranslated sequences sort alpha-cardiac and beta-cytoplasmic actin messenger RNAs to different cytoplasmic compartments. *J. Cell Biol.*, **123**, 165–172.

68 Ross, A.F., Oleynikov, Y., Kislauskis, E.H., Taneja, K.L. and Singer, R.H. (1997) Characterization of a beta-actin mRNA zipcode-binding protein. *Mol. Cell Biol.*, **17**, 2158–2165.

69 Farina, K.L., Huttelmaier, S., Musunuru, K., Darnell, R. and Singer, R.H. (2003) Two ZBP1 KH domains facilitate beta-actin mRNA localization, granule formation, and cytoskeletal attachment. *J. Cell Biol.*, **160**, 77–87.

70 Mowry, K.L. and Melton, D.A. (1992) Vegetal messenger RNA localization directed by a 340-nt RNA sequence element in Xenopus oocytes. *Science*, **255**, 991–994.

71 Schwartz, S.P., Aisenthal, L., Elisha, Z., Oberman, F. and Yisraeli, J.K. (1992) A 69-kDa RNA-binding protein from Xenopus oocytes recognizes a common motif in two vegetally localized maternal mRNAs. *Proc. Natl. Acad. Sci. USA*, **89**, 11895–11899.

72 Czaplinski, K., Kocher, T., Schelder, M., Segref, A., Wilm, M. and Mattaj, I.W. (2005) Identification of 40LoVe, a Xenopus hnRNP D family protein involved in localizing a TGF-betarelated mRNA during oogenesis. *Dev. Cell*, **8**, 505–515.

73 Czaplinski, K. and Mattaj, I.W. (2006) 40LoVe interacts with Vg1RBP/Vera and hnRNP I in binding the Vg1₁-localization element. *RNA*, **12**, 213–222.

74 Hoek, K.S., Kidd, G.J., Carson, J.H. and Smith, R. (1998) hnRNP A2 selectively binds the cytoplasmic transport sequence

of myelin basic protein mRNA. *Biochemistry*, **37**, 7021–7029.

75 Long, R.M., Singer, R.H., Meng, X., Gonzalez, I., Nasmyth, K. and Jansen, R.P. (1997) Mating type switching in yeast controlled by asymmetric localization of ASH1 mRNA. *Science*, **277**, 383–387.

76 Beach, D.L., Salmon, E.D. and Bloom, K. (1999) Localization and anchoring of mRNA in budding yeast. *Curr. Biol.*, **9**, 569–578.

77 Huttelmaier, S., Zenklusen, D., Lederer, M., Dictenberg, J., Lorenz, M., Meng, X., Bassell, G.J., Condeelis, J. and Singer, R.H. (2005) Spatial regulation of beta-actin translation by Src-dependent phosphorylation of ZBP1. *Nature*, **438**, 512–515.

78 Gu, W., Deng, Y., Zenklusen, D. and Singer, R.H. (2004) A new yeast PUF family protein, Puf6p, represses ASH1 mRNA translation and is required for its localization. *Genes Dev.*, **18**, 1452–1465.

79 Irie, K., Tadauchi, T., Takizawa, P.A., Vale, R.D., Matsumoto, K. and Herskowitz, I. (2002) The Khd1 protein, which has three KH RNA-binding motifs, is required for proper localization of ASH1 mRNA in yeast. *EMBO J.*, **21**, 1158–1167.

80 Paquin, N., Menade, M., Poirier, G., Donato, D., Drouet, E. and Chartrand, P. (2007) Local activation of yeast ASH1 mRNA translation through phosphorylation of Khd1p by the casein kinase Yck1p. *Mol. Cell.*, **26**, 795–809.

81 Gu, W., Pan, F., Zhang, H., Bassell, G.J. and Singer, R.H. (2002) A predominantly nuclear protein affecting cytoplasmic localization of beta-actin mRNA in fibroblasts and neurons. *J. Cell Biol.*, **156**, 41–51.

82 Sundell, C.L. and Singer, R.H. (1991) Requirement of microfilaments in sorting of actin messenger RNA. *Science*, **253**, 1275–1277.

83 Latham, V.M., Yu, E.H., Tullio, A.N., Adelstein, R.S. and Singer, R.H. (2001) A Rho-dependent signalling pathway operating through myosin localizes beta-actin mRNA in fibroblasts. *Curr. Biol.*, **11**, 1010–1016.

84 Liu, G., Grant, W.M., Persky, D., Latham, V.M. Jr, Singer, R.H. and Condeelis, J. (2002) Interactions of elongation factor 1alpha with F-actin and beta-actin mRNA: implications for anchoring mRNA in cell protrusions. *Mol. Biol. Cell*, **13**, 579–592.

85 Carson, J.H., Worboys, K., Ainger, K. and Barbarese, E. (1997) Translocation of myelin basic protein mRNA in oligodendrocytes requires microtubules and kinesin. *Cell Motil. Cytoskel.*, **38**, 318–328.

86 Kanai, Y., Dohmae, N. and Hirokawa, N. (2004) Kinesin transports RNA: isolation and characterization of an RNA-transporting granule. *Neuron*, **43**, 513–525.

87 Atlas, R., Behar, L., Elliott, E. and Ginzburg, I. (2004) The insulin-like growth factor mRNA binding-protein IMP-1 and the Ras-regulatory protein G3BP associate with tau mRNA and HuD protein in differentiated P19 neuronal cells. *J. Neurochem.*, **89**, 613–626.

88 Nilson, L.A. and Schüpbach, T. (1999) EGF receptor signaling in *Drosophila* oogenesis. *Curr. Top. Dev. Biol.*, **44**, 203–243.

89 Gonzalez-Reyes, A. and St Johnston, D.S. (1998) Patterning of the follicle cell epithelium along the anterior-posterior axis during *Drosophila* oogenesis. *Development*, **125**, 2837–2846.

90 van Eeden, F. and St Johnston, D. (1999) The polarization of the anterior-posterior and dorsal-ventral axes during *Drosophila* oogenesis. *Curr. Opin. Genet. Dev.*, **9**, 396–404.

91 MacDougall, N., Clark, A., MacDougall, E. and Davis, I. (2003) Drosophila *gurken* (TGFalpha) mRNA localizes as particles that move within the oocyte in two dynein-dependent steps. *Dev. Cell.*, **4**, 307–319.

92 Goodrich, J.S., Clouse, K.N. and Schupbach, T. (2004) Hrb27C, Sqd and Otu cooperatively regulate *gurken* RNA localization and mediate nurse cell

chromosome dispersion in *Drosophila* oogenesis. *Development*, **131**, 1949–1958.

93 Norvell, A., Kelley, R.L., Wehr, K. and Schupbach, T. (1999) Specific isoforms of squid, a Drosophila hnRNP, perform distinct roles in *gurken* localization during oogenesis. *Genes Dev.*, **13**, 864–876.

94 Weil, T.T., Forrest, K.M. and Gavis, E.R. (2006) Localization of *bicoid* mRNA in late oocytes is maintained by continual active transport. *Dev. Cell*, **11**, 251–262.

95 Hachet, O. and Ephrussi, A. (2001) *Drosophila* Y14 shuttles to the posterior of the oocyte and is required for *oskar* mRNA transport. *Curr. Biol.*, **11**, 1666–1674.

96 Hachet, O. and Ephrussi, A. (2004) Splicing of *oskar* RNA in the nucleus is coupled to its cytoplasmic localization. *Nature*, **428**, 959–963.

97 Forrest, K.M. and Gavis, E.R. (2003) Live imaging of endogenous RNA reveals a diffusion and entrapment mechanism for *nanos* mRNA localization in *Drosophila*. *Curr. Biol.*, **13**, 1159–1168.

98 Kalifa, Y., Huang, T., Rosen, L.N., Chatterjee, S. and Gavis, E.R. (2006) Glorund, a *Drosophila* hnRNP F/H homolog, is an ovarian repressor of *nanos* translation. *Dev. Cell*, **10**, 291–301.

99 Crucs, S., Chatterjee, S. and Gavis, E.R. (2000) Overlapping but distinct RNA elements control repression and activation of *nanos* translation. *Mol. Cell*, **5**, 457–467.

100 Forrest, K.M., Clark, I.E., Jain, R.A. and Gavis, E.R. (2004) Temporal complexity within a translational control element in

the *nanos* mRNA. *Development*, **131**, 5849–5857.

101 Bashirullah, A., Halsell, S.R., Cooperstock, R.L., Kloc, M., Karaiskakis, A., Fisher, W.W., Fu, W., Hamilton, J.K., Etkin, L.D. and Lipshitz, H.D. (1999) Joint action of two RNA degradation pathways controls the timing of maternal transcript elimination at the midblastula transition in *Drosophila melanogaster*. *EMBO J.*, **18**, 2610–2620.

102 Ding, D., Parkhurst, S.M., Halsell, S.R. and Lipshitz, H.D. (1993) Dynamic *Hsp83* RNA localization during *Drosophila* oogenesis and embryogenesis. *Mol. Cell. Biol.*, **13**, 3773–3781.

103 Davis, I. and Ish-Horowicz, D. (1991) Apical localization of pair-rule transcripts requires 3′ sequences and limits protein diffusion in the *Drosophila* blastoderm embryo. *Cell*, **67**, 927–940.

104 Lall, S., Francis-Lang, H., Flament, A., Norvell, A., Schupbach, T. and Ish-Horowicz, D. (1999) Squid hnRNP protein promotes apical cytoplasmic transport and localization of *Drosophila* pair-rule transcripts. *Cell*, **98**, 171–180.

105 Wilkie, G.S. and Davis, I. (2001) *Drosophila wingless* and pair-rule transcripts localize apically by dynein-mediated transport of RNA particles. *Cell*, **105**, 209–219.

106 Delanoue, R. and Davis, I. (2005) Dynein anchors its mRNA cargo after apical transport in the *Drosophila* blastoderm embryo. *Cell*, **122**, 97–106.

9
Protein Dynamics and Interactions

Ted A. Laurence and Shimon Weiss

9.1
Introduction

9.1.1
The Single-molecule Approach to Protein Dynamics and Interactions

Proteins are amazingly versatile, serving as motors, enzymes, messengers, and structural elements in all living things. They operate in the still less familiar nanometer-scale (mesoscopic) world, with dimensions smaller than the macroscopic world, but larger than the world of quantum mechanics. Although proteins are small, many concepts from the macroscopic world may be used. Comparisons of protein machines to objects of classical mechanics (rotors, motors, structural elements) are often made, and many experiments have verified the validity of such comparisons [1–4].

The primary difference between the world of macroscopic machines and nanometer-scale machines is that random thermal motion within the machines cannot be averaged out to thermodynamic quantities. The random structural fluctuations of proteins are often intrinsic to their operation. Also, protein function often involves repeated chemical reactions involving single molecules of substrates. Along with the deterministic component, there is a stochastic, or random nature to the timing of the protein function. Non-equilibrium statistical mechanics methods must be used to describe the dynamics of these machines.

To probe nanometer scale fluctuations experimentally requires methods that either synchronize, or in the case of single-molecule spectroscopy, isolate molecules. Non-equilibrium methods, such as fast mixing, nanosecond laser heating, time-resolved fluorescence, femtosecond spectroscopy can create a temporary, non-equilibrium state. Observing the molecules in the non-equilibrium state as they relax back to equilibrium, probes the fluctuations. An alternative method used to create a non-equilibrium state is the spontaneous fluctuations present in the system even at equilibrium; fluorescence correlation spectroscopy (FCS) is an example [5].

Unfortunately, once equilibrium is reached, no more dynamical information is available.

Single-molecule spectroscopy provides a new way to understand protein dynamics: isolate the protein and observe that one molecule for a long period of time. This is a revolutionary concept; its simplicity is one of its strengths, allowing its use in many different fields. By watching proteins in action one at a time, we are able to obtain the previously hidden dynamical information necessary to understand the mechanisms and limitations of protein machines. The goal of our description of protein machines is to move beyond cartoon depiction of protein action. Rather than simply determining what the moving parts are, we want to know how much friction there is between the moving parts, how much power is supplied by a chemical reaction, and how the binding of a another protein affects the action of the protein fluctuations. We want to know the order and timing of events.

There are three major benefits of performing measurements at the single molecule level: the abilities to (i) determine distributions of subpoupulations, (ii) measure long time scale dynamics that are unsynchronized with initial conditions, and (iii) measure the relative timing of coordinated events, especially those unsynchronized with initial conditions.

9.1.1.1 Distributions of Subpoupulations
Biological systems are in general not homogeneous. An ensemble of proteins may be in different states. These states may differ in structure, may differ in binding to other proteins or nucleic acids, or in the progression of a reaction. Single-molecule methods provide new information on the distributions of these states. Based on signals from single proteins, molecules can be sorted into the various states, permitting further study of the properties of these states in isolation.

9.1.1.2 Dynamics of Unsynchronized Trajectories
Biological processes are not random, but thermal fluctuations do add a random element to the timing. Random fluctuations always occur, but there is a non-random sequence to events; see for example studies monitoring the motion of molecular motors such as DNA polymerase [6] and myosin [7]. Due to this, synchronization at the ensemble level is lost very quickly. With single-molecule spectroscopy, it is possible to answer questions about dynamics of multiple, successive events even in the presence of randomizing factors.

9.1.1.3 Order of Events/States
Questions related to the order of events and motion on the energy landscape in between molecular states, are also related to unsynchronized dynamics. Many biological processes are known to involve binding of two different partners. There are very few ways to determine the order of binding events. For instance, does *A* bind *B* before or after binding *C*, or do they all bind at the same time? Also, many enzymes undergo an ordered series of conformational changes during their catalytic cycle. Single-molecule spectroscopy allows the unraveling of these binding and/or conformational sequences.

We focus this chapter on measuring protein dynamics using fluorescence. For some (mechanical-based) questions, single-molecule force experiments (laser and magnetic tweezers, AFM) are more appropriate; however, fluorescence-based single-molecule experiments have several unique benefits. Fluorescence is minimally invasive, more generally applicable, and many of the techniques can be extended to work inside living cells, and possibly inside living tissues and even small organisms.

We plan to provide a researcher outside the single-molecule field with an understanding of what questions regarding protein dynamics and interactions are ideally answered using single-molecule fluorescence methods. In order to illustrate the use of single-molecule spectroscopy for monitoring protein dynamics and interactions, we have chosen two example areas: protein-folding and DNA-processing enzymes. We plan to help the researcher understand the bewildering array of single-molecule methodologies, what motivates their development and how to choose the methodology best suited to a particular question.

9.1.2
Example Biological Systems

The ability of single-molecule detection to separate signals from different conformations of a molecule (e.g. folded and unfolded) and to quantify their respective proportions under conditions of their coexistence can be exploited to study several problems that would otherwise hardly be addressable at the ensemble level. Rather than attempt to survey all of the single-molecule literature regarding protein dynamics and interactions, we have chosen to use a few examples from our research to illustrate how these experiments are motivated and designed. We hope that, in describing our difficulties, mistakes, and successes, newcomers to this exciting field will have a better knowledge of what awaits them.

Proteins are not like clay that can be molded into any shape; they contain structural information in their sequence of amino acids, and fold into precise structures. The overriding question in the protein folding field is how proteins fold based on nothing more than the sequence. In protein folding, single-molecule fluorescence studies give access to the structure and conformational changes in the denatured subensemble, polypeptide chain collapse under a variety of solvent conditions, and thermodynamic parameters of the unfolding process. The protein folding field has already benefited from the abilities of single-molecule spectroscopy to distinguish between the multiple species present and to monitor the folding and unfolding of single proteins over extended periods.

We will also illustrate the application of single-molecule spectroscopy to research into motion and interactions of DNA processing enzymes, particularly RNA polymerase (RNAP). In order to understand how DNA processing enzymes start, perform, regulate, and finish their tasks, it is necessary to elucidate the specific dynamic details of ordering and movement. Already, many questions have been answered or are being answered using single-molecule spectroscopy. For instance, how long does the RNA polymerase remain attached to a promoter? What is the drag on the polymerases – how much molecular friction is there between components?

The dynamic information from single-molecule spectroscopy can complement the fine structural detail obtained using crystallography and NMR for structure determination.

We will review in detail a few recent case studies which illustrate the power of single-molecule approaches, and briefly survey the growing amount of work using these methods. For each example, we will examine how the authors of the papers used many of the principles and procedures outlined in Sections 9.2 and 9.3 to obtain the information sought.

9.2
Fluorescence Spectroscopy as a Tool for Dynamic Measurements of Molecular Conformation and Interactions

There are ways to perform finer measurements of distance, measurements of faster timescales, and less invasive ways to measure protein dynamics, but fluorescence makes up for all of these in sensitivity. Fluorescence can be measured from single molecules, and this has caused increased interest in the methods available to extract the information needed from fluorescence techniques.

9.2.1
Jablonski Diagram (Intensity, Spectrum, Lifetime, Polarization)

The Jablonski diagrams shown in Figure 9.1 is a representation of the energy levels of a fluorescent molecule, or fluorophore [9]. The ground state S_0 is shown as a thick black line. Also part of S_0 is a series of vibronic energy levels. These differ from the ground state only in vibrational energy of the fluorophore. The first excited state S_1 differs from S_0 by an energy $E_{S1}-E_{S0}$. The fluorophores used in single-molecule studies operate in the visible region, so the energy is 1.8–3.1 eV, or using $E = h\nu_a$ hc/λ_a, λ_a is between 400 and 700 nm. A photon is absorbed by the fluorophore with energy E, which excites the molecule from S_0 to S_1. The rate at which the fluorophore is excited depends on the intensity of the incident light and the absorption cross-section of the fluorophore at the wavelength of the incident light. Typically, absorption cross-sections are quoted as molar extinction coefficients (in $M^{-1}cm^{-1}$).

The excitation of the fluorophore from S_0 to S_1 with rate k_e can be to any of the vibronic energy levels of S_1. The vibrational energy is quickly dissipated (~ 1 ps), and the fluorophore remains in the lowest vibronic energy level of S_1. At this point, the fluorophore waits in the excited state until one of four processes occurs. First, the fluorophore may emit a photon; this process has a rate k_r. Second, the fluorophore may de-excite non-radiatively, with rate k_{nr}. Third, the fluorophore may undergo intersystem crossing to a triplet state with rate k_{ISC}, which may also emit a red-shifted phosphorescence photon with rate k_{Ph}. Fourth, the fluorophore may undergo photobleaching with rate k_{bl}, which chemically modifies the fluorophore so that it no longer fluoresces. The fluorescence rate k_r is maximized for good fluorophores, while the other processes are minimized.

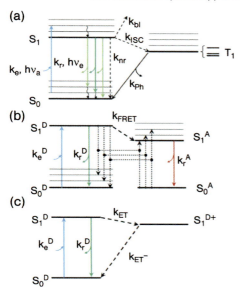

Figure 9.1 Kinetics of fluorescence processes can be summarized in Jablonski diagrams. (A) Upon absorption of a photon of energy hv_a close to the resonance energy E_{S1}–E_{S0}, a molecule in a vibronic sublevel of the ground singlet state S_0 is promoted to a vibronic sublevel of the lowest excited singlet state S_1. Non-radiative, fast relaxation brings the molecule down to the lowest S_1 sublevel in picoseconds. Emission of a photon of energy $hv_e < hv_a$ (radiative rate k_r) can take place within nanoseconds and bring the molecule back to one of the vibronic sublevels of the ground state. Alternatively, collisional quenching may bring the molecule back to its ground state without photon emission (non-radiative rate k_{nr}). A third type of process present in organic dye molecules is intersystem crossing to the first excited triplet state T_1 (rate k_{ISC}). Relaxation from this excited state back to the ground state is spin-forbidden and thus the lifetime of this state ($1/k_{Ph}$ is in the order of microseconds to milliseconds). Relaxation to the ground state takes place either by photon emission (phosphorescence) or non-radiative relaxation. (B) Fluorescence resonance energy transfer involves two molecules: a donor D and an acceptor A whose absorption spectrum overlaps the emission spectrum of the donor. Excitation of the acceptor to the lowest singlet excited state is a process identical to that described for single-molecule fluorescence (A). In the presence of a nearby acceptor molecule (within a few nm), donor fluorescence emission is largely quenched by energy transfer to the acceptor by dipole–dipole interaction with a rate $k_{FRET} \sim R^{-6}$, where R is the D–A distance. The acceptor and donor exhibit fluorescent emission following the rules outlined in A and omitted in this diagram for simplicity. (C) Photo-induced electron transfer effectively oxidizes the donor molecule with a rate $k_{ET} \sim \exp(-\beta R)$, preventing its radiative relaxation. Upon reduction, the molecule relaxes non-radiatively to its ground state. In this scheme, the electron acceptor does not fluoresce and is therefore not represented. Reproduced with permission from [8].

In general, the rate k_r is similar to k_{nr}, but k_{ISC}, k_{Ph}, and k_{bl} are orders of magnitude slower. The lifetime of S_1 is the reciprocal of the sum of rates of all de-excitation pathways,

$$\tau = 1/(k_r + k_{nr} + k_{ISC} + k_{bl}) \approx 1/(k_r + k_{nr}). \tag{9.1}$$

Typical lifetimes of single molecule fluorophores are 1–4 ns. In order to see emission from the fluorophores, it is important to maximize k_r versus k_{nr}. This is quantified as the quantum yield of fluorescence,

$$Q_r = k_r/(k_r + k_{nr} + k_{ISC} + k_{bl}) \approx k_r/(k_r + k_{nr}). \tag{9.2}$$

One important feature of fluorescence is the Stokes shift, where the emission is red-shifted compared to the absorption. This is caused by the vibronic energy levels. In either S_0 or S_1, the fluorophore remains within an energy $k_B T$ of the lowest vibronic state. Within ~ 1 ps, any excitation into another vibronic state quickly relaxes down to the lowest vibronic states. However, the density of vibronic states is much greater above the lowest vibronic states. This means that it is much easier to excite the fluorophore from S_0 to a high vibronic of S_1 than to the lowest vibronic state of S_1. Also, the state S_1 tends to fluoresce such that the fluorophore de-excites into a higher vibronic state of S_0. The vibronic energy levels increase the energy of the photon required to efficiently excite the fluorophore, and decrease the energy of the emitted photons. This shift in the excitation and emission photons, or Stokes shift, allows the emission photons to be efficiently separated from the excitation photons (using a filter). This is one of the most important features of fluorescence that allows for single-molecule spectroscopy by exclusion of background laser scattering.

At the single-molecule level, the process of fluorescence involves repeated cycling of the fluorophore through the energy levels shown in Figure 9.1. The jumps between states occur at random time intervals, with rates as outlined above. The waiting times to jumps follow Poisson statistics (exponential distributions), with rates as outlined above. If at time t = 0, the fluorophore is in the ground state, then the time to a jump to S_1 has a probability of $P(t) = k_e \exp(-k_e t)$. At time t' after the excitation to S_1, the fluorophore usually decays back to the ground state with a probability of $P'(t') = \exp(-t'/\tau)/\tau$, where τ was defined above. The probability that the fluorophore emits a photon upon de-excitation is the quantum yield Q. At this point the fluorophore repeats the cycle. It is in this repeated cycling that the slow processes such as inter-system crossing and photobleaching become important. After many cycles, on average $(k_r + k_{nr} + k_{ISC} + k_{bl})/k_{ISC}$, the fluorophore will undergo intersystem crossing to T1. Since k_{Ph} is much slower than fluorescence, the fluorophore remains in the triplet state for a comparatively long time. If the excitation is high, as is the case in single-molecule studies, these excursions to the triplet state will be interspersed among the emitted fluorescence photons as dark periods. Also, eventually the fluorophore will photobleach after on average $(k_r + k_{nr} + k_{ISC} + k_{bl})/k_{bl}$ cycles, and stop emitting altogether.

9.2.2
Point Emission-Localization Measurements

Fluorophores are in general very small (<0.5 nm), much smaller than the wavelength of light they emit, and therefore may be considered as point sources of light. Repeated excitation of the fluorophore causes emission from the same point. This is a simple observation, but it has important consequences. The position of the fluorophore may be determined within the limitations imposed by resolution of the detection method, mobility of the emitter and binning (integration) time. This localization accuracy can be very high (~ 1 nm) [3], allowing accurate tracking of the fluorophore- and what it is attached to. Also, point emission from the fluorophores allows single-molecule spectroscopy since isolation of molecules is possible. Precise positioning of the fluorophore requires isolation.

Another important consequence is that, if two spectrally separable fluorophores (different colors) are used, the relative position of those molecules attached to two fluorophores may be obtained. The relative position and association can be measured using co-localization, for molecules immobilized on surfaces or in a matrix, and cross-correlation measurements, for freely diffusing molecules. The different colors allow each fluorophore to be isolated even when in close proximity [10]. More recently, it has been shown that even if the fluorophores are of the same color, but can be turned on or off at different times, the same high accuracy resolution and even imaging can be achieved [11].

9.2.3
Fluorescence Polarization-Measures Rotational Movement and Freedom of Movement

Fluorophores do not absorb or emit radiation uniformly in all directions. They act as electric dipoles, which preferentially absorb and emit radiation that is aligned with this dipole. The orientation of the fluorophore plays a role in both absorption and emission processes, due to their dipolar nature. The excitation rate k_e (Figure 9.1A) depends on the incident power, absorption cross-section σ and relative orientation of the incident electromagnetic field E and the absorption dipole moment μ_{abs}:

$$k_e = \sigma |\vec{\mu}_{abs} \cdot \vec{E}|^2 \qquad (9.3)$$

As mentioned, the emitted intensity is not only proportional to the population of the excited state S_1, but also to the detection efficiency which can be chosen to be polarization sensitive. Polarization-sensitive (time-resolved) measurements can thus yield information on the (time-dependent) orientation of the fluorophore, and have been used to study DNA and protein conformations at the single-molecule level as will be reviewed later [12–15].

9.2.4
Fluorescence Resonance Energy Transfer-nm-scale Ruler

Fluorescence resonance energy transfer (FRET) is one of the primary tools used in single molecule spectroscopy, allowing measurements of distances between 2 and 8 nm at the single-molecule level [16–19]. Some prefer to call it Förster resonance energy transfer, after the scientist who first described it, primarily because FRET involves the *non-radiative* transfer of energy from one fluorophore to another. They feel that since the process is non-radiative, the appellation "fluorescence" is inappropriate. However, in order for the non-radiative transfer to occur, the fluorescence emission spectrum of a *donor* (D) fluorophore must overlap the fluorescence absorption spectrum of an *acceptor* (A) fluorophore. In other words, the fluorescence of the donor and acceptor must be in resonance for the energy transfer to occur. Hence, we will continue to use the appellation "fluorescence".

When two fluorophores are close enousgh together, the excitation of one fluorophore may be transferred to the other. Figure 9.1B shows how the FRET process appears in a Jablonski diagram. Fluorophore D, with higher energy excitation levels, is brought close to fluorophore A, with lower energy levels. If there are a large number of emissions pathways from S_1^D to the manifold of vibronic energy levels of S_0^D that match energy of excitation pathways from S_0^A to the vibronic energy levels of S_1^A, then efficient FRET can occur. FRET adds another de-excitation pathway for S_1^D with a distance dependent rate,

$$k_{FRET} = \frac{1}{\tau_0}\left(\frac{R_0}{R}\right)^6 \tag{9.4}$$

where τ_0 is the donor fluorescence lifetime in the absence of an acceptor, R is the distance between D and A, and the Förster radius R_0 is the distance at which 50% of D excitations are transferred A.

A

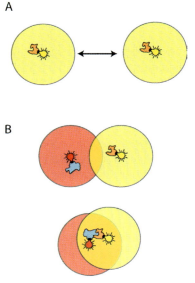

B

Figure 9.2 Effects of diffraction on localization accuracy of single fluorescent molecules. (A) Even though fluorophores are small (<0.5 nm), diffraction causes emission from the fluorophores to be imaged as larger spots with sizes comparable to the wavelength of light. The diffraction-limited spots are shown as larger yellow or red circles around the fluorophores. Even though the spot size is near the wavelength of light, the center of each spot may be determined with very good accuracy, as long as each emitting molecule is well separated as in A. (B) If emission from two molecules is spectrally separable, the relative centers of the two spots may be determined to very high accuracy, even if they overlap. Emission from two fluorophores may also be separated even if they are of same color, but turned on or off at different times.

FRET operates through a dipole–dipole interaction mechanism, accounting for the distance dependence in Equation 9.4. The angular dependence is absorbed in the Förster radius R_0 using the orientation factor κ^2, which depends on the relative orientation of the donor and acceptor transition dipoles.

The FRET efficiency E is the quantum efficiency of the FRET process. We calculate E similarly to Q_r,

$$E = k_{\text{FRET}}/(k_{\text{FRET}} + k_r + k_{\text{nr}} + k_{\text{ISC}} + k_{\text{bl}}) \approx k_{\text{FRET}}/(k_{\text{FRET}} + k_r + k_{\text{nr}}). \quad (9.5)$$

During the repeated cycling through the Jablonski diagram during single molecule measurements, E is the probability that a D excitation will be transferred to A. Even without a change in R, some excitations will cause D emission, and others will be transferred to A, causing A emission. Using Equations 9.1 and 9.4, Equation 9.5 can be rewritten as

$$E = 1/\left(1 + \left(\frac{R}{R_0}\right)^6\right) \quad (9.6)$$

The distance dependence of E allows its use as measure of the distance between D and A near R_0 (typically 5–7 nm for dyes used in single-molecule fluorescence spectroscopy).

9.2.5
Single-molecule Electron Transfer-Ångstrøm-scale Ruler

Electron transfer between a donor and an acceptor molecule occurs at a rate k_{ET} that depends on two main factors: the coupling between the reactant and product *electronic* wave functions V_R^2, and the Franck–Condon weighted density of states FC [20, 21]. The latter is usually constant for macromolecules and therefore one is left with the electronic coupling term, which depends exponentially on the distance between donor and acceptor:

$$V_R^2 = V_0^2 \exp(-\beta R) \quad (9.7)$$

where β has been measured for different systems to vary from 1.0 to 1.4 Å^{-1} in proteins. The electron transfer rate k_{ET} can be accessed by measuring the donor fluorescence emission, or the donor fluorescence lifetime, assuming that no other process influences these two observables. Fluorescence quenching by electron transfer can be used to monitor minute conformational changes in biopolymers as it requires close contact between the donor and acceptor molecules. This approach has been used at the single-molecule level to study dye-labeled polypeptides containing a tryptophan residue [22], and a flavin reductase enzyme in which the fluorescence of an isoalloxazine is modulated by photo-induced electron transfer to a tyrosine [23].

This brief overview of some of the photophysical characteristics of fluorophores used in single-molecule spectroscopy has illustrated theoretically how sensitive the absorption and emission can be to the local environment of the fluorophore. Therefore, any modification or fluctuation of fluorescence intensity or lifetime (the two main observables) observed in an experiment, should be carefully analyzed to exclude genuine photophysical effects that have nothing to do with potential intra- or intermolecular changes of the macromolecule to which the fluorophore is attached. Simple controls such as, for instance, studying the excitation power dependence of the observed effect, analysis of the photophysics of the dye alone or of singly-labeled macromolecules in the case of FRET studies, are necessary to ascertain the physical origin of the observable variation (see e.g. [24]).

9.3
Single-molecule Data Acquisition and Analysis Methods

The number of fluorescence-based single-molecule methodologies can be daunting for newcomers to the field. Nearly every time a new question is approached, a new methodology is also developed. In order to understand why this is so, an understanding of the motivating factors in the development of these methods must be derived. Depending on the question of interest, the proper method can usually be found. Or, can be developed as a combination of previous methods (and given a new name!).

There is only one chance to look at each molecule. In order to maximize the amount of information obtained from that molecule, experimental schemes of increasing complexity have been devised. Also, single-molecule measurements require the same types of controls as ensemble experiments. Many of these controls must be performed simultaneously with the single-molecule experiments in order to be meaningful.

Single molecule detection and analysis of protein dynamics and interactions requires three components: an observable for answering the question of interest, sufficient signal, and isolation of the signal from any other signals. In this section, we will discuss how to satisfy these three requirements, and how to acquire and analyze the data once the molecules are prepared.

9.3.1
Choosing a Labeling Configuration: What is the Observable?

The choice of observable must be chosen specifically to answer each new question. Example labeling configurations are shown in Figure 9.3. Single-molecule methods have the largest impact in measuring dynamic changes in the conformation and interactions of proteins. For observables monitoring protein dynamics and interactions, the diagrams and suggestions suggested in a previous review still apply [25]. In choosing the observable, the first choice to be made is to determine whether the conformation of an individual protein will be observed, or whether the configuration of protein interactions will be observed. Examples of the former include protein folding measurements and conformational changes induced by protein interactions such as changes in DNA structure upon binding of CAP [26]. Examples of the latter include the motion of RNA polymerase on DNA and antibody–antigen interactions.

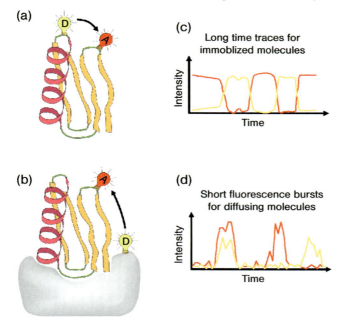

Figure 9.3 Two basic labeling configurations used for monitoring protein dynamics and interactions. Both are based on using FRET to dynamically monitor distances between two sites of proteins. (A) Monitoring internal protein dynamics is accomplished by site-specifically labeling one protein with both donor and acceptor. As the protein changes conformation, folds or unfolds, changes in the distance between D and A are monitored in real time using FRET. (B) Dynamic monitoring of protein interactions is accomplished by site-specifically labeling two interacting molecules, one with D and one with A. FRET is used to monitor distance changes between the two sites, and the localization as described in Figure 9.2 is used for detecting association of the molecules. (C) Fluorescence intensities of D (yellow) and A (red) are monitored as a function of time for a single molecule. If the molecule is immobilized, changes in distance between D and A are seen as anti-correlated changes in the intensity of D and A (when A goes up, D goes down). (D) For freely diffusing molecules, observation times are typically too short to see changes in structure. Small snapshots of D and A can be seen, allowing the formation of histograms of distance distributions of all molecules. Adapted from [25].

The most general way to measure conformational changes in a protein, DNA, RNA, etc., either internal fluctuations or externally-induced changes, is to monitor the distance between two points in the protein using FRET (Figure 9.3B). For appropriately chosen labeling sites, single-molecule measurements will allow the conformational state to be monitored for extended periods (Figure 9.3C). Distances shorter than 3–4 nm can be monitored using electron transfer. Fluorescence polarization may also be used as an observable; in this case, rather than monitoring distance between two points, the mobility of the fluorophore or the rotational movement of the protein is monitored.

We can use some of these same observables for monitoring changes in the configuration of interactions (Figure 9.3B). FRET and ET may be used to dynamically monitor the distance between two proteins or a protein and DNA. Fluorescence polarization may also be used to observe changes in binding, although expected changes in anisotropy are usually less dramatic than changes in FRET or ET. One additional property may be used to monitor changes in interactions: co-localization of proteins. If a protein is bound to DNA or another protein and each component is labeled, the signal from the various fluorophores are co-localized, and will either be in close proximity (if the system is not moving) or will diffuse together (in solution-based experiments). These signals are extremely important for monitoring the presence of various components over time.

Once the choice of observable is made, the more difficult task of actually producing the labeled configuration is next [27–29]. In labeling the molecules, three things must be borne in mind. First, the specificity and efficiency of the labeling must be sufficient to remove ambiguity in the single-molecule measurements. The most common method for labeling proteins is a cysteine-based labeling. Labeling of DNA is generally less difficult due to automated synthesis and may often be purchased ready labeled. Second, does the presence of the fluorophore, or changes made to the protein or DNA to facilitate labeling affect its function? Third, does the system actually show the changes or movement expected at the single-molecule level? Even in cases where the final question requires immobilization, we find that solution-based techniques are often preferable in showing that the observables actually work.

After choosing labeling sites, it is necessary to decide on which fluorophores will be used. This is an important decision as good performance from fluorophores is required to obtain useful information from single-molecule studies. In addition to more traditional dyes such as tetramethylrhodamine (TMR), many excellent fluorophores have become available in recent years. Series of fluorophores from Atto-tec and Molecular Probes (Alexa fluor) are in general excellent choices. In choosing the fluorophores, it is generally best to start with fluorophores used successfully elsewhere. However, issues to consider are extinction coefficient (how strong the absorption of light is), quantum efficiency (how many excitations lead to fluorescence), resistance to photobleaching, sources of background for the laser wavelengths chosen, and, when monitoring multiple colors, the extent of overlap of emission spectra, and the Forster radii of the fluorophores.

9.3.2
Should a Freely-diffusing or Immobilized Format be used?

Next to the labeling configuration, the most important decision to make in a single-molecule measurement is whether to carry out the measurements on freely diffusing molecules in solution or immobilized molecules. The determining factor for which type of measurement to use is the length of the observation time required to answer the question at hand.

For example, watching a single protein fold and unfold requires immobilization since the waiting times between folding and unfolding events can be in the ms–s time scale. However, studying the sub-ms structural fluctuations of unfolded proteins does not require immobilization.

Immobilization of the system is of similar difficulty to fluorescent labeling, requiring an attachment point on one of the molecules in the system. Successful immobilization strategies include immobilization on coated glass surfaces, trapping in immobilized vesicles, and immobilization in a gel matrix.

The primary benefit of immobilization is the ability to monitor one molecule for an extended period. This allows both the monitoring of equilibrium fluctuations as well as the response of a single molecule to triggered reactions or other non-equilibrium events. In cases where solutions can be exchanged (surface immobilization), it is also possible to wash away unused reagents during the preparation of the sample. The drawback is the additional work of immobilizing the molecules, and the possibility that the immobilization strategy will affect the function of the molecules.

In experiments using freely-diffusing molecules, fluorescence is observed briefly (~ 1 ms) from each molecule as it diffuses through a small observation volume. Many molecules can be observed with this approach during a relatively short experiment, and can be sorted into several subpopulations. No immobilization strategy is needed, so the only changes to the molecules are the fluorophores, thus minimizing potential problems. Sample preparation for each experiment is much simpler and does not require any steps for immobilization. The main drawback is the short observation time. Equilibrium measurements in solution provide "snap shots" of the states of many molecules. By invoking the ergodic theorem, stating that the behavior of an ensemble of molecules is equivalent to the behavior of a single molecule over very long time periods, we can determine the states accessible to each molecule.

Even in cases where immobilization is necessary, it is often desirable to perform preliminary experiments on freely-diffusing molecules. This helps fine tune experimental conditions for single-molecule measurements, verify that fluorophores do not affect protein function, and can often answer parts of the question or alternative issues related to the primary question. Comparing the results of the equilibrium distributions obtained from solution-based measurements and the smaller number of long time trajectories obtained, immobilized measurements can be used to demonstrate ergodicity in the system.

As discussed in the next section, small observation volumes are used in single-molecule spectroscopy; these small volumes allow freely-diffusing molecules to diffuse out of the observation volumes quickly (typically near 1 ms). If longer observation times are necessary to answer the experimenter's question, it will be necessary to immobilize the molecule.

9.3.3
What Excitation/Optical Isolation Format should be used?

In order to obtain single-molecule detection, fluorescence from individual molecules must be isolated from all other fluorescence and scattering sources. Sufficiently efficient excitation and detection of a molecule is necessary to detect typically several thousands of photons from a fluorescent molecule per second. A small optical detection volume must be obtained, minimizing background from Raman scattering,

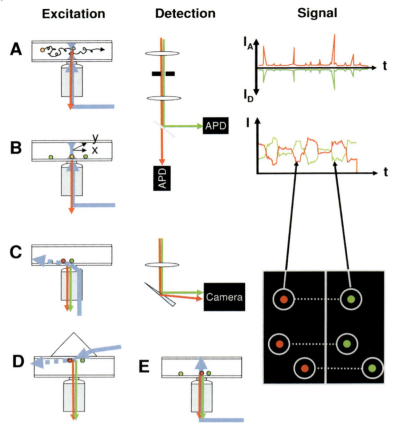

Figure 9.4 Experimental geometries used in single-molecule fluorescence spectroscopy. Two main types of geometries can be used for single-molecule fluorescence spectroscopy: confocal and wide-field. In the *confocal geometry* (A, B), a collimated laser beam is sent into the back focal plane of a high numerical aperture objective lens, which focuses the excitation light into a diffraction limited volume (or point spread function – PSF) in the sample. Fluorescence emitted by molecules present in this volume is collected by the same objective and transmitted through dichroic mirrors, lenses and color filters to one or several point detectors (avalanche photodiodes – APD). An important aspect of this geometry is the presence of a pinhole in the detection path, whose size is chosen such as to let only light originating from the region of the excitation PSF reach the detectors. Freely-diffusing molecules (A) will yield signals comprised of bursts of various size and duration (but typically less than a few ms), as indicated schematically on the right-hand side (RHS). Immobile molecules (B) will first need to be localized using a scanning device (indicated as two perpendicular arrows x and y), before recording can commence. Typical time traces are comprised of one or more fluctuating intensity levels until the molecule eventually bleaches after a few seconds, as indicated schematically on the RHS. The wide-field geometry (C–E) can be used in two different modes: (C, D) total internal reflection (TIR) or (E) epifluorescence. In TIR, a laser beam is shaped in such a way that a collimated beam reaches the glass–buffer interface at a critical angle $\theta = \sin^{-1}(n_{buffer}/n_{glass})$, where n designates the index of refraction. This creates an evanescent wave (decay length of a few 100 nm) in the sample (dashed arrow), which only excites the fluorescence of molecules in the vicinity of the surface, resulting in very low background. TIR

and intrinsic fluorescence from other molecular species in the sample. The concentration of the sample must be diluted so that, on average, less than one molecule is within this optical detection volume at a time. Under these conditions, individual molecules can be observed.

There are two general methods of performing single-molecule spectroscopy. In the first, a small point detection volume is defined using confocal microscopy. Typical sizes for the detection volume are just below a femtoliter. The fluorescence is observed using high sensitivity single photon detectors. In the second, a wide, thin volume of solution is illuminated at a water–glass interface. Both formats are commonly used. As we will see, the critical issue for deciding on the excitation format is whether the process under observation is irreversible, and for how long the process needs to be observed (Figure 9.4).

9.3.3.1 Optical Isolation of a Single Point

Confocal   For single molecule spectroscopy at a single point, a confocal excitation–detection format is used. A laser excitation is focused using a high numerical aperture (N.A.) microscope objective to a diffraction limited spot (size is $\sim\lambda/$N.A.). In order to achieve the best focus, the laser beam must be collimated with a Gaussian beam shape. Spatial filtering using a pinhole or single mode fiber optic is used to clean up the beam. Fluorescent molecules within the focus are efficiently excited from the ground state S_0 to the excited state S_1. The fluorescence produced by transitions from S_1 to S_0 may be emitted in all directions. For a high N.A. objective, a large fraction (up to 30%) of the fluorescence is collected by the same objective that focuses the laser beam. This fluorescence travels the same optical path as the focused laser beam, but in reverse. In order to separate these paths, a dichroic mirror (DM) is used. The DM reflects wavelength of the laser excitation, but transmits the wavelengths of the fluorescence. Here, we take advantage of the Stokes shift described above, which causes a separation between the excitation and emission bands of fluorophores.

Although the excitation of fluorophores is most efficient at the focus of the laser beam, there is also a large amount of excitation of fluorescence outside the focus. The resulting fluorescence from outside the focus is also collected by the objective, and creates a large background. In order to carry out single-molecule spectroscopy, optical

◄──

Figure 9.4 *(Continued)*

can be obtained either with illumination through the objective (C) or by coupling the laser through a prism (D), both methods having their advantages and inconveniences (for details, see [30]). In epifluorescence (E), a laser beam focused at the back focal plane of the objective or a standard arc lamp source is used to illuminate the whole sample depth, possibly generating additional background signal. A wide-field detector (camera) is used in all three cases, allowing the recording of several single molecule signals in parallel, although with a potentially reduced time resolution than that achievable with point detectors. The image on the RHS represents the case of a dual-color experiment, where both spectral channels are imaged simultaneously on the same camera (signals from the same molecule are connected by dotted line). Individual intensity trajectories can be extracted from films, resulting in similar information as that obtained with the confocal geometry. Reproduced with permission from [8].

isolation of individual molecules is necessary. The background fluorescence from out-of-focus regions prevents optical isolation, unless a pinhole is used on the detection path. The fluorescence, after passing through DM, is focused by another lens onto a pinhole, which excludes fluorescence from regions outside the focus. By choosing a pinhole of similar size to the focus (multiplied by the magnification of the objective), fluorescence from regions down to 0.2–1.0 fl can be isolated; or, on surfaces, fluorescence from areas down to $\sim 0.1\,\mu m^2$ can be isolated. This excitation/optical isolation format may be used either at glass–water interfaces, in aqueous solutions, or in living cells. The primary drawback of this approach is that only one point is being observed at a time. This can slow down acquisition considerably, especially in cases where triggered chemical reactions are occurring, and the behavior of each molecule during the entire chemical reaction needs to be monitored.

Two-photon excitation of the fluorophores is the non-linear absorption of two photons, exciting the fluorophore from S_0 to S_1. Fast, powerful IR lasers are required to excite in this way, but there are advantages to this technique. The non-linearity of the excitation prevents excitation outside the focus; the excitation rate is proportional to the square of the laser intensity, which rapidly drops to negligible levels outside the focus. Drawbacks of this method include intense laser intensities and photobleaching.

In order to obtain information about the sample in different regions, either to form an image or to search for immobilized single molecules, the sample is generally scanned using a piezo-actuated microscope stage. This changes the position of the confocal detection volume in the sample, allowing measurements of different regions of the sample.

High concentration   A second limitation of the confocal geometry is the size. In order to isolate single molecules with a 1.0-fl volume, sub-nanomolar concentrations must be used. For intramolecular studies, where the internal dynamics of a protein are being studied, this is not a problem. For example, protein folding studies are not adversely affected by diluting the sample down to sub-nanomolar concentrations. Even if the effects of molecular crowding are being studied, this is not a problem, since unlabeled protein may be present at much higher concentrations. However, for studies of functional proteins that interact with other labeled components, dilution can adversely affect the function. At the low concentrations, the labeled proteins may not even be bound to each other.

This creates a need for a smaller detection volume than can be achieved using standard confocal optics. Several methods have been developed, but these have not yet become commonplace. These are important for *in vitro* measurements at physiological concentrations, as well as potential cell-based and *in vivo* measurements. For such measurements, *in vivo* labeling strategies will also be needed.

One simple method of achieving this is to use total internal reflection (TIR) with confocal detection, reducing the volume 10-fold [31]. More recently, zero mode waveguides [32], stimulated emission depletion (STED) [33], and supercritical angle fluorescence (SAF) [34] have all been demonstrated. Zero mode waveguides have the smallest volumes, but require the presence of metallic coated surfaces. STED provides very small volumes that can be focused in the far-field, even inside cells.

It does require expensive laser and other components. SAF requires a water–glass interface and special objectives, but provides very small volumes.

Although these methods are critically important for the single-molecule analysis of interactions, they still need further development to be widely applicable.

9.3.3.2 Multiple Points

TIR In order to look at many molecules at one time, an alternative optical isolation method is commonly used. Total internal reflection is 100% reflection obtained at a dielectric interface when the approaching light beam is contained inside a medium with a higher index of refraction than that of the medium on the other side of the interface and the angle of incidence is larger than a critical angle. Light incident on a glass–water interface can undergo total internal reflection for >60° from the normal to the interface. When a light beam is reflected under these conditions, it is 100% reflected, but an evanescent (exponentially decaying) beam does penetrate the lower index of the refraction medium, in this case the water. This evanescent wave can excite fluorophores near the surface of the glass (within ∼150 nm), but does not excite molecules which are deeper in the water. This provides optical isolation without a pinhole. Total internal reflection can be obtained over a wide area ($50 \times 50\,\mu m^2$ or more), allowing excitation and optical isolation of many single molecules simultaneously. This is a great help in saving experimental time. The main drawback of this technique is that the molecules are required to be at the glass–water interface.

Multi-confocal spectroscopy Multiple spot techniques have been demonstrated or proposed in a few papers that provide a glimpse of how the single point measurements could be extended to multiple simultaneous points. Two-point cross-correlation techniques have been used to monitor systems undergoing flow [35]. The zero-mode waveguides can be patterned in arrays that may be simultaneously monitored [32]. Cubic lattices of foci have been proposed for rapid multi-confocal analysis [36].

9.3.3.3 How many Excitation Lasers?

If the proposed study monitors only fluorophores of one color, this question is easy to answer – a single laser should be used. However, most studies of protein dynamics, and especially interactions, will require more than one color and one fluorophore to address the scientific questions. For instance, consider a labeled RNA polymerase attached to a labeled DNA fragment. In order to verify that RNAP is actually interacting with the DNA, it is necessary to see both the RNAP and the DNA in the same optically isolated volume. This requires two colors in order to distinguish between the two components.

The most straightforward method of obtaining multiple colors is to excite two fluorophores of different colors with two lasers, one for each color. If the optical isolation provided for both wavelengths overlap, then co-localization of fluorescence for both colors indicates binding of the two molecules. The problem is that, especially in the confocal case, the optical excitations do not overlap perfectly, and much time can be spent trying to overlap them perfectly. Alternatively, the lasers can be focused at different spots, and flow used to put molecules through both laser foci [37]. If fine measurements of centroids of fluorescence are to be used to attain precise positioning of the molecules, this methodology will not work well (because of the chromatic aberrations mentioned above). When extending this concept to more than two colors, all lasers will need to be focused to the same focal volume.

One excitation, many colors   If possible, it is simpler to excite all colors simultaneously with a single laser. This could be accomplished in at least two ways. First, FRET between a donor and acceptor dye may be used to detect the presence of the acceptor-labeled component as long as it is bound to the donor-labeled component. Unfortunately, in order for this method to work, donor and acceptor must be relatively close to each other (within ~ 2–$8\,nm$).

Second, alternative dyes may be used. Examples include quantum dots [10] and hybrid fluorophores that have extra large Stokes shifts [38]. Proof-of-principle of these methods has been shown to work, and they will be of great interest in monitoring interactions.

Multiple excitations – alternating laser excitations   Since it is possible to replace multiple excitation lasers with a single excitation laser, it would seem that use of multiple laser excitations would fade with time. However, the recent development of alternating laser excitation (ALEX) promises to stop this from happening [26]. ALEX allows the combination of FRET information with the co-localization information available from molecules being in the same optically isolated volume.

In previous multiple laser excitation applications, both lasers were on continuously. Emission from each color is separated and monitored individually. In the example of RNAP and DNA, if the fluorophores are close together and FRET occurs from one fluorophore on the protein (donor) to the fluorophore on the DNA (acceptor) at high efficiency, it is not possible to distinguish between acceptor emission from FRET and acceptor emission from direct excitation by the second (red) laser. Additionally, the donor emission is quenched, and the observer may incorrectly conclude that the RNAP and DNA are not bound, since only acceptor emission can be observed.

Conversely, if the RNAP and DNA are bound, but outside the range of FRET, a single excitation format would result in only donor emission. The observer would incorrectly conclude that the RNAP and DNA are unbound, because only donor emission can be observed.

ALEX allows the correct conclusion to be drawn in both cases by using time-division and wavelength division multiplexing. Only one laser is on at a given time, but the donor excitation and acceptor excitation lasers are alternated more rapidly than the system diffuses through the optically isolated volume or the dynamics of the system. When the donor excitation laser is on, any FRET that occurs can be easily detected. When the acceptor excitation laser is on, the presence of the acceptor fluorophore is interrogated, even if there is no FRET.

ALEX essentially replaces the multiple excitation methods. There is much to be gained, with very little to lose. The only issue is whether the alternation period of the lasers interferes with observations of dynamics which is on that time scale. However, ALEX has been demonstrated with alternation periods in the microsecond, nanosecond, and millisecond range, so the alternation period can be chosen not to interfere with the dynamics in question [39].

9.3.3.4 Pulsed Laser Excitation

With a pulsed laser, the laser energy is produced in the form of a train of energy packets (or pulses) rather than a continuous stream of energy. For short pulses (< 1 ns) the pulse may be used as a trigger to identify when the fluorescence excitation occurred. Pulsed lasers are used in three ways: (i) to allow measurement of fluorescence lifetime [40]; (ii) to allow two-photon excitation [41]; and (iii) to allow for stimulated emission depletion (STED) [42]. If these issues are not important to the question at hand, then continuous wave lasers are sufficient.

9.3.4
What Detection Format should be used?

The fluorescence obtained from the sample, after passing through the pinhole or simply passing through the dichroic mirror, must be separated into various detection channels. For a two-color system, the fluorescence must be separated by a second dichroic mirror into bands specific for the two colors. Additional bandpass filters are used to exclude stray laser reflection and other background that make it through the imperfect dichroic mirrors. Splitting the donor and acceptor emission by polarization as well requires four detectors [43, 44]. In applications using FRET, monitoring fluorescence polarization or anisotropy is desirable since it allows acquisition of information on the orientation factor κ^2 (described above). Additional colors require more dichroic mirrors and bandpass filters to define detection channels [45–47].

The type of detector used generally depends on the optical isolation method. For confocal, single point methods, single photon counting avalanche photodiode (APD) detectors are typically used. A fluorescence photon impinging on the detector will trigger an electronic pulse at the output of the detector. The electronic pulses may be counted in fixed time intervals to provide a time trace of the number of photon counting as a function of time. Even better, each electronic pulse may be timed with high accuracy (depending on the counting/timing electronics, ~ 10 ns typically, and down to 4 ps with a dedicated circuit), providing a list of all detected photons with their arrival times. If a pulsed laser excitation is used, the time difference between the arrival of each photon and the laser pulse could be obtained with an accuracy of several picoseconds (time-correlated single photon counting, TCSPC) constituting a measurement of the fluorescence lifetime. The main drawback of APD detection is that only one point can be monitored at a time.

For TIR applications with a wide field of view, the APD detectors are not appropriate, since no spatial information is available. Instead, an intensified CCD camera is used, which acquires film of a large field-of-view (FOV) near video rate (30-ms frames), but may be significantly faster for smaller FOVs ($\sim 1\,kHz$). As mentioned above, such sensitive CCD cameras allow single-molecule detection to be undertaken on a wide area. The main drawback is that the time resolution is currently limited to the ms time range. This does not allow time-correlation information below the millisecond time scale (see data analysis below) or fluorescence lifetime information to be obtained. Gated CCD cameras are available to obtain fluorescence lifetime information, but a large proportion of signal is lost using gating techniques [48].

The combination of wide-area detection and photon counting would remove the timing drawbacks of TIR methods, or a allow detection with high time resolution from many confocal spots simultaneously. Such a detector is under development [49, 50].

9.3.5
Data Reduction and Analysis Methods

The most general way to view single-molecule data is as a stream of photon-detection events. All of the single-molecule data reduction and analysis methods branch off from this point. Data reduction is the process by which the raw data is transformed into time traces, histograms, and correlations, and sorted into single-molecule events that can then be analyzed. Data analysis is the process by which the reduced data is interpreted to provide quantitative results.

9.3.5.1 Photon Streams and Films
When detected, a photon carries several pieces of information. There is the arrival time of the photon, its polarization, energy or wavelength, and spatial position at the image plane. With current detection schemes, not all of this information can be obtained simultaneously.

For confocal microscopy with APD detection, only the arrival time of the photon is directly recorded. The spatial position of the photon is only used to exclude out-of-focus light at the pinhole. In order to obtain other information about wavelength and polarization, detection channels are defined using optical filters and dichroic and polarizing beamsplitters that correspond to specific polarization and wavelength ranges. Spatial information about the sample is obtained by scanning the sample, the beam [51], or even the objective (Microtime 200, Picoquant, Berlin, Germany). The information obtained from such detection schemes is a list of photons, containing the arrival time and detection channel. In fluorescence lifetime measurements using TCSPC, two arrival times are obtained for each photon detected: one measured with respect to the beginning of the measurement (at a relatively "low" temporal resolution of $\sim 10\,ns$), and one measured as a time difference with respect to the laser pulse (typically at higher time resolution, of ~ 4–$50\,ps$). Some data acquisition techniques acquire data as numbers of photon counts over fixed time widths (photon binning). The photon timing technique is in general preferred (because of higher information content), unless photon count rates are so high that the size of the data files becomes unmanageable.

For CCD-based detection with TIR excitation, the most important information available is the spatial position of the detected photon. This is determined by which pixel detects the photon. Temporal information is also detected, but readout of the images limits the time resolution that can be obtained to the frame rate of the camera. Spectral information is obtained by using beamsplitters and filters forming different images of different spectral regions on different sections of the CCD. CCD images have somewhat more noise than the APD images due to intensification and readout [52]. The data obtained are a time series of images, with pixels containing different numbers of counts. These counts do not correspond directly to numbers of photon counts (but are proportional to the number of photons detected through the quantum efficiency of the camera).

9.3.5.2 Time Traces

Probably the most basic data reduction procedure is the production of time traces which are simply the number of photons detected during fixed time widths as a function of time (photon binning). For example, a 10-ms resolution time trace is formed from photon timing data by going through the list of photon arrival times and counting the number of photons that were detected in each channel in 10-ms increments. Examples of time traces are shown in Figure 9.3C and D. Time traces are used for analyzing long time trajectories of immobilized molecules as discussed below. They are also used as online diagnostics; as data is acquired, a simple time trace is very useful for making sure the set-up is aligned properly, and that nothing is amiss with the sample.

A time series of CCD images is already a time trace, and does not need to be transformed into time traces as do photon timing data. However, to form a time trace for an individual molecule, the pixels that contain the emission from that molecule are added together, and the intensity from that molecule can be formed into a single time trace as a function of time. For this analysis, a region of pixels needs to be identified for each molecule.

9.3.5.3 Single-molecule Identification

Most single-molecule identification is performed using simple threshold search algorithms. These searches can be performed spatially for immobilized molecules or temporally for diffusing molecules. A threshold level is set, and any signals above that background level are tagged as potential single-molecule events. Sometimes, more sophisticated filters are used to discriminate between background and signal, but the basic idea of a threshold level above which a signal indicates a single molecule remains.

A single-molecule region of interest is chosen in an image by finding groups of pixels above a threshold value. This can be done visually, with subsequent fitting to accurately find the position of the molecule [3]. Automated searching and fitting may also be used to find and track positions of molecules [53].

For diffusing molecules, a single-molecule event is the traversal of a molecule through the optically isolated detection volume. This produces a "burst" of fluorescence photons over a limited time. The selection of single molecules in diffusion begins with the identification of these bursts. Burst search algorithms are essentially a region-of-interest selection as described above, but in the time domain (see for example [54]). Rather than identification of full transits of molecules diffusing through the optically isolated detection volume it is possible to threshold-analyze time traces on a bin-by-bin basis, without a burst search [55, 56].

Once a single-molecule signal is identified, time traces, time correlations, and various quantities describing the molecule can be calculated and analyzed to provide information on the molecule.

9.3.5.4   Histogram-based Analysis (Including Correlation Analysis)

Before going on to describe the analysis of identified single molecules, we will describe another way to analyze single-molecule data. This involves the formation of histograms and correlations from the data without first identifying regions containing single molecules. Techniques in this category include fluorescence correlation spectroscopy (FCS) [5], photon counting histogram (PCH) [57], fluorescence intensity distribution analysis (FIDA) [58], photon arrival-time interval distribution (PAID) [59], and time-integrated fluorescence cumulant analysis (TIFCA) [60] with all subsequent additions and variations. The main benefit of these techniques is that the effects of the single-molecule identification threshold are eliminated, allowing for easier mathematical modeling. In addition, there is less potential for biasing the results that may occur during the single-molecule identification process.

In PCH and FIDA and their variations, a fixed time bin width is used for analysis of photons. The number of photons counts per channel is calculated over the time bin. A one- or two-dimensional histogram is formed, where the axis or axes are the number of photons counted. Each point in the histogram counts the number of time bins in the experiment with the specified number of photon counts. Fitting functions have been obtained that allow for the concentration, molecular brightness, and background to be obtained for one or more fluorescent species. Extensions of FIDA and PCH to two channels, to include temporal or fluorescence lifetime information have been demonstrated.

In FCS, a time-correlation function is formed. The temporal cross-correlation function is defined as $C_{AB}(\tau) \equiv \langle I_A(t) I_B(t + \tau) \rangle / \langle I_A(t) \rangle \langle I_B(t + \tau) \rangle$, where $I_A(t)$ and $I_B(t)$ are the detected intensities for channels A and B, and t and τ are continuous time and time lag variables. The correlation function decay is used to determine the time scales of fluorescence fluctuations, which is key for monitoring molecular dynamics. Often, correlations are calculated using hardware correlators, and only the correlations are kept. For experiments at the single-molecule level, signal amplitudes are not high enough to make this beneficial. Recording photon events allows flexibility to perform other types of analysis, and, with proper algorithms [61–63], the correlations can be calculated quickly. Extensions of FCS include the use of cross-correlations [64]. PAID extends FCS to include information about brightness, similar to the information obtained from PCH and FIDA [59].

Generally, these methods are very useful for obtaining quantitative, unbiased values for brightness and concentrations of fluorescent analytes. This is particularly true for determining stoichiometry of binding partners (for example see [65]).

However, these methods are limited to low numbers of dimensions (two or three). This precludes their use in analyzing many multi-dimensional single-molecule observables.

9.3.5.5 Analysis of Histograms of Single Molecules

The analysis of histograms of single molecules can be very similar to the histogram analysis just described. The axes of the histograms may even be the same. The difference is what is counted in the histogram. Whereas the histograms above counted time intervals or photon pairs, here we count identified molecules.

For each molecule, several quantities are calculated, revealing information about the molecule. For example, for alternating laser experiments, two important quantities may be calculated: Alternating excitation recovers distinct emission signatures (Figure 9.3A) by calculating two fluorescence ratios: the FRET efficiency E which reports on donor–acceptor distance [66–68], and the distance-independent ratio S which reports on the donor–acceptor stoichiometry:

$$E = F_{D_{exc}}^{A_{em}} / (F_{D_{exc}}^{A_{em}} + \gamma F_{D_{exc}}^{D_{em}})$$
$$S = F_{D_{exc}} / (F_{D_{exc}} + F_{A_{exc}})$$

where $F_{D_{exc}}^{D_{em}}$ is the D-excitation-based D-emission, $F_{D_{exc}}^{A_{em}}$ is the D-excitation-based A-emission, $F_{D_{exc}}$ is a sum of D-excitation based emissions, $F_{A_{exc}}$ is a sum of A-excitation based emissions (Figure 9.2B), and γ is a detection-correction factor [68–70]; all emissions refer to single molecules. In addition to E and S, each burst can be tabulated by the number of its photon counts (giving rise to a histogram of burst size distribution), its duration (giving rise to a histogram of burst duration distribution), its polarization, and its lifetime – already defining a six-dimensional space. Similar quantities and multi-dimensional histograms can be calculated using multiparameter fluorescence detection (MFD) [43].

9.3.5.6 Single-molecule Sorting

Stoichiometry S provides information even without close proximity between fluorophores; it allows thermodynamic and kinetic analysis of interactions, identification of interaction stoichiometry, and study of local environment (as detected by changes in fluorophore brightness). Combination of E and S on two-dimensional histograms (Figure 9.3A) allows virtual molecule sorting [71]; we define this analysis as Fluorescence-Aided Molecule Sorting (FAMS), and its implementation using alternating-laser excitation as ALEX-FAMS. When applied on diffusing molecules, ALEX-FAMS is a homogeneous, "mix-and-read" assay, in which interacting species are combined and optical readouts report *simultaneously* on their association status and conformational status.

This sorting procedure allows the observer to determine to the number of subpoupulations and their mean distance as well as the distance distribution. The distance distribution is time averaged over the $\sim$ 1-ms observation time for each molecule, and there is shot noise. Hence the distance distribution has limited resolution [72].

Single-molecule sorting may also be used as just that – a way to sort molecules into different subpopulations for further study. Generally, the molecules switch states frequently, so a physical sorting into different solutions is impossible (unlike cell sorting). However, it does allow a brief time of $\sim 1\,\text{ms}$ to study the subpopulation using other methods. In this way, we have used single-molecule sorting to facilitate the measurement of fluorescence lifetime curves of individual subpopulations, and to analyze them in order to determine distance distributions that fluctuate on nanosecond time scales using time-resolved FRET [44].

9.3.5.7 Trajectory Analysis of Single Molecules
The most striking data from single-molecule experiments show real-time conformational changes and transitions as a function of time. For transitions or changes on a relatively slow time scale ($\sim 10\,\text{ms}$ or longer), there are a sufficient number of photons to produce a strong signal on time scales over which the system does not change. At each of these time points, the state of the system may be determined and followed as a function of time. All transitions or changes are observable and determined. This simple analysis of time traces is commonly used for immobilized molecules, and has produced striking results. Hidden Markov modeling [73] and Bayesian models [74–76] may be used to obtain more precise results. For marginal cases, where the signals are not so clear, these more advanced methods are necessary to obtain proper results. Even in cases with strong signal to noise, these methods are able to obtain more precise results with better defined error estimates.

9.3.6
Modeling and Simulations of Single-molecule Experiments

In order to properly understand the limitations of the experiments and determine whether the analysis applied actually provides the information expected, modeling and simulation of single-molecule experiments are necessary.

The modeling of nano-scale machines requires simultaneous modeling of deterministic and random forces, as well as coupling to chemical reactions. Examples of such modeling for molecular motors are reviewed in [4]. Polymer dynamics have been modeled using molecular dynamics (MD) simulations [77], simple bead-spring models [44] for analysis of single molecule data.

In addition to modeling the dynamics of interest, it is also necessary to model the single-molecule fluorescence experiments themselves. Modeling of single-molecule fluorescence measurements requires coupling of such models to photophysical simulations that include effects such as FRET, triplet states, saturation, etc. Modeling of all of these features has been lacking, but some initial attempts may be seen in [44, 72, 78].

9.4
Examples

In the remaining sections of this chapter we will discuss examples of single-molecule experiments that relate to two classes of problems: intra-chain conformations i.e.

protein folding and macromolecular interactions i.e. DNA processing enzymes (protein-DNA interactions).

9.4.1
Single-molecule Fluorescence Studies of Protein Folding and Conformations

Proteins contain structural information in their sequence and fold spontaneously into specific structures. In protein folding studies, folded, unfolded, and partially folded species may be simultaneously present and rapidly inter-converting, obscuring the properties of individual species. Due to the ability of single-molecule spectroscopy to sort proteins into different conformational subspecies, single-molecule fluorescence methods are beginning to impact the protein- folding field in a substantial manner [8]. The properties of the unfolded, and folded, and intermediate states can be studied individually even when both are simultaneously present. In immobilized experiments, the timing and dynamics of the folding and unfolding processes may be observed directly over extended periods.

9.4.1.1 Observables for Protein Folding
In order to apply single-molecule spectroscopy to protein folding, appropriate observables and labeling schemes must be chosen. One of the main differences between folded, unfolded, and intermediate states is the size of the overall protein. FRET, with its ability to measure the distance between two points, is a primary tool for single-molecule measurements of protein folding. Since folding of small proteins is accomplished by each protein independently, all labels and immobilization attachment points must be on the same protein in order to provide a useful observable. So far, almost all single-molecule protein folding studies use a scheme where each protein has one donor and one acceptor attached (Figure 9.3). This has been adequate for the simple two-state folders studied so far. However, more complex situations can be imagined. For example, recently developed three-color single molecule methods may find use in determining the ordering in folding of larger, more complex proteins: how does the folding of domain 1 (measured by FRET from donor D to acceptor 1) affect the folding of domain 2 (measured by FRET from donor D to acceptor 2)?

For studies of short-range dynamics involved in protein folding, another possible observable is electron transfer or fluorescent quenching of a single fluorophore. Changes in quantum efficiency of a fluorophore would then indicate contact between two sites of a protein. Using correlation analysis, even very fast dynamics may be monitored in this way.

9.4.1.2 Labeling Schemes for Protein Folding
For protein folding studies at the single-molecule level, proteins must be site-specifically labeled by donor and acceptor fluorophores, and, if immobilized, attached at a specific point. Such extensive external modification requires careful planning and extensive work. Several clever strategies have been used to accomplish this complex labeling.

In each of the examples discussed below, the labeling positions were chosen so that FRET efficiency is high in the folded state, with D and A held very close to each other, whereas, in the unfolded state, they are further away. For each protein detected, the FRET efficiency *E* is calculated as a ratio between the number of photons detected from the acceptor versus the number of photons detected from both donor and acceptor channels. When *E* is high (close to 1), the distance *R* between *D* and *A* is small, indicating a folded conformation of the protein. When the protein is unfolded, the protein is in a more expanded conformation with increased *R*, and decreased *E*.

In the first study of protein folding at the single-molecule level, Hochstrasser and collaborators took advantage of the unique properties of GCN4, a transcription factor from yeast [79]. The protein is a cooperatively folded coiled-coil dimer with a disulfide bond between the C termini of identical subunits. This arrangements allows for a simple labeling scheme: a donor fluorophore (rhodamine 6G) was attached to the N terminus of one sample, and an acceptor (Texas Red) was attached to the N terminus of the other. These were mixed, and allowed to fold, forming disulfide bonds between the C termini. After purification, a sample containing heterodimers labeled with one D and one A remained, and was used for single-molecule experiments.

In our first study of protein folding with Chymotrypsin Inhibitor 2 [80], amino acid positions 1 and 40 were chosen since they were fairly close in the folded structure, but were well separated in the polypeptide chain. They were also not inside any major secondary structural elements in the folded protein. The protein was synthesized in two pieces using solid phase synthesis, and one piece was labeled first. The protein was then assembled in refolding conditions, and spliced together using a ligation reaction. Then, the other label was attached, and several purification steps were required to obtain the final sample.

Some subsequent protein folding papers have used a less demanding scheme for obtaining dual-labeled samples. Proteins, including CspTm [81] and CI2 [80], are expressed as two-cysteine mutants. One fluorophore is used to label the protein, and the singly-labeled protein is separated from unlabeled and dual-labeled protein. The other fluorophore is added, and dual-labeled proteins were then separated from single-labeled proteins. The problem with this scheme is that, even though the protein is known to have one donor and one acceptor, these fluorophores may be at either position. Careful controls are then required to ensure that any effects seen are not due to the presence of two different labeled populations.

Our group introduced two other ways to site-specifically label a protein at two sites. In Jager *et al.* [28], the procedure of labeling is the same as that described in the previous paragraph, with one critical change: in the first labeling step, one of the labeling sites is hidden using a protein that binds to the CI2. This prevents labeling at that site, and, upon purification, allows the remaining site to be labeled site specifically with the other fluorophore. This method may be employed in many situations, as long as there is a protein that binds to and blocks a region of the protein of interest. In Jager *et al.* [29], a short glutamine tag was added to the N terminus and enzyme ligation was used to specifically tag CI2 at that position. This allowed site-specific labeling in two positions

9.4.1.3 Equilibrium Unfolding Studies on Simple Model Two-state Folders

In order to properly study protein folding, it is important to have access to information on many time scales, from nanoseconds to seconds or longer. Since confocal-based single-molecule measurements currently have detectors with much better time resolution, most, if not all, single-molecule protein folding studies have used this optical isolation technique. In the first protein-folding measurements described in this section and the next, a focused, continuous-wave (cw; non-pulsed) laser was used to excite the donor fluorophore. The fluorescence emission from single molecules was collected and refocused onto a detection pinhole used to select the small observation volume. The fluorescence emission is split using a dichroic mirror and two bandpass filters into two detection channels, one for the donor and one for the acceptor.

For the studies on freely diffusing molecules, time traces of donor and acceptor emission were formed with 100-μs–1-ms time resolution, and intense, single-molecule signals (or bursts) above the background were selected. For each selected single molecule, the ratio $E = I_A/(I_A + \gamma I_D)$ was calculated, where γ is the ratio of the quantum efficiencies and detection efficiencies of the fluorophores. In the initial studies, this ratio was not precisely calibrated, although it was typically near 1. Histograms are formed from these E ratios, showing how many proteins have a specific value of E. In most of these studies, there was a large subpopulation of proteins with a photobleached or inactive acceptor. For these proteins, only donor emission is observed, and E is 0 (assuming that leakage of donor emission into the acceptor channel is subtracted). Since this donor-only subpopulation is indistinguishable from a donor- and acceptor-labeled protein with dimensions too large for FRET to occur, this limits the usable range of E to be well above 0 (generally 0.3 and above).

In the first paper to demonstrate single-molecule measurements of proteins undergoing folding and unfolding, the Hochstrasser group studied immobilized GCN4 using FRET [79]; they constructed similar E histograms, where molecules were searched in an image (rather than a burst in a time trace). Extended time trajectories lasting several seconds were recorded for single molecules, allowing for dynamical information to be accessed. Correlations calculated on these time trajectories were used to provide information on the temporal dynamics. Adding denaturant to the protein increases the fraction of proteins that are unfolded. Upon addition of 7 M of denaturant, all GCN4 molecules were unfolded, increasing on average the distance between D and A, and lowering E. In this completely denatured state, a variety of distances was seen, covering a large range in E. This was attributed to the sticking of the unfolded protein to the surface. This prevented the authors from making a strong correspondence between the results of the study with the results from previous, ensemble level studies. Even so, the general pattern of increasing the number of unfolded proteins with denaturant was seen at the single-molecule level.

In order to study proteins undergoing folding and unfolding in a less perturbative environment, it was necessary to keep the protein (especially unfolded proteins) away from glass surfaces. In [80], we introduced a method of studying protein folding on freely diffusing molecules, and demonstrated it using the extensively studied two state model protein Chymotrypsin Inhibitor 2 (CI2). The ability to monitor long time trajectories ($> \sim 1$ ms) was sacrificed, but the FRET efficiency E was calculated for each protein as it diffused through the optically isolated detection volume. The distribution at low denaturant contained a single peak at high E indicating a compact, folded conformation. The distribution at high denaturant also contained a single peak, but now at low E, indicating a more extended, unfolded conformation. At the midpoint of the sigmoidal curve, when 50% of the protein is unfolded, both peaks were present in the E distribution, indicated two interconverting subpopulations in solution. A good correspondence between the denaturation curves from ensemble and single-molecule measurements was found (Figure 9.5).

Simple proteins often have two states as extensively documented by many experiments over the years. We chose a simple, a two-state model protein folder precisely to demonstrate the validity of single-molecule level measurements. Even so, for many researchers we have talked to, the initial response to the single-molecule experiments showing folded and unfolded subpopulations is satisfaction that there really are two well-separated, interconverting subpopulations. The more direct

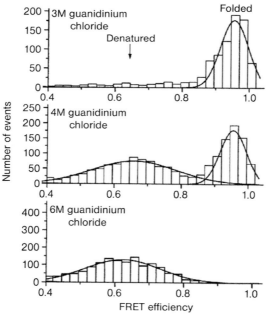

Figure 9.5 FRET efficiency distributions of chymotrypsin inhibitor 2 (CI2) at three different denaturant concentrations. Top: 3 M GdmCl (native conditions), middle: 4 M GdmCl (close to the midpoint of unfolding), bottom: 6 M GdmCl (strongly denaturing conditions). The bimodal distribution of the FRET-efficiency clearly indicates the two-state nature of the unfolding process. Reproduced with permission from [80].

observations available using single-molecule spectroscopy have a power that should not be underestimated.

The group that studied GCN4 also pursued studies of single molecules freely diffusing in solution [82], and they found distributions in line with expectations. In their case, the interconversion time between folded and unfolded states was shorter than the observation time, so they were able to use correlation analysis to probe the time scales of interconversion. For CI2, the time scales for unfolding and refolding were longer than the observation time so that folded and unfolded populations were well separated.

These initial experiments on protein folding helped form our opinion that new experiments in single-molecule fluorescence spectroscopy should in general be first approached with measurements on freely diffusing molecules. Such measurements are easier to initiate and optimize than immobilized experiments. Problems with the biological system working at the single-molecule level and with fluorophores attached are easier to address without the additional complications of immobilization. Long time trajectories are sacrificed initially, but can be added later.

We made two additional observations in our initial experiments with CI2. First, the width of the distribution in energy transfer efficiency E is wider than what we predicted based on shot noise. Second, the average distance between donor and acceptor of the unfolded subpopulation decreases with decreasing denaturant; this result was not very strong, but was outside the noise levels. Schuler *et al.* performed similar single-molecule experiments on another protein, cold shock protein from *Thermotoga maritima* (CspTm) [81]. Most importantly, they compared the width of the distribution in E found in their CspTm experiments with controls using polyproline (approximating a rigid rod). They found that the widths of the E distributions from CspTm were identical to widths found using polyproline, indicating that the wide E distributions must originate from experimental noise rather than from structural changes in the protein, contradicting our initial speculations. Based on this and the measurement time per protein of ($\sim 100\,\mu s$), they were able to conclude that the fluctuations of unfolded CspTm must be averaged out in less than $100\,\mu s$. This in turn allowed them to conclude that the free energy barrier between folded and unfolded proteins must be greater than $2\,k_B T$.

Schuler *et al.* also saw a decrease in the size of the unfolded protein with decreasing denaturant similar to that seen with CI2. The decrease in size was more dramatic in the case of CspTm. They explained this shift in terms of a continuous expansion of the unfolded protein with increasing denaturant, although this expansion does not fit completely with the expectation that the unfolded state would continue to expand with increasing denaturant. The next two sections describe two subsequent papers that focused on elucidating the nature of the unfolded state in more native-like conditions.

9.4.1.4 Single-molecule Protein Folding under Non-equilibrium Conditions

Lipman *et al.* [83] coupled a laminar flow mixer with the single-molecule measurements in solution to trigger refolding of CspTm. They were able to monitor E for both the folded and unfolded subpopulations after diluting the denaturant to native

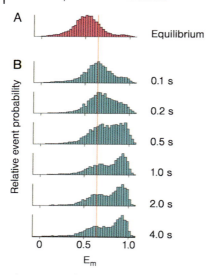

Figure 9.6 Probing single-molecule protein folding kinetics under non-equilibrium conditions. (A) Top: view of the mixing region of the microfluidic laminar mixing device, with the denaturant concentration indicated by a false color code. The laser beam (light blue) and collected fluorescence (yellow) are shown 100 μm from the center inlet. Bottom: cross-section of the mixing region. (B) Histograms of measured FRET-efficiency as a function of the mixing time. Within the first 100 ms of mixing, the expanded chain collapsed into a more compact coil structure, indicated by the slight increase in mean *E*. The collapsed denatured species reconfigure into the compact folded state on a timescale of a few seconds. Note that the mean *E* of the collapsed denatured subpopulation does not vary with time (red line) nor final denaturant concentration (data not shown). Reproduced with permission from [83].

conditions. They found that the unfolded population becomes more compact rapidly (within 100 ms), matching the size of unfolded protein in their previous measurements. This validates the previous detection of change in size of the unfolded protein at lower denaturant concentrations, and allowed the study of this more compact unfolded state at denaturant concentrations down to 0 M. Subsequent to this fast compaction of the unfolded protein, the population of folded proteins increases as the population of unfolded proteins decreases. Rather than a shift in the position of the peak in the *E* distribution that accompanies the fast compaction of the unfolded state, the folding process is seen as changes in the relative populations of two states (Figure 9.6).

Such non-equilibrium studies hold great promise in elucidating the nature of the protein folding process.

9.4.1.5 Monitoring Conformational Dynamics using Fluorescence Lifetime
In the protein-folding experiments described above, the only information recorded from the molecules was the intensity of the signal from the donor and acceptor. What is obtained from such measurements is a distance measurement between donor and

acceptor time-averaged over the measurement time of 100 μs to 1 ms. As demonstrated by Schuler *et al.* [81], this is insufficient time resolution to detect any structural fluctuations of the unfolded protein. For a folded protein, there is in general one structure, with only small fluctuations from this structure. There may be changes to the structure due to interactions with other proteins, DNA, or small molecules, but these are very different from the unfolded protein. The unfolded state of a protein is an ensemble of many random conformations of the protein chain that are interconverting rapidly due to thermal fluctuations.

Measurements of fluorescence lifetime of FRET have been used for many years at the ensemble level to monitor distributions of distances within unfolded and folded proteins [84]. Using time-correlated single photon counting (TCSPC), the time between a laser pulse and the fluorescence photon is timed to 4–50-ps accuracy. Fluorophores generally exhibit single exponential lifetime fluorescence decays, which are shortened by the process of FRET. Shorter lifetimes indicate higher FRET efficiency, and longer lifetimes indicate lower FRET efficiency. If the donor exhibits a single-exponential fluorescence lifetime decay with FRET, there is a single distance between the donor and acceptor. Multi-exponential decays for the donor undergoing FRET indicate multiple distances. By analyzing such decays, it is possible to resolve distance distributions down to the nanosecond time scale.

In Laurence *et al.* [44], we introduced several methodological improvements to single-molecule protein-folding studies that provide additional information on fluorescence lifetime and polarization. In this study, we were able to combine nearly all of the observables available for a single FRET pair, and use them advantageously. Following Seidel and collaborators, we increased the number of detection channels to four, further dividing the donor and acceptor fluorescence emission by polarization, and we used TCSPC to measure the fluorescence lifetime of the donor and acceptor. Lastly, we introduced alternating laser excitation (ALEX) using two interlaced, pulsed lasers, one laser exciting the donor, and one laser exciting the acceptor. For the acceptor, there are lifetime curves both for excitation by the donor laser (via FRET) and by the acceptor laser (via direct excitation). Using the ALEX procedure allows us to distinguish between subpopulations with donor-only, acceptor-only, folded and unfolded proteins. After this procedure of "single-molecule sorting," we select all bursts emanating from the unfolded protein, and form fluorescence lifetime decays for that subpopulation. We then globally fit all of the lifetime curves, obtaining information on distance distributions along with the polarization information.

Using this analysis, we determined the mean and width of the distance distributions of dsDNA, ssDNA with various salt concentrations, and two unfolded proteins. For random polymer fluctuations there is a limit to the possible widths of the distributions. Distribution widths for the nucleic acids were below or at this limit, but, interestingly, the unfolded proteins exhibited larger fluctuations than possible by random polymer fluctuations. These "excess" fluctuations increased as the denaturant decreased. We proposed that these fluctuations are due to transient formation of residual structure, although transient compaction of the unfolded state is another possible explanation [85] (Figure 9.7).

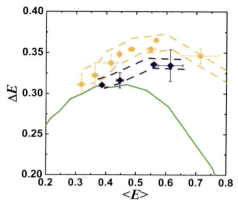

Figure 9.7 Effects of denaturant concentration on distance distributions of unfolded CI2 (blue) and ACBP (orange). The width of the E distribution ΔE is plotted versus the mean of the distribution $\langle E \rangle$. These values are extracted using time-resolved FRET. Nanosecond ALEX-based single-molecule sorting allows the exclusion of signal from folded CI2 and ACBP. Note that all of the measurements for the unfolded proteins are at or above the Gaussian chain limit (green line). For each data set (CI2 or ACBP), lowering the denaturant concentration results in measurements moving toward the upper-right, indicating more compact states with increasing fluctuations. Reproduced with permission from [44].

There is one important methodological achievement that should be mentioned with respect to ALEX-FRET. In [86], a methodology was developed to determine accurate, unbiased energy transfer efficiency E values that proved in many ways to be more reliable than ensemble FRET. In the nsALEX paper, a second methodology was developed using single-molecule selection and fluorescence lifetime fitting to obtain accurate E values along with the width of the E distributions down to the nanosecond time scale. The values obtained by both methods for the same samples matched very well, even though the manner in which the values were obtained were quite different. In the first measurements of FRET at the single-molecule level, it seemed that single-pair FRET would only be useful for measuring dynamic changes in distance. Now, however, the accuracy and reproducibility of FRET efficiency values now rivals, and, in some cases, surpasses that of ensemble methods (Figure 9.8).

9.4.1.6 Single-pair FRET Studies on Immobilized Proteins

As mentioned earlier, in order to monitor repeated folding and unfolding events it is necessary to immobilize the proteins. This allows observations over several seconds. In order to circumvent the difficulties of surface immobilization, two methods have been used. First, encapsulation of the proteins in lipid vesicles, and attachment of those vesicles to surfaces have been used to monitor such transitions. In this way, the protein that is folding and unfolding is not held in contact with any surface, and only comes into contact with the lipid surface of the vesicle, which is relatively inert. Folding and unfolding were clearly seen, which allowed the timing of these processes

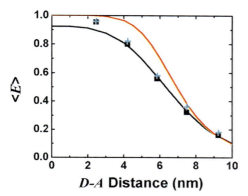

Figure 9.8 Accurate E measurements may be performed at the single-molecule level. $\langle E \rangle$ measured by nsALEX versus distance between attachment points for D and A for dsDNA with 7-, 12-, 17-, 22-, and 27-base pair separations (black squares). $\langle E \rangle$ calculated from calibrated single-molecule intensity ratios rather than fluorescence lifetime information are shown as cyan crosses. FRET model ($E = 1/1 + (R/R_0)^6$) with measured $R_0 = 69$ Å and $\langle \kappa^2 \rangle = 2/3$ (solid red) and simulation accounting for linkers and the measured slower rotational diffusion ($\tau_r^D = 3.0$ ns and $\tau_r^A = 1.3$ ns; solid black) are shown. Simulations were adjusted to $R_0 = 62$ Å to match gray squares. Reproduced with permission from [44].

to be timed. By comparing the observed folding and unfolding trajectories for proteins exhibiting complex [88] and simple two-state [87] folding pathways, the roughness of the free energy surfaces can be probed using single-molecule spectroscopy (Figure 9.9).

A second approach to facilitating immobilized protein-folding studies of single molecules was to passivate the surface in a better way. Nienhaus and colleagues introduced surface coatings made of ultrathin networks of isocyanate-terminated "star-shaped" polyethylene oxide (PEO) molecules, cross-linked at their ends via urea groups to minimize the intertwining of a denatured polypeptide chain with the PEO-polymer. Immobilizing proteins on this surface allowed single-molecule protein-folding trajectories to be monitored [89]. Repeated changes in denaturant concentrations demonstrated that folding and unfolding were completely reversible on the specialized surface.

9.4.1.7 Probing Biomolecular Dynamics via Fluorescence Quenching and Electron Transfer

Fluorescence quenching by electron transfer or other processes promises to provide a method of measuring contacts between two sites at short distances at the single-molecule level. Correlations calculated from these fluctuating signals would reveal the time scales of the fluctuations directly. Folding and unfolded of DNA hairpins were monitored using FCS on a donor-quencher system [90]. An ensemble method was used to provide information on time scales of unfolded protein and nucleic acid fluctuations [91, 92]. Electron transfer from native fluorophores was used to monitor the fluctuations of an enzyme at the Angstrom scale [93]. The greatest weakness in

(a)

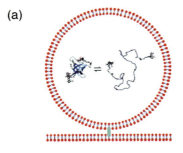

(b)

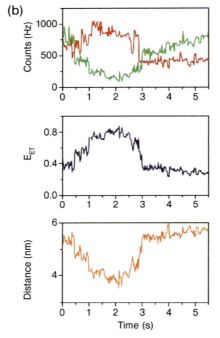

(c)

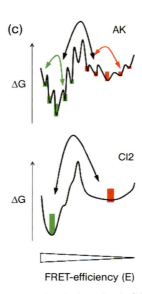

Figure 9.9 Probing conformational dynamics within single, surface-immobilized molecules. (A) Encapsulation of single molecules in lipid vesicles. Immobilization of the trapped molecule is achieved indirectly by tethering the biotinylated vesicle to a streptavidin-coated polymer coating. As the diameter of the encapsulating vesicle (∼100 nm) is considerably larger than the diameter of the trapped molecule, the trapped molecule can diffuse freely, thus minimizing a perturbation of the energy landscape due to confinement and non-productive interactions of the trapped protein within the vesicle interior.

(B) Representative single-molecule folding trajectories of encapsulated adenylate kinase (AK) at denaturant concentrations around the midpoint of unfolding, conditions at which the native and denatured subpopulations are equally populated. The top panel shows time traces of the D- and A-emission (green and red, respectively), employing 20-msec bin times. D- and A-emission are anticorrelated, indicative of conformational transitions within the polypeptide. Notice however that the D-emission continues to increase after bleaching of the acceptor (t ∼ 3 s), suggesting a variable these methods is that only one channel is monitored, preventing the use of ratiometric variables that are so beneficial in FRET measurements. However, for monitoring such short distances at the single-molecule level, there is no current alternative to designing these more challenging experiments.

9.4.2
Single-molecule Measurements of DNA-processing Enzymes

The methodologies we have discussed in this chapter so far have been mainly applicable to DNA–protein interactions rather than protein–protein interactions. One reason is their importance: maintaining the fidelity of DNA replication is necessary for cell survival, and transcription is a primary point at which gene expression is regulated. Another reason is practical: these methods require the labeling of two macromolecules in order to observe both movement and interactions. Compared to proteins, DNA is relatively easy to label site specifically. Having DNA as one of the labeled macromolecules lessens the work involved in getting the biochemical system working.

We focus this section on the example of our measurements on RNA polymerase. The methods can be generalized to other DNA-processing enzymes. Transcription is the process of copying a gene from a DNA template to RNA, which in turn serves as the template for protein synthesis. It is one of the main steps used in regulating the expression levels of proteins, and thus maintaining the proper level of protein. RNA polymerase (RNAP) is the protein machine that copies DNA into RNA. In collaboration with the Ebright group, we set out to investigate the initiation of transcription.

Our initial measurements of RNAP were attempted with a surface-immobilized format. These measurements proved to be an exercise in frustration. A donor-labeled RNAP was supposed to bind to an acceptor-labeled DNA, and, upon adding the required nucleotides, transcription would proceed in a controlled manner. The

---

Figure 9.9 (*Continued*)

environment (e.g. diffusion within the vesicle). Two types of fluctuations are visible in the E-trajectory: fast, step-wise fluctuations that cannot be resolved and slow, continuous transitions that can take > 1 s to complete. The middle panel depicts a FRET-efficiency trajectory calculated from the signals in the top panel, whereas the interprobe distance trajectory (calculated according to Förster theory) is shown in the bottom panel. (C) Schematic of a one-dimensional energy landscape for AK and CI2 at denaturant concentrations close to the midpoint of folding, obtained by averaging of the folding landscape over many degrees of freedom and projection onto the FRET-efficiency axis. Both energy landscapes exhibit two global free energy minima separated by a free energy barrier, as suggested by the bimodal FRET-efficiency distributions. The landscape of AK is characterized by local free energy barriers and traps, resulting in fluorescence time-trajectories that can start and end at any FRET-efficiency value (the population weight of the folded species is indicated by green bars, while the unfolded conformers are color coded red). The landscape of CI2 is smoother, and time trajectories exhibit fluctuation between two relatively constant levels of FRET-efficiencies. A reproduced with permission from [87]. (Copyright © 2004 American Chemical Society). C reproduced with permission from [8]. (Copyright © 2002 National Academy of Sciences USA).

(a) Leading-edge FRET

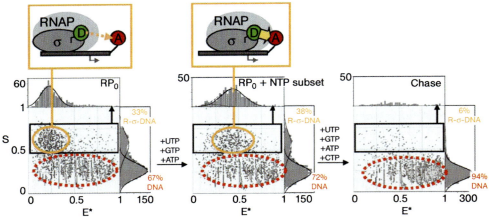

(b) Trailing-edge FRET

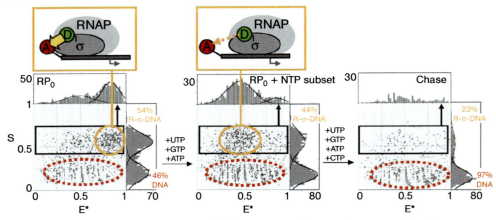

Figure 9.10 ALEX-FRET on single diffusing complexes allows simultaneous monitoring of movement of RNAP on DNA and association of transcription factor σ^{70} with RNAP core. Two-dimensional *E-S* histograms of single molecule fluorescence bursts are shown. *E* is the FRET efficiency (low for larger D–A distance, high for smaller D–A distance), and *S* is the stoichiometric ratio (S = 1 for D-only, S = 0 for A-only, and S = 0.5 for both D and A present). (A) Subpopulations identified using ALEX-FRET may be followed as a reaction progresses. Reproduced with permission from [95].

RNAP was attached to the surface, and the DNA was not. Initial encouraging results showing activity were found to be photophysical in origin. At the time we were unable to determine whether RNAP was functioning properly on the surface. An additional problem was that the single-molecule FRET methodology, as then constituted, was unable to tell the difference between a molecule with only the donor attached and a molecule with the donor bound to a molecule with the acceptor, but too far away to see

Distance between −10/−35 spacer DNA and downstream DNA

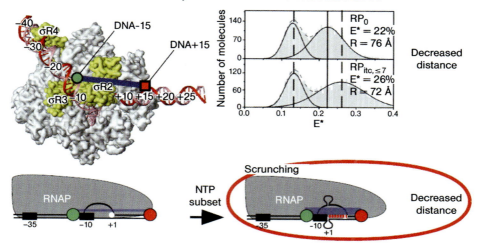

Figure 9.11 Direct evidence that DNA is "scrunched" during initiation of transcription by RNA polymerase. Upon addition of nucleotides (which allow transcription to progress), the distance between D and A on DNA is seen to decrease. The peak in the E histogram shows an increase in E, corresponding to a shorter distance between D and A. Other measurements using LE-FRET and TE-FRET (as in Figure 9.10) show that the back end of RNAP does not move on the DNA during initiation of transcription, whereas the front end of RNAP does move with respect to DNA. Reproduced with permission from [1].

FRET. If the donor-labeled RNAP moved away from the acceptor-labeled DNA it would be impossible to tell whether RNAP was attached to the DNA or not. We knew that cross-correlation methods with two lasers would allow us to determine whether both donor and acceptor were present within the optically isolated detection volume. However, this would obscure FRET measurements since we would be unable to tell which photons from the acceptor were due to FRET and which were due to direct excitation from the acceptor excitation laser. We concluded that a quick alternation between a laser exciting the donor and a laser exciting the acceptor while recording the arrival time of every photon would solve the problem. This initiated the development of alternating laser excitation (ALEX) for single-molecule spectroscopy [26]. The development of this methodology rapidly accelerated our efforts in this area. We will first discuss our results in the area of DNA–protein interactions, and then discuss briefly other important single-molecule work.

9.4.2.1 **RNAP – Retention of Sigma**

In *E. coli*, RNAP initiates transcription at specific (promoter) DNA sequences after binding σ transcription factors. The primary transcription factor of exponentially growing *E. coli* is σ^{70}. Initial studies indicated that after initiation of transcription, the σ transcription factor dissociated from RNAP and DNA. A model for initiation of

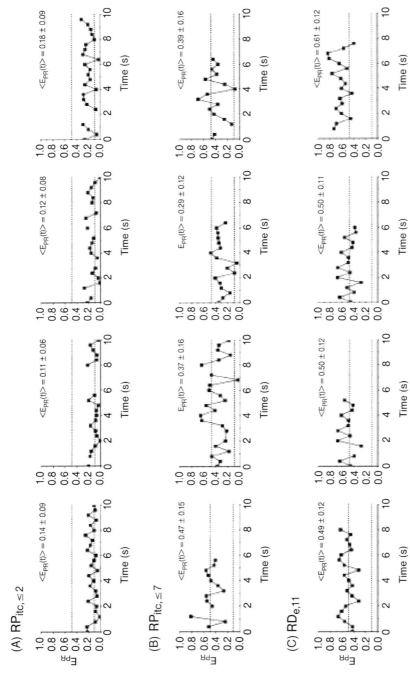

Figure 9.12 Single-molecule time traces of *E* are plotted as a function of time, for different representative RNAP–DNA complexes. Only time points with active D and A are shown. The average *E* values obtained for RNAP allowed to progress up to 2 bases $RP_{itc,\leq2}$ and $RP_{e,11}$ (0.11 and 0.5, respectively) are represented by two horizontal lines to guide the eye. (A) $RP_{itc,\leq2}$ static distribution with $\langle E \rangle = 0.11$. (B) $RP_{itc,\leq7}$, scattered distribution with a majority of time points distributed around a relatively high *E*(t) value (distribution centered around $\langle E \rangle = 0.3$ for the active molecules), consistent with the forward translocation of the leading edge of RNAP relative to downstream DNA during abortive initiation. (C) $RP_{e,11}$, static distribution with $\langle E \rangle = 0.5$. Reproduced with permission from [96].

transcription was suggested that required release of σ^{70} before continuation of transcription. There were however, steps in the purification process that could have taken off weakly bound σ^{70} even if there were no obligatory release. Because of this, there was an alternative model proposed that release of σ^{70} was not necessary, but the interactions were only weakened.

Recently, FRET measurements between the donor on σ^{70} and acceptor on DNA were used to determine whether there is obligatory release of σ^{70}. Ensemble-level experiments were performed first [94], followed up by single-molecule measurements using the same system [95]. This provides a good opportunity to compare directly the information content of single-molecule and ensemble-level measurements.

In both papers, a labeling scheme was used to monitor via FRET the movement of RNAP. With trailing edge FRET, a donor label on σ^{70} toward the back of the RNAP–σ^{70} complex undergoes FRET to an acceptor behind the RNAP (upstream of the promoter). FRET efficiency starts high, and decreases as transcription progresses. With leading edge FRET, a donor label on σ^{70} toward the front of the RNAP–σ^{70} complex undergoes FRET to an acceptor in front of the RNAP (downstream of the promoter). FRET efficiency starts low, and increases as transcription progresses. In [94], ensemble FRET measurements were used to establish that a significant fraction of the σ^{70} remained attached to RNAP even to mature elongation complexes. This contradicted some earlier models that maintained that the release of σ^{70} was necessary to continue transcription. In this paper, the leading-edge FRET measurements were critical, because they were able to differentiate between release and non-release of σ^{70}; if FRET increased with the progress of transcription, then σ^{70} was still attached. However, the trailing-edge FRET measurements predicted a decrease in FRET efficiency with or without release of σ^{70}.

A primary advantage of single-molecule FRET measurements (with ALEX) over ensemble FRET methods is the ability to detect association of molecules even when outside the range of FRET. This allows the detection of non-release of σ^{70} even with the trailing edge FRET measurements. We were able to determine the fraction of complexes that retained σ^{70} as a function of elongation.

This was an example of the power of single-molecule spectroscopy to determine ordering of events, directly visualizing the heterogeneity of the system under study. Even in this case, where ensemble measurements are able to answer some of the basic questions, the single-molecule measurements provided significant, quantitative information that was otherwise unavailable. These tools have now been applied to a problem whose answer has remained elusive for over 25 years as we see next.

9.4.2.2 RNAP – Abortive Initiation

During the initiation of transcription, it takes several attempts to get started, to "prime the transcription pump". Abortive products (short RNA transcripts) are formed in these initiation attempts, giving the name "abortive initiation" to the process. Interestingly, DNA footprinting experiments indicating that during abortive initiation, the RNAP was protecting the upstream boundary of the promoter (the upstream DNA element preceding the transcription start site, the binding site for the core enzyme), even as it was synthesizing the ~ 9–11 nucleotide abortive RNA transcripts. Three models were proposed to explain these observations. In the first model, the DNA is pulled into the RNAP during initiation of transcription, "scrunching" the DNA (model 1). In the second model, there is a flexible element in the RNAP that needs to stretch to another part of DNA in order to continue transcription (inchworming). The third model explains the footprinting results dynamically. The RNAP transiently moves downstream on the DNA until the abortive transcript releases, and then backs up once the abortive transcript is released, spending more time at the retracted position (Figure 9.11).

By using single-molecule leading- and trailing-edge FRET ALEX experiments on freely-diffusing molecules, we have been able to determine conclusively that the DNA scrunching model is correct [1]. The single-molecule methods were able to obtain these results in a simple, direct manner.

Similar solution-based measurements of DNA–protein interactions were performed on HIV reverse transcriptase (with fluorescence lifetime information, but no ALEX), revealing heterogeneity in the reverse transcriptase/primer–template complexes [43]. These measurements demonstrate the use of fluorescence lifetime information to obtain structural information regarding DNA–protein interactions.

Observation of abortive initiation with immobilized molecule   As seen in the previous section, single-molecule spectroscopy on freely-diffusing molecules has been able to answer several questions in the dynamics of DNA-processing machines. Even so, single-molecule spectroscopy using an immobilized format is necessary to watch an entire process over an extended period. We were able to adapt previously successful immobilization strategies to RNAP, to watch movement of RNAP with respect to the DNA as a function of time [96]. The immobilization strategy was the same as that used for Helicase experiments described by Professor Ha elsewhere in this book [97]. The measurements on Helicase are an important example of measurements of immobilized DNA-processing enzymes. We were also able to extend the ALEX techniques to the TIR excitation/CCD imaging approach to single-molecule spectroscopy. In this way, the movement of RNAP with respect to the DNA was monitored during abortive initiation. The time scales could not be determined due to limitations of the CCD frame rate. However, the system was clearly working on the surface (Figure 9.12).

We would like to note that comparison of the results using immobilized molecules with those from freely-diffusing molecules helped immensely in getting these experiments to work.

9.4.2.3 Future Directions

There are many possible methodological directions for this research; we mention two examples. First, the interactions of additional proteins, or additional distances, may be monitored with three-color methods [45–47]. Second, intracellular labeling techniques may allow monitoring of transcription *in vivo*. If combined with the high concentration methods outlined earlier, an extremely powerful methodology will have been developed for monitoring protein dynamics and interactions in their native environment.

9.5
Conclusion

Single-molecule fluorescence spectroscopy has come far in the last several years, and promises to continue to reveal a large amount of new information about biological systems that would otherwise remain hidden. Some of the more established single-molecule fluorescence methodologies are now being used by those whose main expertise is in biology rather than in optics. We expect this trend to accelerate, and hope we have encouraged those interested in single-molecule fluorescence spectroscopy to enter this exciting field.

Acknowledgments

This work was performed under the auspices of the U.S. Department of Energy by Lawrence Livermore National Laboratory under Contract DE-AC52-07NA27344 for T.A.L. and NIH grant GM069709-01, DOE grant 02ER63339 and DOE grant 04ER63938 to S.W.

References

1 Kapanidis, A.N., Margeat, E., Ho, S.O., Kortkhonjia, E., Weiss, S. and Ebright, R.H. (2006) Initial transcription by RNA polymerase proceeds through a DNA scrunching mechanism. *Science* (in press).

2 Noji, H., Yasuda, R., Yoshida, M. and Kinosita, K. (1997) Direct observation of the rotation of F-1-ATPase. *Nature*, **386**, 299–302.

3 Yildiz, A., Forkey, J.N., McKinney, S.A., Ha, T., Goldman, Y.E. and Selvin, P.R. (2003) Myosin V walks hand-over-hand: Single fluorophore imaging with 1.5-nm localization. *Science*, **300**, 2061–2065.

4 Karplus, M. and Gao, Y.Q. (2004) Biomolecular motors: the F-1-ATPase paradigm. *Curr. Opin. Struct. Biol.*, **14**, 250–259.

5 Magde, D., Elson, E. and Webb, W.W. (1972) Thermodynamic fluctuations in a reacting system: measurement by fluorescence correlation spectroscopy. *Phys. Rev. Lett. (USA)*, **29**, 705–708.

6 Wuite, G.J., Smith, S.B., Young, M., Keller, D. and Bustamante, C. (2000) Single-molecule studies of the effect of template tension on T7 DNA polymerase activity. *Nature*, **404**, 103–106.

7 Finer, J.T., Simmons, R.M. and Spudich, J.A. (1994) Single myosin molecule mechanics: piconewton forces and nanometre steps. *Nature*, **368**, 113–119.

8 Michalet, X., Weiss, S. and Jager, M. (2006) Single-molecule fluorescence studies of protein folding and conformational dynamics. *Chem. Rev.*, **106**, 1785–1813.

9 Lakowicz, J.R. (1999) *Principles of Fluorescence Spectroscopy*, 2nd edn, Kluwer Academic/Plenum, New York.

10 Lacoste, T.D., Michalet, X., Pinaud, F., Chemla, D.S., Alivisatos, A.P. and Weiss, S. (2000) Ultrahigh-resolution multicolor

colocalization of single fluorescent probes. *Proc. Natl Acad. Sci. USA*, **97**, 9461–9466.

11 Betzig, E., Patterson, G.H., Sougrat, R., Lindwasser, O.W., Olenych, S., Bonifacino, J.S., Davidson, M.W., Lippincott-Schwartz, J. and Hess, H.F. (2006) Imaging intracellular fluorescent proteins at nanometer resolution. *Science*, **313**, 1642–1645.

12 Ha, T., Laurence, T.A., Chemla, D.S. and Weiss, S. (1999) Polarization spectroscopy of single fluorescent molecules. *J. Phys. Chem. B*, **103**, 6839–6850.

13 Forkey, J.N., Quinlan, M.E. and Goldman, Y.E. (2000) Protein structural dynamics by single-molecule fluorescence polarization. *Prog. Biophys. Mol. Biol.*, **74**, 1–35.

14 Kinosita, K., Adachi, K. and Itoh, H. (2004) Rotation of F1-ATPase: how an ATP-driven molecular machine may work. *Ann. Rev. Biophys. Biomol. Struct.*, **33**, 245–268.

15 Rosenberg, S.A., Quinlan, M.E., Forkey, J.N. and Goldman, Y.E. (2005) Rotational motions of macro- molecules by single-molecule fluorescence microscopy. *Acc. Chem. Res.*, **38**, 583–593.

16 Forster, T. (1946) Energiewanderung und fluoreszenz. *Die Naturwissenschaften*, **6**, 166–175.

17 Forster, T. (1948) Zwischenmolekulare energiewanderung und fluoreszenz. *Ann. Physik*, **2**, 55–75.

18 Stryer, L. (1978) Fluorescence energy transfer as a spectroscopic ruler. *Annu. Rev. Biochem.*, **47**, 819–846.

19 Ha, T., Enderle, T., Ogletree, D.F., Chemla, D.S., Selvin, P.R. and Weiss, S. (1996) Probing the interaction between two single molecules: Fluorescence resonance energy transfer between a single donor and a single acceptor. *Proc. Natl Acad. Sci. USA*, **93**, 6264–6268.

20 Marcus, R.A. and Sutin, N. (1985) Electron transfers in chemistry and biology. *Biochem. Biophys. Acta*, **811**, 265–322.

21 Moser, C.C., Keske, J.M., Warncke, K., Farid, R.S. and Dutton, P.L. (1992) Nature of biological electron transfer. *Nature*, **355**, 796–802.

22 Neuweiler, H., Schulz, A., Böhmer, M., Enderlein, J. and Sauer, M. (2003) Measurement of submicrosecond intramolecular contact formation in peptides at the single-molecule level. *J. Am. Chem. Soc.*, **125**, 5324–5330.

23 Wang, H., Luo, G., Karnchanaphanurach, P., Louie, T.-M., Cova, S., Xun, L. and Xie, X.S. (2003) Protein conformational dynamics probed by single-molecule electron transfer. *Science*, **302**, 262–266.

24 Eggeling, C., Volkmer, A. and Seidel, C.A.M. (2005) Molecular photobleaching kinetics of rhodamine 6G by one- and two-photon induced confocal fluorescence microscopy. *Chem. Phys. Chem.*, **6**, 791–804.

25 Weiss, S. (1999) Fluorescence spectroscopy of single biomolecules. *Science*, **283**, 1676–1683.

26 Kapanidis, A.N., Lee, N.K., Laurence, T.A., Doose, S., Margeat, E. and Weiss, S. (2004) Fluorescence-aided molecule sorting: analysis of structure and interactions by alternating-laser excitation of single molecules. *Proc. Natl Acad. Sci. USA*, **101**, 8936–8941.

27 Kapanidis, A.N. and Weiss, S. (2002) Fluorescent probes and bioconjugation chemistries for single-molecule fluorescence analysis of biomolecules. *J. Chem. Phys.*, **117**, 10953–10964.

28 Jager, M., Michalet, X. and Weiss, S. (2005) Protein–protein interactions as a tool for site-specific labeling of proteins. *Protein Sci.*, **14**, 2059–2068.

29 Jager, M., Nir, E. and Weiss, S. (2006) Site-specific labeling of proteins for single-molecule FRET by combining chemical and enzymatic modification. *Protein Sci.*, **15**, 640–646.

30 Axelrod, D. (2001) Total Internal Reflection Fluorescence microscopy in cell biology. *Traffic*, **2**, 764–774.

31 Starr, T.E. and Thompson, N.L. (2001) Total internal reflection with fluorescence correlation spectroscopy: Combined surface reaction and solution diffusion. *Biophys. J.*, **80**, 1575–1584.

32 Levene, M.J., Korlach, J., Turner, S.W., Foquet, M., Craighead, H.G. and Webb, W.W. (2003) Zero-mode waveguides for single-molecule analysis at high concentrations. *Science*, **299**, 682–686.

33 Kastrup, L., Blom, H., Eggeling, C. and Hell, S.W. (2005) Fluorescence fluctuation spectroscopy in subdiffraction focal volumes. *Phys. Rev. Lett. (USA)*, **94**, 178104.

34 Ruckstuhl, T. and Verdes, D. (2004) Supercritical angle fluorescence (SAF) microscopy. *Optics Express*, **12**, 4246–4254.

35 Brinkmeier, M., Dorre, K., Stephan, J. and Eigen, M. (1999) Two beam cross correlation: A method to characterize transport phenomena in micrometer-sized structures. *Anal. Chem.*, **71**, 609–616.

36 Betzig, E. (2005) Excitation strategies for optical lattice microscopy. *Optics Express*, **13**, 3021–3036.

37 Neely, L.A., Patel, S., Garver, J., Gallo, M., Hackett, M., McLaughlin, S., Nadel, M., Harris, J., Gullans, S. and Rooke, J. (2006) A single-molecule method for the quantitation of microRNA gene expression. *Nat. Methods*, **3**, 41–46.

38 Hwang, L.C., Gosch, M., Lasser, T. and Wohland, T. (2006) Simultaneous multicolor fluorescence cross-correlation spectroscopy to detect higher order molecular interactions using single wavelength laser excitation. *Biophys. J.*, **91**, 715–727.

39 Kapanidis, A.N., Laurence, T.A., Lee, N.K., Margeat, E., Kong, X.X. and Weiss, S. (2005) Alternating-laser excitation of single molecules. *Acc. Chem. Res.*, **38**, 523–533.

40 Tellinghuisen, J., Goodwin, P.M., Ambrose, W.P., Martin, J.C. and Keller, R.A. (1994) Analysis of fluorescence lifetime data for single rhodamine molecules in flowing sample streams. *Anal. Chem.*, **66**, 64–72.

41 Schwille, P., Haupts, U., Maiti, S. and Webb, W.W. (1999) Molecular dynamics in living cells observed by fluorescence correlation spectroscopy with one- and two-photon excitation. *Biophys. J.*, **77**, 2251–2265.

42 Klar, T.A., Jakobs, S., Dyba, M., Genre, A. and Hell, S.W. (2000) Fluorescence microscopy with diffraction resolution barrier broken by stimulated emission. *Proc. Natl Acad. Sci. USA*, **97**, 8206–8210.

43 Roth well, P.J., Berger, S., Kirsch, O., Felecia, S., Antonio, M., Whorl, B.M., Restle, T., Goody, R.S. and Seidel, C.A. (2003) Multiparameter single-molecule fluorescence spectroscopy reveals heterogeneity of HIV-1 reverse transcriptase:primer/template complexes. *Proc. Natl Acad. Sci. USA*, **100**, 1655–1660.

44 Laurence, T.A., Kong, X., Jager, M. and Weiss, S. (2005) Probing structural heterogeneities and fluctuations of nucleic acids and denatured proteins. *Proc. Natl Acad. Sci. USA*, **102**, 17348–17353.

45 Clamme, J.P. and Deniz, A.A. (2005) Three-color single-molecule fluorescence resonance energy transfer. *Chemphyschem.*, **6**, 74–77.

46 Haustein, E., Jahnz, M. and Schwille, P. (2003) Triple FRET: A tool for studying long-range molecular interactions. *Chemphyschem.*, **4**, 745–748.

47 Lee, N.K., Kapanidis, A.N., Koh, H.R., Korlann, Y., Ho, S.O., Kim, Y., Gassman, N., Kim, S.K. and Weiss, S. (2006) Three-color alternating-laser excitation of single molecules: monitoring multiple interactions and distances. Biophys. J. (in press).

48 Webb, S.E.D., Gu, Y., Leveque-Fort, S., Siegel, J., Cole, M.J., Dowling, K., Jones, R., French, P.M.W., Neil, M.A.A., Juskaitis, R., Sucharov, L.O.D., Wilson, T. and Lever, M.J. (2002) A wide-field time-domain fluorescence lifetime imaging microscope with optical sectioning. *Rev. Scient. Instr.*, **73**, 1898–1907.

49 Michalet, X., Siegmund, O.H.W., Vallerga, J.V., Jelinsky, P., Millaud, J.E. and Weiss, S. (2006) A space- and time-resolved single photon counting detector for fluorescence microscopy and spectroscopy. *Proc. SPIE*, **6092**, 60920M.

50 Michalet, X., Siegmund, O.H.W., Vallerga, J.V., Jelinsky, P., Millaud, J.E. and Weiss, S. (2006) Photon-counting H33D detector for biological fluorescence imaging. *Nuclear Instr. Methods Phys. Res. A.* (in press).

51 Farrer, R.A., Previte, M.J.R., Olson, C.E., Peyser, L.A., Fourkas, J.T. and So, P.T.C. (1999) Single-molecule detection with a two-photon fluorescence microscope with fast-scanning capabilities and polarization sensitivity. *Optics Lett.*, **24**, 1832–1834.

52 Michalet, X. and Weiss, S. (2002) Single-molecule spectroscopy and microscopy. *Comp. Rend. Phys.*, **3**, 619–644.

53 Schmidt, T., Schutz, G.J., Baumgartner, W., Gruber, H.J. and Schindler, H. (1995) Characterization of photophysics and mobility of single molecules in a fluid lipid-membrane. *J. Phys. Chem.*, **99**, 17662–17668.

54 Fries, J.R., Brand, L., Eggeling, C., Kollner, M. and Seidel, C.A.M. (1998) Quantitative identification of different single molecules by selective time-resolved confocal fluorescence spectroscopy. *J. Phys. Chem. A*, **102**, 6601–6613.

55 Dahan, M., Deniz, A.A., Ha, T., Chemla, D.S., Schultz, P.G. and Weiss, S. (1999) Ratiometric measurement and identification of single diffusing molecules. *Chem. Phys.*, **247**, 85–106.

56 Li, H.T., Ying, L.M., Green, J.J., Balasubramanian, S. and Klenerman, D. (2003) Ultrasensitive coincidence fluorescence detection of single DNA molecules. *Anal. Chem.*, **75**, 1664–1670.

57 Chen, Y., Müller, J.D., So, P.T. and Gratton, E. (1999) The photon counting histogram in fluorescence fluctuation spectroscopy. *Biophys. J.*, **77**, 553–567.

58 Kask, P., Palo, K., Ullmann, D. and Gall, K. (1999) Fluorescence-intensity distribution analysis and its application in biomolecular detection technology. *Proc. Natl Acad. Sci. USA*, **96**, 13756–13761.

59 Laurence, T.A., Kapanidis, A.N., Kong, X.X., Chemla, D.S. and Weiss, S. (2004) Photon arrival-time interval distribution (PAID): A novel tool for analyzing molecular interactions. *J. Phys. Chem. B*, **108**, 3051–3067.

60 Wu, B. and Muller, J.D. (2005) Time-integrated fluorescence cumulant analysis in fluorescence fluctuation spectroscopy. *Biophys. J.*, **89**, 2721–2735.

61 Laurence, T.A., Fore, S. and Huser, T. (2006) Fast, flexible algorithm for calculating photon correlations. *Optics Lett.*, **31**, 829–831.

62 Magatti, D. and Ferri, F. (2003) 25 ns software correlator for photon and fluorescence correlation spectroscopy. *Rev. Sci. Instr.*, **74**, 1135–1144.

63 Wahl, M., Gregor, I., Patting, M. and Enderlein, J. (2003) Fast calculation of fluorescence correlation data with asynchronous time-correlated single-photon counting. *Opt. Express*, **11**, 3583–3591.

64 Schwille, P., Meyer-Almes, F.J. and Rigler, R. (1997) Dual-color fluorescence cross-correlation spectroscopy for multicomponent diffusional analysis in solution [see comments]. *Biophys. J.*, **72**, 1878–1886.

65 Margeat, E., Poujol, N., Boulahtouf, A., Chen, Y., Muller, J.D., Gratton, E., Cavailles, V. and Royer, C.A. (2001) The human estrogen receptor alpha dimer binds a single SRC-1 coactivator molecule with an affinity dictated by agonist structure. *J. Mol. Biol.*, **306**, 433–442.

66 Dahan, M., Deniz, A.A., Ha, T., Chemla, D.S., Schultz, P.G. and Weiss, S. (1999) Ratiometric measurement and identification of single diffusing molecules. *Chem. Phys.*, **247**, 85–106.

67 Ha, T., Ting, A.Y., Liang, J., Deniz, A.A., Chemla, D.S., Schultz, P.G. and Weiss, S. (1999) Temporal fluctuations of fluorescence resonance energy transfer between two dyes conjugated to a single protein. *Chem. Phys.*, **247**, 107–118.

68 Deniz, A.A., Dahan, M., Grunwell, J.R., Ha, T., Faulhaber, A.E., Chemla, D.S., Weiss, S. and Schultz, P.G. (1999) Single-pair fluorescence resonance energy transfer on freely diffusing molecules:

observation of Förster distance dependence and subpopulations. *Proc. Natl Acad. Sci. USA*, **96**, 3670–3675.

69 Kapanidis, A.N., Lee, N.K., Laurence, T., Doose, S., Margeat, E. and Weiss, S. (2004) Fluorescence-aided molecule sorting: analysis of structure and interactions using alternating-laser of single molecules. *Proc. Natl Acad. Sci. USA*, **101**, 8936–8941.

70 Ha, T., Ting, A.Y., Liang, J., Caldwell, W.B., Deniz, A.A., Chemla, D.S., Schultz, P.G. and Weiss, S. (1999) Single-molecule fluorescence spectroscopy of enzyme conformational dynamics and cleavage mechanism. *Proc. Natl Acad. Sci. USA*, **96**, 893–898.

71 Eigen, M. and Rigler, R. (1994) Sorting single molecules – application to diagnostics and evolutionary biotechnology. *Proc. Natl Acad. Sci. USA*, **91**, 5740–5747.

72 Nir, E., Michalet, X., Hamadani, K., Laurence, T.A., Neuhauser, D., Kovchegov, Y. and Weiss, S. (2006) Shot-noise limited single-molecule FRET histogram: comparison between theory and experiments. J. Phys. Chem. (in press).

73 Andrec, M., Levy, R.M. and Talaga, D.S. (2003) Direct determination of kinetic rates from single-molecule photon arrival trajectories using hidden Markov models. *J. Phys. Chem. A*, **107**, 7454–7464.

74 McHale, K., Berglund, A.J. and Mabuchi, H. (2004) Bayesian estimation for species identification in single-molecule fluorescence microscopy. *Biophys. J.*, **86**, 3409–3422.

75 Witkoskie, J.B. and Cao, J.S. (2004) Single molecule kinetics. I. Theoretical analysis of indicators. *J. Chem. Phys.*, **121**, 6361–6372.

76 Witkoskie, J.B. and Cao, J.S. (2004) Single molecule kinetics. II. Numerical Bayesian approach. *J. Chem. Phys.*, **121**, 6373–6379.

77 Schuler, B., Lipman, E.A., Steinbach, P.J., Kumke, M. and Eaton, W.A. (2005) Polyproline and the "spectroscopic ruler" revisited with single-molecule fluorescence. *Proc. Natl Acad. Sci. USA*, **102**, 2754–2759.

78 Wohland, T., Rigler, R. and Vogel, H. (2001) The standard deviation in fluorescence correlation spectroscopy. *Biophys. J.*, **80**, 2987–2999.

79 Jia, Y.W., Talaga, D.S., Lau, W.L., Lu, H.S.M., DeGrado, W.F. and Hochstrasser, R.M. (1999) Folding dynamics of single GCN4 peptides by fluorescence resonant energy transfer confocal microscopy. *Chem. Phys.*, **247**, 69–83.

80 Deniz, A.A., Laurence, T.A., Beligere, G.S., Dahan, M., Martin, A.B., Chemla, D.S., Dawson, P.E., Schultz, P.G. and Weiss, S. (2000) Single-molecule protein folding: diffusion fluorescence resonance energy transfer studies of the denaturation of chymotrypsin inhibitor 2. *Proc. Natl Acad. Sci. USA*, **97**, 5179–5184.

81 Schuler, B., Lipman, E.A. and Eaton, W.A. (2002) Probing the free-energy surface for protein folding with single-molecule fluorescence spectroscopy. *Nature*, **419**, 743–747.

82 Talaga, D.S., Lau, W.L., Roder, H., Tang, J.Y., Jia, Y.W., DeGrado, W.F. and Hochstrasser, R.M. (2000) Dynamics and folding of single two-stranded coiled-coil peptides studied by fluorescent energy transfer confocal microscopy. *Proc. Natl Acad. Sci. USA*, **97**, 13021–13026.

83 Lipman, E.A., Schuler, B., Bakajin, O. and Eaton, W.A. (2003) Single-molecule measurement of protein folding kinetics. *Science*, **301**, 1233–1235.

84 Haas, E., Wilchek, M., Katchalski-Katzir, E. and Steinberg, I.Z. (1975) Distribution of end-to-end distances of oligopeptides in solution as estimated by energy transfer. *Proc. Natl Acad. Sci. USA*, **72**, 1807–1811.

85 Bilsel, O. and Matthews, C.R. (2006) Molecular dimensions and their distributions in early folding intermediates. *Curr. Opin. Struct. Biol.*, **16**, 86–93.

86 Lee, N.K., Kapanidis, A.N., Wang, Y., Michalet, X., Mukhopadhyay, J., Ebright, R.H. and Weiss, S. (2005) Accurate FRET

measurements within single diffusing biomolecules using alternating-laser excitation. *Biophys. J.*, **88**, 2939–2953.

87 Rhoades, E., Cohen, M., Schuler, B. and Haran, G. (2004) Two-state folding observed in individual protein molecules. *J. Am. Chem. Soc.*, **126**, 14686–14687.

88 Rhoades, E., Gussakovsky, E. and Haran, G. (2003) Watching proteins fold one molecule at a time. *Proc. Natl Acad. Sci. USA*, **100**, 3197–3202.

89 Kuzmenkina, E.V., Heyes, C.D. and Nienhaus, G.U. (2005) Single-molecule Forster resonance energy transfer study of protein dynamics under denaturing conditions. *Proc. Natl Acad. Sci. USA*, **102**, 15471–15476.

90 Bonnet, G., Krichevsky, O. and Libchaber, A. (1998) Kinetics of conformational fluctuations in DNA hairpin-loops. *Proc. Natl Acad. Sci. USA*, **95**, 8602–8606.

91 Lapidus, L.J., Steinbach, P.J., Eaton, W.A., Szabo, A. and Hofrichter, J. (2002) Effects of chain stiffness on the dynamics of loop formation in polypeptides. Appendix: Testing a 1-dimensional diffusion model for peptide, dynamics. *J. Phys. Chem. B*, **106**, 11628–11640.

92 Wang, X. and Nau, W.M. (2004) Kinetics of end-to-end collision in short single-stranded nucleic acids. *J. Am. Chem. Soc.*, **126**, 808–813.

93 Yang, H., Luo, G.B., Karnchanaphanurach, P., Louie, T.M., Rech, I., Cova, S., Xun, L.Y. and Xie, X.S. (2003) Protein conformational dynamics probed by single-molecule electron transfer. *Science*, **302**, 262–266.

94 Mukhopadhyay, J., Kapanidis, A.N., Mekler, V., Kortkhonjia, E., Ebright, Y.W. and Ebright, R.H. (2001) Translocation of sigma(70) with RNA polymerase during transcription: fluorescence resonance energy transfer assay for movement relative to DNA. *Cell*, **106**, 453–463.

95 Kapanidis, A.N., Margeat, E., Laurence, T.A., Doose, S., Ho, S.O., Mukhopadhyay, J., Kortkhonjia, E., Mekler, V., Ebright, R.H. and Weiss, S. (2005) Retention of transcription initiation factor sigma70 in transcription elongation: single-molecule analysis. *Mol. Cell*, **20**, 347–356.

96 Margeat, E., Kapanidis, A.N., Tinnefeld, P., Wang, Y., Mukhopadhyay, J., Ebright, R.H. and Weiss, S. (2006) Direct observation of abortive initiation and promoter escape within single immobilized transcription complexes. *Biophys. J.*, **90**, 1419–1431.

97 Ha, T., Rasnik, I., Cheng, W., Babcock, H.P., Gauss, G.H., Lohman, T.M. and Chu, S. (2002) Initiation and re-initiation of DNA unwinding by the *Escherichia coli* Rep helicase. *Nature*, **419**, 638–641.

10
Two Rotary Motors of ATP Synthase

Ryota Iino and Hiroyuki Noji

10.1
Introduction

10.1.1
ATP Synthase: a Significant and Ubiquitous Enzyme in the Cell

Adenosine triphosphate (ATP) is one of the most important compounds for living organisms and serves as an energy carrier to drive many biological reactions that support the activity of life. For example, in many enzymes, an energetically unfavorable chemical reaction is coupled to the energetically favorable ATP hydrolysis reaction. ATP also powers many biological molecular machines such as molecular motors, ion pumps, molecular chaperones, and proteasomes. ATP synthase is responsible for the generation of ATP. It synthesizes ATP from adenosine diphosphate (ADP) and inorganic phosphate (Pi). ATP synthase is one of the most abundant proteins found in nature from bacteria to plants and animals; however, its reaction mechanism is unique and different from those of the other enzymes. ATP synthase uses physical rotation of its own subunits for catalysis. Prior to the discovery of ATP synthase as a rotary motor, the bacterial flagella motor – a large complex of many proteins – was the only known rotary motor.

ATP synthase is composed of two rotary motors, namely, F_1 and F_o, and it is also termed F_oF_1-ATP synthase (Figure 10.1, left) [1–3]. F_1 is a water-soluble component and rotates upon ATP hydrolysis. When isolated, F_1 catalyzes only the ATP hydrolysis reaction and is often termed F_1-ATPase (Figure 10.1, lower right). Bacterial F_1 has the subunit structure, $\alpha_3\beta_3\gamma\delta\epsilon$. The α and β subunits have noncatalytic and catalytic nucleotide-binding sites, respectively, and form a ring-shape structure. The γ subunit rotates in the $\alpha_3\beta_3$ ring, and the δ and ϵ subunits connect the $\alpha_3\beta_3$ ring and γ to F_o, respectively. Bacterial F_o has the subunit structure, ab_2c_{10-15}, and is embedded in the cell membrane (Figure 10.1, upper right). Although direct observation has not yet been possible, it is widely accepted that the central c_{10-15} ring rotates against the peripheral ab_2 complex, with the proton flow across the membrane driven by

Single Molecule Dynamics in Life Science. Edited by T. Yanagida and Y. Ishii
Copyright © 2009 WILEY-VCH Verlag GmbH & Co. KGaA, Weinheim
ISBN: 978-3-527-31288-7

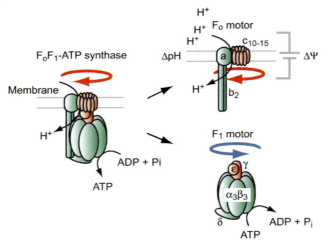

Figure 10.1 Schematic structure of ATP synthase. ATP synthase (left) is composed of two rotary motors. F_o (upper right) is embedded in the cell membrane and rotates with the translocation of proton, and is driven by proton motive force. The proton motive force is formed by the difference in proton concentration (ΔpH) and membrane potential ($\Delta\Psi$) across the membrane. F_1 (lower right) rotates with ATP hydrolysis. Note that the rotational direction of F_o is opposite to that of F_1. The rotor and stator subunits are shown in red and green, respectively.

the electrochemical potential gradient of the protons (or Na ions in several species). This potential gradient, which is often termed the proton motive force, is generated by respiratory chain or photosynthesis [4]. The rotational direction of F_o is opposite to that of F_1 driven by ATP hydrolysis. Generally, the proton motive force is large *in vivo*, and F_o generates a larger torque than F_1 and forcibly rotates F_1 in the reverse direction, thereby catalyzing the reverse chemical reaction, i.e. ATP synthesis.

10.1.2
Boyer's Proposal and Walker's Crystal Structure

Approximately three decades ago, the rotational catalysis of F_1 (and ATP synthase) was first proposed by Paul Boyer, as a result of his studies on "binding change mechanism" for ATP synthase [5]. His model had two central features. First, the energy required for ATP synthesis is mainly used to promote the binding of substrates (ADP and Pi) and the release of the tightly bound product (ATP). Second, the three catalytic β subunits of ATP synthase participate in the reaction in a highly cooperative manner. According to these features, each of the catalytic sites alternately changes the affinity to nucleotides and Pi, coupled with energy input/output. To substantiate this model Boyer proposed that the asymmetric central γδε subunit of ATP synthase rotates against the surrounding catalytic β subunits to cause sequential change. To date, in addition to the γ subunit, rotation of the ε subunit has also been observed.

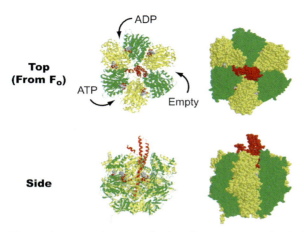

Figure 10.2 A crystal structure for the $\alpha_3\beta_3\gamma$ subcomplex of F_1. The α, β, and γ subunits are shown in yellow, green, and red, respectively. The catalytic sites are located at the interface between the α and β subunits (arrows); however, most of the residues responsible for catalysis are associated with the β subunit. The noncatalytic nucleotide-binding site on the α subunit is located at the opposite interface to the β subunit. Three α subunits bind ANPPNP (non-hydrolyzing ATP analog). Each of the three β subunit binds ANPPNP, ADP, or none.

Boyer's novel idea did not gain much support until John Walker and colleagues revealed the crystal structure of the $\alpha_3\beta_3\gamma$ subcomplex of F_1 from bovine mitochondria in 1994 (Figure 10.2) [6]. As mentioned above, in this crystal structure, the α and β subunits are alternately arranged and form a ring. A significant feature of this structure was the conformations of the three β subunits (Figure 10.3). Each of the

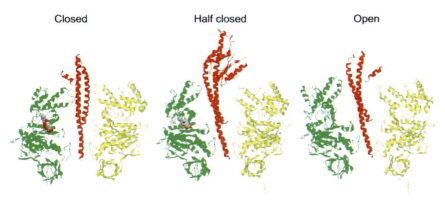

Figure 10.3 Three different conformations of the β subunit. One of the three α and β subunits and the γ subunit are shown in yellow, green, and red, respectively. In the 1994 structure, one β subunit that does not bind nucleotide is shown in "open" conformation (right), and two β subunits that bind nucleotide are shown in "closed" conformation (left). In the 2001 structure, one β subunit that corresponds to the β subunit without nucleotide in the 1994 structure, binds ADP and sulfate and is shown in "half closed" conformation (center).

three β subunits exhibited different "open" and "closed" conformations and bound different nucleotides – AMPPNP (a non-hydrolyzing ATP analog), ADP, or none. Later, an intermediate "half-closed" conformation that binds ADP and sulfate (mimicking Pi) was also reported [7]. The catalytic sites are located at the interfaces of the α and β subunits; however, most of the residues responsible for catalysis are associated with the β subunit. In contrast to the β subunits, the three α subunits exhibited almost the same conformations. Furthermore, the amino- and carboxyl-terminal residues of the γ subunit formed an asymmetric coiled-coil structure and extended into the central cavity of the $\alpha_3\beta_3$ ring. These structural features strongly supported Boyer's proposal.

Once rotation of F_1 was considered to be plausible, many researchers attempted to prove it. Although the results of chemical cross-linking exchange experiments and fluorescence polarization measurements proved the relative motion of the γ subunit against the three β subunits and were consistent with the rotation hypothesis, these results were not the direct evidence of unidirectional rotation [8, 9]. Finally, in 1997, single-molecule imaging by Yoshida and Kinosita's group directly demonstrated unidirectional successive rotation driven by ATP hydrolysis [10]. The direction of rotation was counterclockwise when viewed from F_o, which is consistent with the expectation from the crystal structure.

10.2
Rotation of ATP Synthase

10.2.1
Single-molecule Imaging of Rotation of F_1 Driven by ATP Hydrolysis

10.2.1.1 Strategy for Visualization of Rotation
The $\alpha_3\beta_3\gamma$ subcomplex of F_1 suffices for ATP hydrolysis, and direct observation of rotation was first carried out using this subcomplex [10] (hereafter, unless specified otherwise, the $\alpha_3\beta_3\gamma$ subcomplex is referred to as F_1). To visualize rotation, Yoshida and Kinosita's group executed the following strategies (Figure 10.4A) [11]. First, rapid lateral and rotational Brownian motion of F_1 (10 nm in diameter) in aqueous solution was suppressed by fixing it on a Ni-nitrilotriacetic acid-modified glass surface via a polyhistidine tag introduced into the amino terminus of the β subunit. Next, the small turning radius (approximately 1 nm) of the central γ subunit was magnified by attaching a fluorescent-labeled actin filament – a large probe having a length of several microns. Using these strategies, the rotation of F_1 from thermophilic *Bacillus* PS3 was first visualized under an optical microscope. Instead of the actin filament, latex beads have often been used as a rotational probe in recent experiments.

10.2.1.2 Large Torque Generated by F_1
Due to the hydrodynamic friction against the large probe, the rotational speed depends on the length of the actin filaments (Figure 9.2B) [12]. Long filaments rotate slowly, while short ones rotate rapidly. However, the torque generated by F_1 calculated

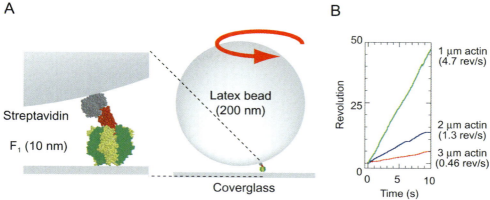

Figure 10.4 The experimental system for single-molecule observation of the rotation of F_1. (A) The $\alpha_3\beta_3$ ring is fixed on the glass surface to suppress the lateral and rotational Brownian motion of the F_1 molecule. A large probe, for example, actin filaments and latex beads, was attached to the γ subunit to visualize rotation. (B) Example of the dependence of the rotational speed on the length of the actin filament at 2 mM ATP.

from angular velocity and the drag coefficient was constant and approximately 40 pNnm. If we assume that the torque is generated at the interface between the β and γ subunits in a radius of 1 nm, it corresponds to a force of 40 pN. This value is considerably larger than that generated by most of the known nucleotide-driven molecular motors such as kinesin ($\sim$6 pN) [13], myosin ($\sim$4 pN) [14], and RNA polymerase ($\sim$14 pN) [15]. The only known nucleotide-driven motor stronger than F_1 is the portal motor of bacteriophage ($\sim$57 pN) which packages DNA inside the virus against a large internal pressure [16].

In addition to ATP, the rotation of F_1 was also supported by other purine nucleotides (guanosine triphosphate and inosine triphosphate) that generated torque comparable to that of ATP but not by pyrimidine nucleotides (cytidine triphosphate and uridine triphosphate) [17]. This result suggests that the mechanical characteristics of rotation are inherent in the F_1 structure, and purine nucleotides but not pyrimidine nucleotides can trigger and maintain the rotation. Truncation of the carboxyl-terminal 21-amino acid residues of the γ subunit decreased the torque by 50% (20 pNnm), indicating that the torque is actually generated by mechanical interaction between the β and γ subunits [18]. The reconstitution of the ϵ subunit to the $\alpha_3\beta_3\gamma$ subcomplex did not affect the torque [19].

10.2.1.3 Steps in Rotation
At low [ATP] in which ATP binding was rate-limiting, the rotation became stepwise (Figure 10.5A) [12]. The step size was 120°, which was consistent with the pseudo-threefold symmetric structure of F_1. The 120° stepping rotation is an intrinsic feature since the angle-resolved imaging of a single fluorophore attached to the γ subunit also

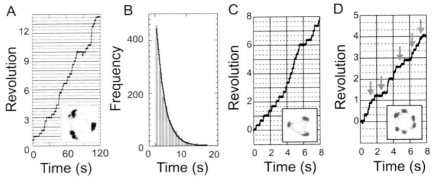

Figure 10.5 Steps in the rotation of F_1. (A) Rotation of F_1 with 120° steps observed at 20 nM ATP. (B) Histogram of the dwell time between 120° steps at 20 nM ATP. The histogram was fitted by a single exponential function (solid line). (C) Rotation of a mutant F_1 (E190D) with 120° steps at 2 mM ATP. (D) Rotation of F_1 (E190D) with 80° and 40° substeps at 2 μM ATP. The dwells at the 80° substep are indicated by arrows.

revealed 120° steps [20]. At low [ATP], the rotational speed is proportional to concentration down to 200 pM, suggesting that the same rotary mechanism is functional even at very low [ATP] [21]. The histogram of the dwell time between the steps was well fitted by a single exponential function, indicating that each step is driven by one ATP molecule (Figure 10.5B). This conclusion was further supported by the comparison of the rotational speed with the turnover rate of ATPase at low [ATP]; the value of the former was in good agreement with the one-third value of the latter.

The large probe limited the maximum rotational speed of F_1 to 6–8 revolutions per second (rps). When the large probe of rotation was replaced with a smaller one such as a colloidal gold particle with a diameter of 40 nm, the viscous load became negligible and faster rotation was observed [22]. At high [ATP], the rotational speed reached 130 rps, which is comparable to that expected from the maximum rate of ATP hydrolysis ($\sim 300\,\mathrm{s}^{-1}$ at 25 °C). At a recording rate of 8000 frames/s, 120° steps were observed even at 2 mM ATP. These steps are not driven by ATP binding since the binding rate (k_{on}) of ATP is around 2–$3 \times 10^7\,\mathrm{M}^{-1}\,\mathrm{s}^{-1}$ (this corresponds to a binding dwell time of 17–25 μs at 2 mM) and they could not be detected at the above-mentioned recording rate. When [ATP] was decreased, 120° steps were resolved into $\sim 90°$ and $\sim 30°$ substeps. While the dwell time before the 90° substep was inversely proportional to [ATP], no change was observed in the dwell time before the 30° substep. This implies that the 90° substep is driven by ATP binding, while the 30° substep is driven by other events. Statistical analysis of the dwell time before the 30° substep indicated that at least two events, each of 1 ms duration, occur before substepping.

When the rotation of an F_1 mutant with a very low maximum ATP hydrolysis rate ($2\sim3\,\mathrm{s}^{-1}$) was driven by high [ATP], 120° steps were observed at the video rate of

30 frames/s [23]. The rotation of this mutant driven by low [ATP] was easily resolved into $\sim 80°$ and $\sim 40°$ substeps (Figure 10.5D). Furthermore, when F_1 rotation was driven by slow hydrolyzing ATP analogs such as ATP-γ-S or Cy3-ATP, the dwells before $\sim 40°$ substeps were longer than those driven by ATP [23, 24]. These results indicate that one of the two events that occur before the 30° substeps is ATP hydrolysis. The slight difference in the angle may be due to the experimental error or a fast 10° substep that may occur during ATP hydrolysis.

The other event that initiates a 30–40° substep is less clear. However, several experiments have suggested some candidates. When F_1 lapses into the so-called "MgADP-inhibited state", in which MgADP that is tightly bound to the catalytic β subunit inhibits catalysis, it pauses the rotation at the position after the 80–90° substep [25]. This is consistent with the assumption that the 30~40° substep is induced by the release of ADP from the β subunit. Alternatively, Pi release may induce this substep since it has been reported that Pi release induces a conformational change in the β subunit [26].

Although the existence of substeps is a striking feature when considering the rotation scheme, it is not essential for a nucleotide-driven rotary motor. V_1, the soluble component of V-ATPase and a relative of F_1, is also an ATP-driven rotary motor and hydrolyzes three ATP molecules per turn [27]. However, in the case of V_1, a buffer exchange experiment indicated that the stepping positions of ATP binding and ATP-γ-S hydrolysis were almost identical [28].

10.2.1.4 A Model of Cooperative Chemo-mechanical Coupling in Rotating F_1

F_1 has three catalytic β subunits. How does each subunit cooperate during ATP hydrolysis and achieve unidirectional rotation? One topic that has been debated for decades is whether F_1 operates under bi- or tri-site mode (Figure 10.6, top and middle) [29, 30]. Simply defined, bi- and tri-site modes are those in which one and two and two and three nucleotides, respectively, alternately occupy the three catalytic sites during rotation. Simultaneous observation of Cy3-ATP binding/dissociation and rotation revealed that after the binding, Cy3-ATP or the hydrolyzed product Cy3-ADP remains bound at least for 240° rotation of the γ subunit [24]. This result supports the tri-site scheme since two additional nucleotides bind before the release of Cy3-ADP. The crystal structure of F_1 showing that three nucleotides bind to the β subunits also supports the tri-site scheme [7].

In contrast, the measurement of fluorescence resonance energy transfer (FRET) between a single pair of donor and acceptor molecule introduced into the β and γ subunits may support the bi-site scheme [31] since the orientation of the γ subunit in the ATP-waiting position estimated by FRET analysis was approximately 40° greater than that in the structure that binds two nucleotides [6]. In addition to bi- and tri-site schemes, an alternative in which ATP binding promotes the release of ADP can also be assumed (Figure 10.6, bottom). Recent computer simulation of the rotation and our preliminary experimental results are consistent with this model [32]. Our current assumption is that Pi release induces the 30–40° rotation and ATP binding induces the concomitant release of ADP. However, further research should be carried out before drawing the final conclusion.

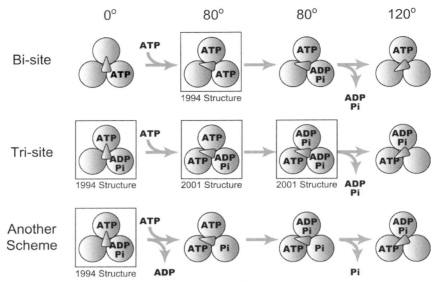

Figure 10.6 Models of chemo-mechanical coupling of F_1. Top, bi-site scheme. Middle, tri-site scheme. Bottom, an alternative scheme. The nucleotide states of F_1 that may correspond to the 1994 and 2001 crystal structures are enclosed in squares.

10.2.2
Single-molecule Manipulation of F_1 Rotation

10.2.2.1 Mechanical Activation of Pausing F_1
As described above, the rotating F_1 occasionally lapses into the MgADP-inhibited state at the position after an 80° substep [25]. Usually, this inhibition lasts for several tens of seconds, and presumably, after the dissociation of the inhibitory MgADP from the catalytic β subunit, F_1 spontaneously restarts its rotation. When the γ subunit of MgADP-inhibited F_1 is pushed and stalled in the forward (rotational) direction by using magnetic tweezers (Figure 10.7A and B), the time constant for activation became shorter [33]. However, the same result was not observed when the γ subunit was pushed and stalled in the backward direction (Figure 10.7C). The probability of activation after stalling was strongly dependent on the angle (Figure 10.7D), and the rate of activation was increased exponentially. The Arrehenius activation energy of the mechanical activation decreased to $-1.3\,k_BT/10°$. This strongly suggests that the mechanical energy input through the γ subunit shifted the energy state of the MgADP-inhibited F_1 to a higher state and lowered the activation energy. The potential of the MgADP-inhibited F_1, which was estimated from the Brownian motion of the beads attached to the γ subunit, was linear at >20° and had a slope of $1.5\,k_BT/10°$. This energy increment with the angle coincides well with the decrement in activation energy, suggesting that 85% of the mechanical energy input was transferred to the catalytic site to weaken the binding energy of MgADP. This result is consistent with

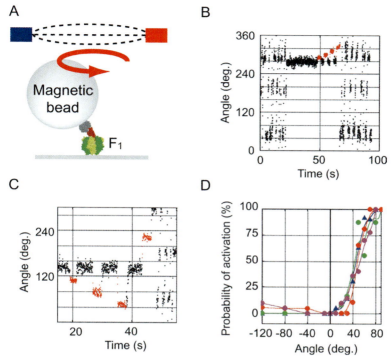

Figure 10.7 Mechanical activation of pausing F_1 by external force. (A) Schematic representation of the experimental set-up. Orientation of the γ subunit was manipulated by rotating the magnetic field. (B) Mechanical activation by pushing in the forward direction. F_1 that lapsed into the MgADP-inhibition state at 280° was pushed at $+10°$, $+20°$, $+30°$, and $+40°$ and stalled for 3 s (red). After being pushed and stalled at $+40°$, F_1 restarted its rotation. (C) Trial of mechanical activation by pulling in the backward direction. F_1 was not activated when pulled and stalled at $-20°$, $-40°$, and $-80°$ but was activated when pushed and stalled at $+80°$. (D) Angle dependence of the probability of activation. MgADP-inhibited F_1 was pushed or pulled and stalled for 3 s in the presence of $0.2\,\mu M$ ATP (blue); $0.2\,\mu M$ ATP and 30 mM Pi (green); $0.2\,\mu M$ ATP and 1 mM ADP (red); and $0.2\,\mu M$ ATP, 30 mM Pi, and 1 mM ADP (purple).

the tight mechanochemical coupling between the rotation of the γ subunit and the chemical reaction on the catalytic β subunits.

The most important result obtained from this experiment is that the angle of the γ subunit determines the rates of ADP dissociation from the catalytic β subunit. This angle dependency is expected to be applicable to ADP release during the rotation of active F_1. Our preliminary results indicate that not only the rate of ADP release but also that of ATP binding and hydrolysis are also angle dependent. The angle dependency of the affinity of ATP is also suggested by the constant torque generation (40 pNnm) by ATP binding, thereby indicating a linear downhill potential with a slope of $1.7\,k_BT/10°$ which corresponds to the affinity being increased by a factor of 5.5 per $10°$. The angle dependency of the rate of each chemical reaction step appears to be the fundamental mechanism that supports the unidirectional rotation of F_1. This property may also be required for the efficient ATP synthesis catalyzed by F_oF_1 in the physiological composition of ATP ($\sim$1 mM), ADP ($\sim$0.1 mM), and Pi ($\sim$1 mM). If the affinity of ATP and ADP is dependent on the angle of the γ subunit and reversed during the rotation, F_1 would be able to bind ADP efficiently even in the presence of high [ATP], when forced to rotate by F_o. This also facilitates the dissociation of synthesized ATP from the catalytic sites.

10.2.2.2 Highly Coupled ATP Synthesis by F₁ Forced to Rotate in the Reverse Direction

The physiological function of F_1 is ATP synthesis and not hydrolysis. As described above, when F_1 hydrolyzes ATP, the rotation of the γ subunit is tightly coupled to the chemical reaction, and three ATP molecules are hydrolyzed per turn (3 ATP/turn). Then, how efficiently does F_1 synthesize ATP when the rotation is reversed? ATP synthesis by the reverse rotation of F_1 was proved in 2004 [34]. In this experiment, many F_1 molecules enclosed in an observation chamber were forced to rotate in the reverse direction by using magnetic tweezers, and the synthesized ATP was detected by a conventional bioluminescence method using the luciferin–luciferase system. However, quantitative estimation of the ratio of mechanochemical coupling was difficult because the number of active F_1 molecules in the chamber was unknown.

The coupling ratio can be directly estimated if the number of ATP molecules synthesized by a single F_1 molecule during the forced rotation can be measured. Using magnetic tweezers, forced rotation of F_1 could be easily achieved. However, the quantitative measurement of ATP generated by a single F_1 molecule was not easy using conventional methods. For example, even if we assume 100% coupling (3 ATP/turn), the rotation of F_1 at 10 Hz for 1 min would result in the generation of only 1800 ATP molecules (3.0×10^{-21} mole). To overcome this problem, we developed a very small reaction chamber having a volume of 1 fl (= $[1\,\mu m]^3$) by using conventional microfabrication methods (Figure 10.8A) [35]. If 1800 ATP molecules are enclosed in this chamber, [ATP] would reach 3 μM, which is sufficiently high to be detected by conventional methods. In the actual experiment, we did not use the bioluminescence assay; instead, we used an enzymatic property of F_1 for the estimation of the number of ATP molecules synthesized after forced rotation using magnetic tweezers. In other words, we estimated [ATP] in a femtoliter chamber from the rotational speed of an enclosed F_1 molecule since the speed (or ATP hydrolysis activity) is proportional to the [ATP] below micromolar level.

After the forced rotation in the reverse direction in the femtoliter chamber, the rotational speed of F_1 driven by ATP hydrolysis actually increased (Figure 10.8B). This result indicates that ATP was actually synthesized by reverse rotation (Figure 10.8B); however, the ratio of coupling catalyzed by the $\alpha_3\beta_3\gamma$ subcomplex was unexpectedly very low and only 10% (0.3 ATP/turn) of that of ATP hydrolysis (3 ATP/turn) (Figure 10.8C) [36]. However, when the ε subunit was reconstituted with the $\alpha_3\beta_3\gamma$ subcomplex, the ratio significantly increased. Some molecules exhibited an almost 100% coupling ratio, and the average value was 77% (2.3 ATP/turn) (Figure 10.8D). The difference in the coupling ratio between each $\alpha_3\beta_3\gamma\varepsilon$ subcomplex was relatively large; this can be attributed to the difference in the volume of chamber and the error in the estimation of the rotational speed.

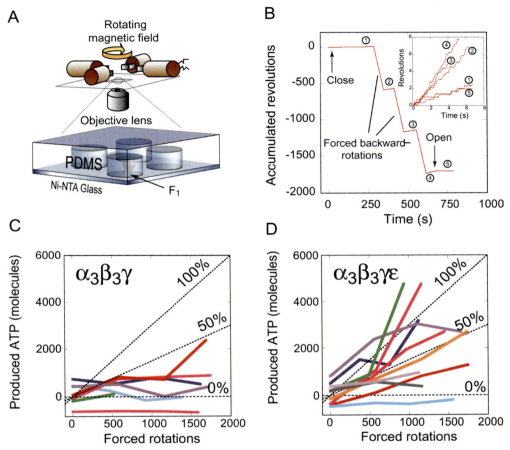

Figure 10.8 ATP synthesis by a single F_1 molecule forced to rotate in the reverse direction. (A) Schematic representation of the experimental set-up. F_1 enclosed in the femtoliter chamber was forced to rotate in the ATP synthesis direction by using magnetic tweezers in the presence of ADP, Pi, and a small amount of ATP. The newly synthesized ATP was accumulated in the chamber. (B) After the forced rotation, the rotational speed by ATP hydrolysis increased (1 to 4). When F_1 was released from the chamber, the rotational speed decreased and returned to the level before the enclosure. The number of synthesized ATP molecules was estimated from the increment in rotational speed. (C and D) The number of ATP molecules synthesized by the $\alpha_3\beta_3\gamma$ (C) and $\alpha_3\beta_3\gamma\varepsilon$ (D) subcomplexes of F_1 after the forced reverse rotation. Each trace represents the data from individual F_1. Dotted lines indicate the slopes where the coupling ratio was expected to be 0% (0 ATP/turn), 50% (1.5 ATP/turn), and 100% (3 ATP/turn).

The above experiment revealed a novel function of the ε subunit as a coupling factor of the ATP synthesis reaction. Then, how does the ε subunit support the highly coupled ATP synthesis? The ε subunit is a small subunit with a molecular mass of 14 kDa and is known to be an endogenous inhibitor of the ATP hydrolysis activity of F_1. In the case of bacterial F_1, its inhibitory effect is reduced in the presence of high

[ATP], and the ε subunit rotates with the γ subunit [19]. In contrast, the ε subunit from the chloroplast or cyanobacteria inhibits ATP hydrolysis activity regardless of [ATP] and inhibited the ATP-driven rotation [37]. In the crystal structure, the ε subunit stabilizes the residues of the γ subunit protruding from the $α_3β_3$ ring and disordered in the $α_3β_3γ$ subcomplex [38]. Since these stabilized residues include the contact interfaces with the β subunits, this may support the highly efficient mechanical coupling between the γ subunit rotation and the conformational change of the β subunits, which results in a highly coupled ATP synthesis reaction.

Furthermore, recent studies on the ε subunit suggested another possibility. Crystal structures of the F_1 subcomplex from different species revealed that at the carboxyl-terminus two α-helices of the ε subunit assume different conformations, namely, "retracted" (Figure 10.9, left) and "partly-extended" (Figure 10.9, right) [38–40]. These results suggested that the ε subunit was able to change its conformation. Conformational change of the ε subunit was supported by chemical cross-linking studies, and the alternative "fully-extended" and "extended-hairpin" conformations were also suggested [41–44]. Cross-linking studies further revealed that F_1 and F_oF_1 with the ε subunit fixed in partly- and fully-extended forms exhibited a significant suppression of ATP hydrolysis activity, but those with the retracted form of the ε subunit did not. In contrast, the extended forms of the ε subunit did not affect the ATP synthesis activity of F_oF_1 driven by the proton motive force. Based on these results, it is suggested that the ε subunit acts as a "switch" that changes the F_oF_1 activity between ATP hydrolysis and synthesis mode [43]. In addition to the cross-linking experiment, the measurement of FRET between the dye molecules introduced into the γ and ε subunits directly revealed that the ε subunit in F_1 actually undergoes nucleotide-dependent, large, reversible conformational change [45]. Furthermore, the isolated ε subunit can bind ATP with relatively low affinity [46, 47], and the bound ATP stabilizes the retracted form of the isolated ε subunit (H. Yagi *et al.*, personal

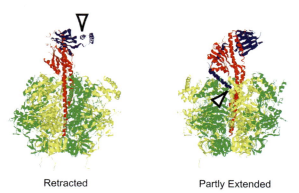

Retracted   Partly Extended

Figure 10.9 Structural models of the ε subunit in F_1. F_1 from *Escherichia coli* (right) and bovine mitochondria (left) are shown in blue. Here, the δ subunit of mitochondrial F_1 is referred to as ε because it is equivalent to the bacterial ε subunit. The positions of carboxyl-terminal residues of the ε subunit are indicated by arrowheads.

communication). The ε subunit in the F_oF_1 complex is also likely to bind ATP since the binding site is outside the contact interfaces between the ε and other subunits.

Considering the above-mentioned results, the conformational dynamics and states of the ε subunit are likely to affect the coupling ratio of ATP synthesis. The rotational direction of the $\gamma\varepsilon c_{10-15}$ complex, i.e. the direction of force applied to the ε subunit, may affect its conformation and function under non-equilibrium conditions. Further insights will be gained by single-molecule FRET measurement of the conformations of the carboxyl-terminal helices of the ε subunit during forced rotation of F_1. With regard to the amino-terminal β-sandwich moiety of the ε subunit in F_oF_1 from *Escherichia coli*, no conformational change was detected between ATP synthesis and hydrolysis conditions [48].

10.2.3
Rotation of F_oF_1 or F_o

10.2.3.1 Steps in the Rotation of F_oF_1 driven by ATP hydrolysis
When the rotation of F_oF_1 from thermophilic *Bacillus* PS3 solubilized with a detergent was driven by ATP hydrolysis, 80° and 40° substeps were observed [49]. F_o from thermophilic *Bacillus* PS3 has a subunit structure of ab_2c_{10}. The rotor c-ring consists of 10 c subunits and has 10 times symmetry [50]. However, the smaller substeps reflecting the structure of the c-ring (36° for c_{10}) were not observed. This indicates that the "friction" between the rotor and stator in F_o is not rate-limiting and does not slow down the ATP-driven rotation.

The number of c subunits in a ring varies among species, and a total of 10, 11, 14, and 15 subunits have been reported [50–54]. Each monomer has a single proton (or sodium ion)-binding site, and the number of c subunits in a ring determines the number of ions transported per turn. Since three ATP molecules are synthesized per turn in the F_1 component, the stoichiometry of ions/ATP becomes 3.3, 3.7, 4.7, and 5.0. Therefore, except for one example, most species exhibit non-integer stoichiometry between ions and ATP. This non-integer ratio or symmetry mismatch between F_1 and F_o may have advantage for the smooth rotation because such mismatch would avoid the formation of deep potential minima. Furthermore, the importance of elastic power transmission between F_1 and F_o was suggested for the smooth rotation under the symmetry mismatch condition [55, 56].

10.2.3.2 Ratchet versus Power Stroke as the Driving Force of F_o Rotation
In the case of F_1, it is widely accepted that the conformational changes in the β subunits coupled to the ATP hydrolysis reaction generate the torque for rotation of the γ subunit. With regard to the rotational mechanism of F_o, there are two controversial models, namely, "power stroke" and "Brownian ratchet" [57–59]. The power stroke model assumes the conformational change of the c subunit in the rotor c-ring. This model was proposed based on the result of structural analysis by NMR [60, 61]. In the c subunit monomer from *E. coli* dissolved in a water/methanol/chloroform mixed solvent, one of the two α-helices of the c subunit swiveled relative to the other at a different pH. Based on this result, Rastogi and

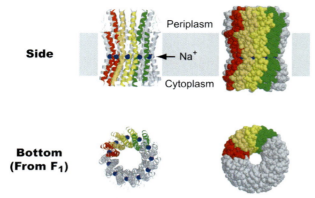

Figure 10.10 Structure of the c-ring of F_o from *Ilyobacter tartaricus*. Among the 11 monomers that form the ring, four are shown in red, orange, yellow, and green. Sodium ions (Na^+) are shown in blue.

Girvin proposed that (de)protonation of the critical asparagine acid residue of the c subunit induces the swiveling of the helix at the interface with the a subunit and generates the torque for rotation of the entire c-ring [60].

In contrast, the Brownian ratchet model does not assume the swiveling of the c subunit as a driving force of rotation. Instead, this model assumes the presence of a barrier for rotation at the interface between the a subunit and c-ring. The c-ring can overcome this barrier and rotate only when the critical asparagine acid residue of the c subunit is deprotonated. In the absence of the proton motive force, the c-ring undergoes rotational Brownian movement [62], and the proton motive force biases the direction of rotation by modulating the frequency of binding and release of protons from the two aqueous phases separated by the membrane.

Recently, the power stroke model that assumes the rotation of the helix in the c subunit has been challenged. The helix of the c subunit monomer from thermophilic *Bacillus* PS3 did not swivel at a different pH [63]. Furthermore, the crystal structure of the c-ring from *Ilyobacter tartaricus* (Figure 10.10) [64] did not support the swiveling of the carboxyl-terminal helix of the c subunit since the ion-binding site was located at the interface of the two monomers and near the outer surface of the ring, and ions would be transferred to and from the a subunit without swiveling. In addition, the crystal structure of the K-ring of V-ATPase from *E. hirae* did not support swiveling of the helix [65]. At present, the Brownian ratchet model is more plausible, although critical evidence is yet to be provided.

10.2.3.3 Rotation of F_oF_1 Driven by the Proton Motive Force

In 2004, the rotation of F_oF_1 driven by the proton motive force was first detected by Diez *et al.*; they measured FRET between single donor and acceptor molecules introduced into the γ subunit of F_1 and the b subunit of F_o [66]. They reconstituted F_oF_1 from *E. coli* into a liposome and applied the proton motive force generated by pH transition and additional electric potential difference. This method revealed a repetitive change in three FRET levels (H, M, and L), indicating the 120° step rotation during ATP synthesis; this result is consistent with the very low friction of the F_o component reported in another study [49]. The order of the three FRET levels (L-H-M-L) was reversed when the rotation was driven by ATP hydrolysis (L-M-H-L), indicating that the rotational direction is reversed in ATP synthesis.

Although the above experiment proved that the rotation of F_oF_1 is driven by the proton motive force, many issues remain to be resolved. For example, the dependence of the rotational speed on the proton motive force and the composition of nucleotides and Pi, and the conditions under which the rotational direction is reversed have not yet been examined. Furthermore, the 80° and 40° substeps found in the rotation driven by ATP hydrolysis remain to be observed, and the reversibility of the reaction scheme between ATP hydrolysis and synthesis has not been investigated.

10.3
Perspectives

As a biological molecular machine, ATP synthase has always been a topic of interest in basic and applied research. Although extensive analysis using single-molecule measurements has revealed various properties of F_1, many issues remain to be resolved; these include the determination of how the three β subunits cooperate in the ATP-driven rotation of F_1, and particularly how the proton motive force drives the rotation of F_o and F_oF_1. Direct observation can provide valuable insights and needs to be achieved as soon as possible. Structural information relating to F_o and the F_oF_1 complex at the atomic scale and computational study of conformational and rotational dynamics are also indispensable. Many more studies will be needed in order to understand this splendid molecular machine.

Acknowledgments

We thank Drs Hiromasa Yagi, Nobuto Kajiwara, and Hideo Akutsu, (Osaka University) for providing their unpublished results. We also thank Drs Masasuke Yoshida (Tokyo Institute of Technology), Katsuya Shimabukuro (Florida State University), Yasuyuki Kato-Yamada (Rikkyo University), and Drs Daichi Okuno, Kazuhito Tabata, Hiromi Imamura, and Hiroshi Ueno in our laboratory for helpful discussions.

References

1 Boyer, P.D. (1997) The ATP synthase – a splendid molecular machine. *Annu. Rev. Biochem.*, **66**, 717–749.

2 Yoshida, M., Muneyuki, E., and Hisabori, T. (2001) ATP synthase – a marvelous rotary engine of the cell. *Nat. Rev. Mol. Cell Biol.*, **2**, 669–677.

3 Capaldi, R.A. and Aggeler, R. (2002) Mechanism of the F(1)F(0)-type ATP synthase, a biological rotary

motor. *Trends Biochem. Sci.*, **27**, 154–160.

4 Mitchell, P. and Moyle, J. (1967) Chemiosmotic hypothesis of oxidative phosphorylation. *Nature*, **213**, 137–139.

5 Boyer, P.D. (1993) The binding change mechanism for ATP synthase–some probabilities and possibilities. *Biochim. Biophys. Acta*, **1140**, 215–250.

6 Abrahams, J.P., Leslie, A.G., Lutter, R., and Walker, J.E. (1994) Structure at 2.8 A resolution of F1-ATPase from bovine heart mitochondria. *Nature*, **370**, 621–628.

7 Menz, R.I., Walker, J.E., and Leslie, A.G. (2001) Structure of bovine mitochondrial F(1)-ATPase with nucleotide bound to all three catalytic sites: implications for the mechanism of rotary catalysis. *Cell* **106**, 331–341.

8 Zhou, Y., Duncan, T.M., Bulygin, V.V., Hutcheon, M.L., and Cross, R.L. (1996) ATP hydrolysis by membrane-bound *Escherichia coli* F0F1 causes rotation of the gamma subunit relative to the beta subunits. *Biochim. Biophys. Acta* **1275**, 96–100.

9 Sabbert, D., Engelbrecht, S., and Junge, W. (1996) Intersubunit rotation in active F-ATPase. *Nature* **381**, 623–625.

10 Noji, H., Yasuda, R., Yoshida, M., and Kinosita, K., Jr. (1997) Direct observation of the rotation of F1-ATPase. *Nature* **386**, 299–302.

11 Adachi, K., Noji, H., and Kinosita, K., Jr. (2003) Single-molecule imaging of rotation of F1-ATPase. *Methods Enzymol.* **361**, 211–227.

12 Yasuda, R., Noji, H., Kinosita, K., Jr., and Yoshida, M. (1998) F1-ATPase is a highly efficient molecular motor that rotates with discrete 120 degree steps. *Cell* **93**, 1117–1124.

13 Svoboda, K. and Block, S.M. (1994) Force and velocity measured for single kinesin molecules. *Cell* **77**, 773–784.

14 Finer, J.T., Simmons, R.M., and Spudich, J.A. (1994) Single myosin molecule mechanics: piconewton forces and nanometre steps. *Nature* **368**, 113–119.

15 Yin, H., Wang, M.D., Svoboda, K., Landick, R., Block, S.M., and Gelles, J. (1995) Transcription against an applied force. *Science* **270**, 1653–1657.

16 Smith, D.E., Tans, S.J., Smith, S.B., Grimes, S., Anderson, D.L., and Bustamante, C. (2001) The bacteriophage straight phi29 portal motor can package DNA against a large internal force. *Nature* **413**, 748–752.

17 Noji, H., Bald, D., Yasuda, R., Itoh, H., Yoshida, M., and Kinosita, K., Jr. (2001) Purine but not pyrimidine nucleotides support rotation of F(1)-ATPase. *J. Biol. Chem.* **276**, 25480–25486.

18 Hossain, M.D., Furuike, S., Maki, Y., Adachi, K., Ali, M.Y., Huq, M., Itoh, H., Yoshida, M., and Kinosita, K., Jr. (2006) The rotor tip inside a bearing of a thermophilic F1-ATPase is dispensable for torque generation. *Biophys. J.* **90**, 4195–4203.

19 Kato-Yamada, Y., Noji, H., Yasuda, R., Kinosita, K., Jr., and Yoshida, M. (1998) Direct observation of the rotation of epsilon subunit in F1-ATPase. *J. Biol. Chem.* **273**, 19375–19377.

20 Adachi, K., Yasuda, R., Noji, H., Itoh, H., Harada, Y., Yoshida, M., and Kinosita, K., Jr. (2000) Stepping rotation of F1-ATPase visualized through angle-resolved single-fluorophore imaging. *Proc. Natl Acad. Sci. USA* **97**, 7243–7247.

21 Sakaki, N., Shimo-Kon, R., Adachi, K., Itoh, H., Furuike, S., Muneyuki, E., Yoshida, M., and Kinosita, K., Jr. (2005) One rotary mechanism for F1-ATPase over ATP concentrations from millimolar down to nanomolar. *Biophys. J.* **88**, 2047–2056.

22 Yasuda, R., Noji, H., Yoshida, M., Kinosita, K., Jr., and Itoh, H. (2001) Resolution of distinct rotational substeps by submillisecond kinetic analysis of F1-ATPase. *Nature* **410**, 898–904.

23 Shimabukuro, K., Yasuda, R., Muneyuki, E., Hara, K.Y., Kinosita, K., Jr., and Yoshida, M. (2003) Catalysis and rotation of F1 motor: Cleavage of ATP at the catalytic site occurs in 1 ms before 40 degree substep

rotation. *Proc. Natl Acad. Sci. USA* **100**, 14731–14736.

24 Nishizaka, T., Oiwa, K., Noji, H., Kimura, S., Muneyuki, E., Yoshida, M., and Kinosita, K., Jr. (2004) Chemomechanical coupling in F1-ATPase revealed by simultaneous observation of nucleotide kinetics and rotation. *Nat. Struct. Mol. Biol.* **11**, 142–148. Epub 2004 Jan 18.

25 Hirono-Hara, Y., Noji, H., Nishiura, M., Muneyuki, E., Hara, K.Y., Yasuda, R., Kinosita, K., Jr., and Yoshida, M. (2001) Pause and rotation of F(1)-ATPase during catalysis. *Proc. Natl Acad. Sci. USA* **98**, 13649–13654.

26 Masaike, T., Muneyuki, E., Noji, H., Kinosita, K., Jr., and Yoshida, M. (2002) F1-ATPase changes its conformations upon phosphate release. *J. Biol. Chem.* **277**, 21643–21649.

27 Imamura, H., Nakano, M., Noji, H., Muneyuki, E., Ohkuma, S., Yoshida, M., and Yokoyama, K. (2003) Evidence for rotation of V1-ATPase. *Proc. Natl Acad. Sci. USA* **100**, 2312–2315.

28 Imamura, H., Takeda, M., Funamoto, S., Shimabukuro, K., Yoshida, M., and Yokoyama, K. (2005) Rotation scheme of V1-motor is different from that of F1-motor. *Proc. Natl Acad. Sci. USA* **102**, 17929–17933.

29 Weber, J. and Senior, A.E. (2001) Bi-site catalysis in F1-ATPase: Does it exist? *J. Biol. Chem.* **276**, 35422–35428.

30 Boyer, P.D. (2002) Catalytic site occupancy during ATP synthase catalysis. *FEBS Lett.* **512**, 29–32.

31 Yasuda, R., Masaike, T., Adachi, K., Noji, H., Itoh, H., and Kinosita, K., Jr. (2003) The ATP-waiting conformation of rotating F1-ATPase revealed by single-pair fluorescence resonance energy transfer. *Proc. Natl Acad. Sci. USA* **100**, 9314–9318.

32 Koga, N. and Takada, S. (2006) Folding-based molecular simulations reveal mechanisms of the rotary motor F1-ATPase. *Proc. Natl Acad. Sci. USA* **103**, 5367–5372.

33 Hirono-Hara, Y., Ishizuka, K., Kinosita, K., Jr., Yoshida, M., and Noji, H. (2005) Activation of pausing F1 motor by external force. *Proc. Natl Acad. Sci. USA* **102**, 4288–4293.

34 Itoh, H., Takahashi, A., Adachi, K., Noji, H., Yasuda, R., Yoshida, M., and Kinosita, K. (2004) Mechanically driven ATP synthesis by F1-ATPase. *Nature* **427**, 465–468.

35 Rondelez, Y., Tresset, G., Tabata, K.V., Arata, H., Fujita, H., Takeuchi, S., and Noji, H. (2005) Microfabricated arrays of femtoliter chambers allow single molecule enzymology. *Nat. Biotechnol.* **23**, 361–365.

36 Rondelez, Y., Tresset, G., Nakashima, T., Kato-Yamada, Y., Fujita, H., Takeuchi, S., and Noji, H. (2005) Highly coupled ATP synthesis by F1-ATPase single molecules. *Nature* **433**, 773–777.

37 Konno, H., Murakami-Fuse, T., Fujii, F., Koyama, F., Ueoka-Nakanishi, H., Pack, C.G., Kinjo, M., and Hisabori, T. (2006) The regulator of the F(1) motor: inhibition of rotation of cyanobacterial F(1)-ATPase by the varepsilon subunit. *EMBO J.* **25**, 4596–4604.

38 Gibbons, C., Montgomery, M.G., Leslie, A.G., and Walker, J.E. (2000) The structure of the central stalk in bovine F(1)-ATPase at 2.4 A resolution. *Nat. Struct. Biol.* **7**, 1055–1061.

39 Rodgers, A.J. and Wilce, M.C. (2000) Structure of the gamma-epsilon complex of ATP synthase. *Nat. Struct. Biol.* **7**, 1051–1054.

40 Hausrath, A.C., Capaldi, R.A., and Matthews, B.W. (2001) The conformation of the epsilon- and gamma-subunits within the *Escherichia coli* F(1) ATPase. *J. Biol. Chem.* **276**, 47227–47232.

41 Kato-Yamada, Y., Yoshida, M., and Hisabori, T. (2000) Movement of the helical domain of the epsilon subunit is required for the activation of thermophilic F1-ATPase. *J. Biol. Chem.* **275**, 35746–35750.

42 Tsunoda, S.P., Rodgers, A.J., Aggeler, R., Wilce, M.C., Yoshida, M., and Capaldi,

R.A. (2001) Large conformational
changes of the epsilon subunit in the
bacterial F1F0 ATP synthase provide a
ratchet action to regulate this rotary motor
enzyme. *Proc. Natl Acad. Sci. USA* **98**,
6560–6564.

43 Suzuki, T., Murakami, T., Iino, R., Suzuki,
J., Ono, S., Shirakihara, Y., and Yoshida, M.
(2003) F0F1-ATPase/synthase is geared to
the synthesis mode by conformational
rearrangement of epsilon subunit in
response to proton motive force and
ADP/ATP balance. *J. Biol. Chem.* **278**,
46840–46846.

44 Bulygin, V.V., Duncan, T.M., and Cross,
R.L. (2004) Rotor/stator interactions of the
epsilon subunit in *Escherichia coli* ATP
synthase and implications for enzyme
regulation. *J. Biol. Chem.* **279**, 35616–
35621.

45 Iino, R., Murakami, T., Iizuka, S., Kato-
Yamada, Y., Suzuki, T., and Yoshida, M.
(2005) Real-time monitoring of
conformational dynamics of the epsilon
subunit in F1-ATPase. *J. Biol. Chem.* **280**,
40130–40134.

46 Kato-Yamada, Y. and Yoshida, M. (2003)
Isolated epsilon subunit of thermophilic
F1-ATPase binds ATP. *J. Biol. Chem.* **278**,
36013–36016.

47 Kato-Yamada, Y. (2005) Isolated epsilon
subunit of *Bacillus subtilis* F1-ATPase binds
ATP. *FEBS Lett.* **579**, 6875–6878.

48 Zimmermann, B., Diez, M., Zarrabi, N.,
Graber, P., and Borsch, M. (2005)
Movements of the epsilon-subunit during
catalysis and activation in single
membrane-bound H(+)-ATP synthase.
EMBO J. **24**, 2053–2063.

49 Ueno, H., Suzuki, T., Kinosita, K., Jr., and
Yoshida, M. (2005) ATP-driven stepwise
rotation of FoF1-ATP synthase. *Proc. Natl
Acad. Sci. USA* **102**, 1333–1338.

50 Mitome, N., Suzuki, T., Hayashi, S., and
Yoshida, M. (2004) Thermophilic ATP
synthase has a decamer c-ring: Indication
of noninteger 10:3 H+/ATP ratio and
permissive elastic coupling. *Proc. Natl
Acad. Sci. USA* **101**, 12159–12164.

51 Stock, D., Leslie, A.G., and Walker, J.E.
(1999) Molecular architecture of the rotary
motor in ATP synthase. *Science* **286**,
1700–1705.

52 Seelert, H., Poetsch, A., Dencher, N.A.,
Engel, A., Stahlberg, H., and Muller, D.J.
(2000) Structural biology: Proton-powered
turbine of a plant motor. *Nature* **405**,
418–419.

53 Stahlberg, H., Muller, D.J., Suda, K.,
Fotiadis, D., Engel, A., Meier, T., Matthey,
U., and Dimroth, P. (2001) Bacterial Na
(+)-ATP synthase has an undecameric
rotor. *EMBO Rep.* **2**, 229–233.

54 Pogoryelov, D., Yu, J., Meier, T., Vonck, J.,
Dimroth, P., and Muller, D.J. (2005)
The c15 ring of the *Spirulina platensis*
F-ATP synthase: F1/F0 symmetry
mismatch is not obligatory. *EMBO Rep.* **6**,
1040–1044.

55 Cherepanov, D.A. and Junge, W. (2001)
Viscoelastic dynamics of actin filaments
coupled to rotary F-ATPase: curvature as an
indicator of the torque. *Biophys. J.* **81**,
1234–1244.

56 Panke, O., Cherepanov, D.A.,
Gumbiowski, K., Engelbrecht, S., and
Junge, W. (2001) Viscoelastic dynamics of
actin filaments coupled to rotary F-ATPase:
angular torque profile of the enzyme.
Biophys. J. **81**, 1220–1233.

57 Junge, W. (1999) ATP synthase and other
motor proteins. *Proc. Natl Acad. Sci. USA*
96, 4735–4737.

58 Fillingame, R.H. and Dmitriev, O.Y. (2002)
Structural model of the transmembrane
Fo rotary sector of H+-transporting ATP
synthase derived by solution NMR and
intersubunit cross-linking *in situ*. *Biochim.
Biophys. Acta* **1565**, 232–245.

59 Oster, G. and Wang, H. (2003) Rotary
protein motors. *Trends Cell Biol.* **13**,
114–121.

60 Rastogi, V.K. and Girvin, M.E. (1999)
Structural changes linked to proton
translocation by subunit c of the ATP
synthase. *Nature* **402**, 263–268.

61 Aksimentiev, A., Balabin, I.A., Fillingame,
R.H., and Schulten, K. (2004) Insights into

the molecular mechanism of rotation in the Fo sector of ATP synthase. *Biophys. J.* **86**, 1332–1344.

62 Kaim, G. and Dimroth, P. (1998) Voltage-generated torque drives the motor of the ATP synthase. *EMBO J.* **17**, 5887–5895.

63 Nakano, T., Ikegami, T., Suzuki, T., Yoshida, M., and Akutsu, H. (2006) A new solution structure of ATP synthase subunit c from thermophilic *Bacillus* PS3, suggesting a local conformational change for H + -translocation. *J. Mol. Biol.* **358**, 132–144.

64 Meier, T., Polzer, P., Diederichs, K., Welte, W., and Dimroth, P. (2005) Structure of the rotor ring of F-Type Na + -ATPase from *Ilyobacter tartaricus. Science* **308**, 659–662.

65 Murata, T., Yamato, I., Kakinuma, Y., Leslie, A.G., and Walker, J.E. (2005) Structure of the rotor of the V-Type Na + -ATPase from *Enterococcus hirae. Science* **308**, 654–659.

66 Diez, M., Zimmermann, B., Borsch, M., Konig, M., Schweinberger, E., Steigmiller, S., Reuter, R., Felekyan, S., Kudryavtsev, V., Seidel, C.A., and Graber, P. (2004) Proton-powered subunit rotation in single membrane-bound F0F1-ATP synthase. *Nat. Struct. Mol. Biol.* **11**, 135–141.

11
Single-molecule FRET Studies of Helicases and Holliday Junctions

Taekjip Ha

11.1
Introduction

Single molecule fluorescence detection is a powerful tool for probing biological events directly without the temporal and population averaging of conventional ensemble studies (reviewed in [1, 2]). We are mainly using one incarnation, single-molecule FRET (fluorescence resonance energy transfer). FRET is a spectroscopic technique for measuring distances in the 30–80-Å range [3, 4]. Excitation energy of the donor is transferred to the acceptor via an induced-dipole, induced-dipole interaction. The efficiency of energy transfer, E, is given by $(1 + (R/R_0)^6)^{-1}$ where R is the distance between the donor and acceptor and R_0 is the distance at which 50% of the energy is transferred and is a function of the properties of the dyes. A small change in distance between the two sites of a biological molecule where donor and acceptor are attached can result in a sizeable change in E. Therefore, structural changes of biological molecules or relative motion between two different molecules can be detected via FRET changes [5]. R_0 contains a contribution from the relative orientation between the two dyes, known as κ^2. It is often assumed that the dipole moments of donor and acceptor are free to rotate in all directions, on a time scale much faster than their radiative lifetime. In this case, a geometrical averaging of the angles yields $\kappa^2 = 2/3$.

Often, conformational changes are very difficult to synchronize or occur too infrequently to detect using ensemble FRET. smFRET (reviews in [1, 2, 6–9]) opens up new opportunities to probe the structural changes of biological molecules in real time. In addition, smFRET readily determines not just average conformations, but also the distribution of distances. SmFRET is also relatively insensitive to incomplete labeling of host molecule with donor and acceptor. Donor-only species simply show up as a FRET = 0 peak while acceptor-only species are excited only very weakly if the probes are selected with a large spectral separation. smFRET, first introduced by us [10], has been now widely adopted by many laboratories around the world to study a variety of biological systems including DNA [11–18], RNA [19–31], proteins [32–40] and large macromolecular complexes [41, 42].

Single Molecule Dynamics in Life Science. Edited by T. Yanagida and Y. Ishii
Copyright © 2009 WILEY-VCH Verlag GmbH & Co. KGaA, Weinheim
ISBN: 978-3-527-31288-7

In this chapter, I will discuss two different biological systems that we have been studying using smFRET over the last few years, the DNA unwinding enzyme, "helicase", and the key intermediate in DNA combination known as "Holliday junction".

11.2
Single-molecule FRET

FRET efficiency is given by $I_A/(I_A + \eta I_D)$ where I_A is the sensitized emission intensity of the acceptor, I_D is the donor intensity, and η is a parameter representing relative detection efficiencies and quantum yields of the two dyes, and can be determined from photobleaching events [43]. Our data using Cy3 and Cy5 show that $\eta \sim 1$. Thus $E \equiv I_A/(I_A + I_D)$ is an excellent approximation for FRET efficiency and all our data are presented in this form. While this FRET value cannot be smaller than 0 or larger than 1 in principle, in practice I_A can be very small for FRET approaching 0 and noise around the background level can make I_A negative after background subtraction, leading to calculated FRET values smaller than 0. For the same reason FRET values larger than 1 can be obtained if it is close to 1. For this reason our FRET histograms sometimes contain data points below 0 or above 1, but this is simply due to a finite signal-to-noise ratio. In almost all of our studies, we do not need absolute distance information because other control experiments can unambiguously assign FRET values to corresponding states. For certain cases, it can be ambiguous, even with control experiments, whether an observed FRET change is due to distance change or the change in dipole orientations of the dyes, hence in κ^2. In such cases we can measure polarization anisotropy of the dyes at the same time as FRET from single molecules so that we can determine whether or not the FRET changes are mere artifacts [44, 45]. Therefore in all of our experiments, we do not need to convert E to distance in order to answer questions of biological interest. The distance information can still play a supporting role to determine whether the signal we obtain is consistent with what is known about the structural properties of the molecules [37, 38].

11.2.1
Non-perturbative Immobilization: BSA and PEG Surfaces, and Vesicle Encapsulation

To study the dynamic changes of individual molecules over extended time periods, molecules need to be localized in space. This is often achieved by surface immobilization (Figure 11.1). An ideal surface would allow specific immobilization of nucleic acids or proteins while rejecting non-specific adsorption. For nucleic acids studies, we prefer to use a quartz slide coated with biotinylated BSA and streptavidin because of the simplicity of this system. We can immobilize DNA and RNA with high specificity (>500 : 1, compared to control experiments that exclude biotin or streptavidin) and were able to faithfully reproduce their bulk solution activities [19, 20, 31]. This is likely because all three surface constituents (BSA, streptavidin and quartz) are

Adapted from Rasnik et al, Acc. Chem. Res. (2005).

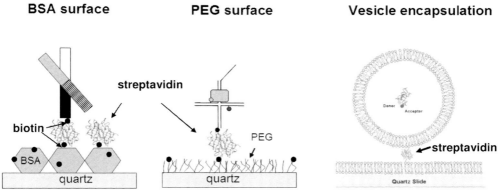

Figure 11.1 Three different methods of immobilizing biomolecules for single-molecule fluorescence studies. Adapted from [45].

negatively charged in neutral pH, repelling nucleic acids. For studies involving proteins, BSA surfaces are too adhesive. Therefore, we developed a PEG (polyethylene glycol)-coated surface that reduces the protein adsorption to an undetectable level [46–48]. Surface passivation using PEG, first introduced by us for single-molecule studies [48], has now been used successfully by several other groups as well as our own [37, 38, 41, 42, 49–56]. If a dense layer of PEG is formed on a quartz surface (we use amino-silane coating followed by conjugation with PEG modified with NHS-ester at one end), it forms a polymer brush that prevents protein adsorption to the underlying surface. We incorporate a small fraction of PEG polymers that are end-modified with biotin to facilitate the immobilization of biotinylated macromolecules. Proteins interact specifically with DNA immobilized to the PEG surface and their bulk solution activities are well reproduced in all the systems we have tested [37–39, 48, 57]. We have also developed a vesicle encapsulation technique that can measure conformational dynamics of biomolecules free of surface tethering [58].

11.3
smFRET Studies of Rep Helicase

Helicases are highly efficient, processive motor proteins, which are powered by the hydrolysis of nucleotide co-factors such as ATP. For instance, RecBCD protein can unwind tens of thousands of base pairs in one run at a speed of $\sim$500 bps per second [59, 60]. Most DNA helicases show directionality in DNA unwinding. Polar binding to ssDNA and the ability to translocate uni-directionally appear to be its origin even though the exact relation between ssDNA translocation and unwinding is not yet clear. Helicases need to couple the conformational changes induced by ATP binding and hydrolysis to directional motion along a DNA track and unwinding of thermodynamically stable, duplex DNA. Recent crystal structures provide valuable insights, but exactly how this is achieved is not known.

We have developed new single-molecule fluorescence assays to study various aspects of a helicases. smFRET approaches have already revealed many surprises in what was considered the simplest of helicases, *E. coli* Rep helicase. For example, we discovered that DNA unwinding catalyzed by Rep stalls if one of the two monomers within the functional unit dissociates and that unwinding reinitiates only if another protein is added to reassemble the functional unit [48]. In addition, we discovered that a Rep monomer undergoes a series of acrobatic movements to reinitiate ssDNA translocation when it encounters a physical blockade [38].

11.3.1
Helicase: Essential Motor Proteins on the Nucleic Acid Highway

Since its original discovery in 1976 as the "DNA unwinding enzyme" [61, 62], it has become clear that helicases participate in virtually all cellular processes that involve nucleic acid. This includes DNA metabolism, such as replication, recombination and repair, as well as RNA processing, such as ribosome assembly, RNA interference, translation initiation and mRNA splicing [63–68]. Helicases are found in all three kingdoms of life and are extremely numerous: 1–2% of eukaryotic genes are helicases. Helicase studies have important biomedical implications. Several severe human genetic diseases (Xeroderma pigmentosum, Werner's, Bloom's, and Rothmund–Thomson syndromes) have been linked to mutations in helicases [69–73]. In particular, Werner's syndrome results in premature aging as well as a high propensity to cancer. Since DNA replication and repair are fundamental to cell growth in all organisms, an understanding of such a basic process as enzyme-catalyzed DNA unwinding will undoubtedly have an impact on our understanding of some cancers that result from defects in replication or repair.

The most fundamental property of all helicases is their "translocation" ability to move along nucleic acids [74]. This translocation is powered by ATP hydrolysis, hence helicases are motor proteins. How this motor works remains a mystery. Even the most elementary issue of the step size of translocation is not yet clear. Also unknown is why some helicases move in one direction on ssDNA while others move in the opposite direction. For example, Rep helicase is a 3′-5′ ssDNA translocase and its DNA unwinding activity requires a 3′ ssDNA tail on the DNA. Unwinding and translocation are two defining features of helicase, but their relationship is poorly understood. Helicase must couple its conformational changes resulting from ATP binding and hydrolysis to its unwinding and translocation action, but exactly how this is achieved is not known. The long-term goal of our efforts is to measure helicase function and structural changes simultaneously to probe the core of the structural–function relationship.

We are using *E. coli* Rep helicase as a model system for this type of fluorescence study since (i) its crystal structure is known allowing the rational choice of labeling sites [75], (ii) its structural homolog PcrA has been crystallized in a number of different states regarding bound DNA and nucleotides [76], (iii) extensive biochemical information is available [63, 77–79], (iv) it can bind to DNA and move on DNA as a monomer [79] thus simplifying the interpretation of FRET data, and (v) none of the five native cysteines are conserved among superfamily I helicase. Although the crystal structures of many other helicases have been determined, most of these do not quite fit the bill. For example, HCV NS3 [80], *E. coli* RecQ [81], and *E. coli* RecBCD [82] have many more cysteines than Rep, some of which are highly conserved. Hexameric ring helicases [83, 84] are more difficult to study and interpretation of single-molecule data would be more complex because of multiple labeling sites.

11.3.2
Single-molecule Techniques Applied to Helicase Studies

Many different single-molecule approaches have been developed to study helicase mechanisms (see Table 11.1 for comparison of their capabilities) [85]. The first two single-molecule studies were performed on the highly processive RecBCD via fluorescence imaging of intercalating dyes on DNA stretched in a microfluidic device [60] and via tethered particle assay [59] with relatively low resolution (100–300 bp). An optical tweezers apparatus was used to measure RecBCD-catalyzed DNA unwinding with 6-bp resolution [86] and NS3-catalzyed unwinding with 2-bp resolution [87]. Magnetic tweezers technique was used to detect UvrD-catalyzed DNA unwinding with 10–30-bp resolution [88]. Branch migration activity of RuvAB was measured via tethered particle assay [89] and magnetic tweezers [90, 91]. We have detected ssDNA translocation, dsDNA unwinding, and conformational changes by Rep with a few bp resolution using single-molecule FRET (smFRET) [37, 38, 48]. Techniques such as tethered particle assay and microfluidics are suitable for studying highly processive enzymes because of their limited spatial resolution. SmFRET and optical tweezers provide in principle the highest possible temporal (milliseconds) and spatial (a few bps) resolution. SmFRET is relatively insensitive to the mechanical noise since it reports on the internal motions within the center of the mass frame of the system under study. However, because the distance range that can be probed by FRET is between approximately 2 and 8 nm, smFRET cannot be used to measure the extended movements of a highly processive helicase. Other single molecule techniques do not have such limitation on the dynamic range but because they measure the motion in the laboratory frame; mechanical noises and thermal drift have to be

Table 11.1 Time and base-pair resolution of *published* single-molecule unwinding assays.

Technique	Base-pair resolution	Time resolution	Helicase studied
Optical trap	2–6 bp	20 ms	RecBCD, NS3
Magnetic tweezers	10~30 bp	0.4–1.0 s	UvrD, RuvAB
Tethered article assay	~100 bp	~0.5 s	RecBCD, RuvAB
Microfluidic assay	~200 bp	~0.3 s	RecBCD
Single-molecule FRET	<5 bp	15 ms	Rep

dealt with. Using the external force as an additional knob is very useful in understanding the mechanochemistry of motor proteins [92], but the functional parameters at the zero force limit can be obtained only through extrapolation because the resolution deteriorates when the force is reduced.

11.3.3
Different Types of smFRET Approaches to Probe Helicase Mechanisms

11.3.3.1 DNA–DNA FRET
In this study [48], we detected, in real time, DNA binding of a Rep protein and Rep-catalyzed DNA unwinding by measuring FRET between dyes attached to defined DNA locations. The DNA was a duplex with a 3′-ssDNA tail denoted as "3′-tailed DNA". The major findings are summarized below and shown diagrammatically in Figure 11.2.

(i) In the unwinding experiment, protein and ATP were simultaneously added to surface-immobilized 3′-tailed DNA labeled at the junction with a donor and an acceptor. FRET, initially near 100%, drops when unwinding initiates. Complete unwinding is seen as the total disappearance of fluorescence because the donor strand which is not directly tethered to the surface diffuses away quickly. The unwinding initiation rate reflects the assembly of the functional helicase unit and exactly matched the rates determined from bulk phase kinetic studies without dye labels or surface tethering [77].

(ii) In the binding experiment, the two dyes were attached to the extremities of the ssDNA tail. Remarkably, the binding of a Rep monomer in the presence of ATP was detected as a burst of FRET fluctuations and the binding saturated at 10 nM protein concentration. In contrast, the unwinding initiation rate was linearly dependent on protein concentration above 20 nM, strongly indicating that the binding of another Rep monomer to a monomer-occupied DNA is the rate-limiting step in unwinding initiation. That is, a Rep dimer is likely the active species for unwinding.

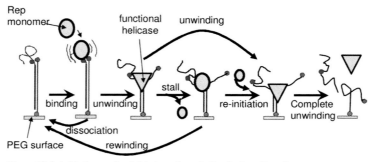

Figure 11.2 Initiation and re-initiation of unwinding by Rep. Based on extensive smFRET data from 3′-tailed DNA tethered to a PEG surface via biotin–streptavidin. Adapted from [48].

(iii) Unwinding of a 40-bp duplex often paused, resulting in either DNA rezipping (FRET recovery) or the restart of unwinding; the latter required free proteins in solution. We interpreted the results as unwinding stalls when a monomer dissociates from the functional dimer. DNA rezips if the remaining monomer also dissociates. If instead another monomer comes to the rescue, they can form a functional dimer and continue to unwind (Figure 11.2).

11.3.3.2 Protein–DNA FRET

In a subsequent study [37], we carried out site-directed mutagenesis to create single-cysteine Rep mutants that were fully functional both *in vivo*, tested through the phage replication assay [93], and *in vitro*, tested by both multiple turnover and single turnover unwinding with and without dye label. We showed that the PEG surface eliminated non-specific binding of labeled protein to the surface and we could detect the moment of single Rep protein binding to an immobilized 3′-tailed DNA as the sudden appearance of a fluorescence signal. By working with sub-nanomolar protein concentrations (K_D is about 5–10 nM [48]) and with over 90% labeling efficiency, single protein binding events were predominant and could be easily distinguished from rare multiple protein binding events which showed higher fluorescence signal and multi-step binding and dissociation (or photobleaching). We measured FRET from a donor on the protein to an acceptor on the DNA junction for each of the eight single cysteine mutants in the absence of ATP. Importantly, we obtained highly similar FRET efficiencies when the donor and the acceptor locations were swapped or when different donors with similar spectral properties were used. Our data confirmed the crystallographic binding orientation of ssDNA relative to Rep and showed that the protein obtains primarily the closed conformation when it is bound to a 3′-tailed DNA in solution, consistent with the PcrA structure bound to a 3′-tailed DNA [76]. This study validated the smFRET approach using the labeled protein and laid a firm foundation for functional studies such as DNA translocation [38] and unwinding powered by ATP hydrolysis.

11.3.3.3 Repetitive Shuttling

In a recent article [38], we showed that the 3′ to 5′ ssDNA translocation of a Rep monomer can be detected as gradual change in FRET between the protein and the DNA. ssDNA is highly flexible so FRET here is averaged over conformations of the intervening ssDNA. Surprisingly, when the protein encountered an insurmountable blockade, either the dsDNA that it cannot unwind as a monomer or a streptavidin attached to a 5′-biotin, it snapped back to near the 3′ end abruptly within a time resolution of 15 ms and repeated the gradual translocation followed by another snapback and so on. This was observed in the data as the "sawtooth pattern" of the smFRET time traces (Figure 11.3A). We termed this novel behavior "repetitive shuttling". Several lines of evidence strongly indicated that a single monomer was doing the repetitive shuttling, not multiple monomers in succession [38], and repetitive shuttling was observed over a wide range of salt concentrations and temperature, and also on DNA with mixed sequences. The repetition period was proportional to the tail length, insensitive to the DNA sequence, and was longer at lower ATP concentrations. We proposed its physical mechanism as blockage-induced

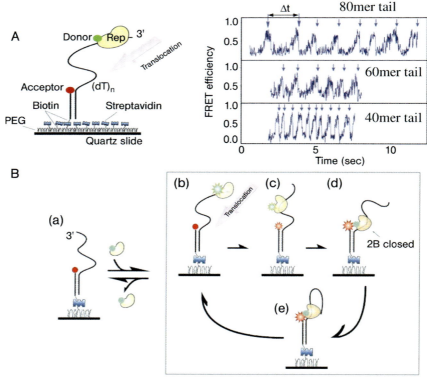

Figure 11.3 (A) Repetitive shuttling of a Rep monomer revealed by smFRET. (B) Physical mechanism of repetitive shuttling involving transient loop formation via binding to the 3′ end.

conformational change of the protein that enhances the affinity on its secondary binding site for a 3′ end, forming a transient ssDNA loop before snapback and repetition of translocation (Figure 11.3B). Supporting evidence includes: (1) periodic, transient FRET increase when the two extremities of the ssDNA are labeled with the FRET pair, (2) no repetitive shuttling when the 5′ end is free, and (3) evidence of 2B domain closing as the protein approaches the blockage . Repetitive shuttling was also observed on ssDNA bounded by a stalled replication fork and an Okazaki fragment analog, a DNA structure highly relevant to the function of Rep *in vivo* [94, 95], and it was shown that Rep can interfere with RecA filament formation on ssDNA. It is possible that one of the *in vivo* functions of Rep is to shuttle repetitively on ssDNA thereby keeping it free of unwanted proteins.

11.3.3.4 ssDNA Flexibility

In a separate study [13], we sought to understand the conformational properties of the primary substrate of helicases by performing a systematic FRET study of ssDNA of various lengths ((dT)$_N$ where N = 10–70) over a wide range of NaCl concentrations (0.025–2 M). FRET increased with decreasing length and with increasing salt. Using a

Monte Carlo simulation of the worm-like chain model we deduced that the persistent length of poly-dT ranges from 1.5 nm in 2 M NaCl to 3 nm in 25 mM NaCl. We were not able to observe any temporal changes in smFRET even with 1 ms time resolution and therefore concluded that ssDNA conformation is rapidly averaged over a time scale much faster than 1 ms. Thus, smFRET detected during ssDNA translocation and DNA unwinding must report on the temporally averaged values over the range of allowed ssDNA conformations.

11.4
SmFRET Studies of Holliday Junction

Homologous recombination is an essential process in maintaining genomic stability and its defects can lead to serious human diseases including cancer [96–99]. To cope with DNA damage encountered during genome duplication, a four-way (Holliday) junction (HJ) is formed by joining two nearly identical DNA molecules. Although conventional ensemble tools such as gel electrophoresis, fluorescence resonance energy transfer (FRET) [100–102], transient electric birefringence [103], chemical probing [104–106], X-ray crystallography [107–111], atomic force microscopy [112–114], small-angle X-ray scattering [115] and NMR [116–121] have provided crucial insights into the HJ lifecycle, many of its dynamic aspects remain poorly understood.

We have developed smFRET assays to study the structural dynamics of the HJ. Many of these properties are difficult to address unambiguously using conventional ensemble techniques. Our studies provide the firm baseline for future studies on the various enzymes that participate in homologous recombination and how their activities are modulated by the intrinsic properties of the HJ.

11.4.1
Structure and Function of HJ

The HJ is a four-way DNA junction in which four helices are connected by the covalent continuity of the four single strands (see Figure 11.4 for conventions used here). The HJ structure has been extensively studied using non-migratable (i.e. sequence prohibits branch migration) HJs built from oligonucleotides [122–124].

In the absence of divalent ions such as Mg^{2+}, the HJ is unfolded and takes on a so-called "open structure", where each of the four helices points to a corner of a square (Figure 11.4A) [101]. In physiologically relevant conditions, for instance with 1 mM Mg^{2+}, the HJ folds into the "antiparallel stacked-X structure" (Figure 11.4B and C) [100, 105]. A pair of helices is stacked against each other (that is, two helices form an essentially continuous helix, much like a regular B-DNA). For instance, helix X in Figure 11.4B can stack with helix R forming a "super-helix" XR while helices H and B can stack to form a super-helix HB. These super-helices can be considered as cylinders that are arranged in an X shape with an inter-helical angle of ~60° [125–127]. An alternative stacking conformer is also possible where helix X stacks on helix R and helix H stacks on helix B (Figure 11.4C).

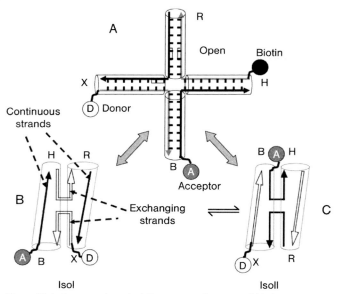

Figure 11.4 Open- and stacked-X structure of HJ. Two folded forms (isoI and isoII) are expected to interconvert via an unstacked intermediate resembling the unfolded open form. Attaching dyes and biotin to various arms can provide information on HJ conformations via FRET.

There are two alternative conformers [100, 105, 106] known as crossover isomers (isoI and isoII, for example). The relative populations of these conformers were determined using time-resolved FRET and NMR [121], and it was also shown that these conformers interchange with each other [128–130]. Equilibrium populations of the two conformers are determined by the junction sequence [105, 106, 121, 130] but not by solution conditions [14, 131]. Therefore, it has not been possible to study the stacking conformer transitions using bulk solution techniques which require synchronization. Below, we will show that conformer transitions can be detected from single HJ molecules and their rates can be determined under a variety of conditions. An important question remains as to whether a non-migratable HJ is a good model system since HJs *in vivo* would mostly consist of homologous sequences that allow spontaneous branch migration in which the branch point can hop forward and backward in a stochastic way. Our single-molecule approaches can measure structural properties even in a migratable junction, allowing us to tackle these issues for the first time.

11.4.2
Conformer Transitions of Non-migratable HJ

We assembled non-migratable HJs from four component strands of 22 nucleotides each such that the four helical arms (B, H, R and X, see Figure 11.4 for nomenclature convention) consist of 11 bp each. Three of the four helices were conjugated at the

5'-termini by various species: for instance, helix X by Cy3 (donor), helix B by Cy5 (acceptor), and helix H by biotin (Figure 11.4). In one stacking conformer (isoI), helix B stacks on helix H, minimizing the inter-fluorophore distance (high FRET), while in the other conformer (isoII), helix B stacks on helix X, maximizing the distance (low FRET). These assignments were independently verified by comparison with gel mobility assays [105]. Three different sequences named junctions 1, 3 and 7 were used. smFRET time records show two-state fluctuations between isoI and isoII (Figure 11.5A). The bias between the two conformers is a function of sequence but not of ionic conditions [14]. The conformer bias observed in smFRET experiments (isoI over isoII for junction 1 shown in Figure 11.5B for instance) exactly matched the bias deduced from the gel mobility assay of unlabeled DNA. In addition, switching the dye positions or biotin position to different arms did not change the transition rates [131]. Therefore surface-immobilization and dye-labeling do not significantly perturb the HJ dynamics.

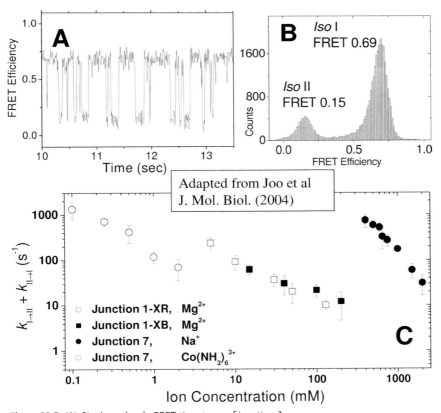

Figure 11.5 (A) Single-molecule FRET time trace of junction 1.
(B) FRET efficiency histogram from 25 junction-1 molecules.
(C) The sum of the two rates (isoI to isoII and isoII to isoI) for a variety of ionic conditions.

The rates of conformer transitions, $k_{I \rightarrow II}$ and $k_{II \rightarrow I}$, averaged over several hundred transitions, strongly depend on the type and concentration of metal ions. Figure 11.5C shows that $(k_{I \rightarrow II} + k_{II \rightarrow I})$ increases with de creasing metal ion concentration for Na^+, Mg^{2+} and hexaminecobalt [131]. These rates were obtained via cross-correlation analysis [31] as described in detail in the introduction to this revised application. Higher valence ions make the transitions slower at the same concentration. These observations suggest that the open structure, stable in the absence of metal ions, may be an intermediate in conformer transitions. Metal ions stabilize the stacked structures and HJ opening (or unfolding) necessary for the conformer transitions, and tend to be less frequent at higher cation concentrations. The characteristics of HJ conformer dynamics such as a relative bias between two conformers are essentially the same in Mg^{2+} concentrations ranging from sub-millimolar (physiological free Mg^{2+} level) to 50 mM, except for changes in the absolute rates.

11.4.2.1 Single-molecule Three-color FRET on HJ

Since regular smFRET cannot obtain information on more than one inter-dye distance at a time, we developed three-color FRET using the HJ as a model system [132]. First, an alternative acceptor (Cy5.5) was attached to the free arm of junction 1 (helix R in Figure 11.4) so that the donor would transfer energy alternatively to Cy5 or Cy5.5 upon conformer transitions (Figure 11.6A). With a judicious choice of fluorescent filters and careful correction for crosstalk between

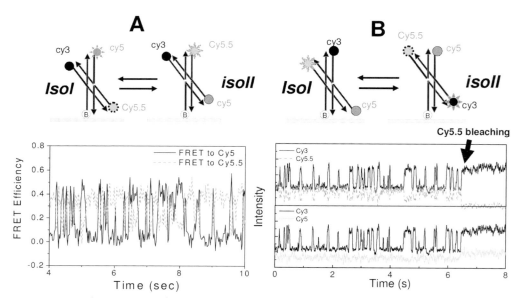

Figure 11.6 First demonstration of three-color FRET. Conformational transitions of Holliday junction were seen as alternating FRET to two distinct acceptors. Adapted from [132].

detection channels, we succeed in measuring Cy5 and Cy5.5 alternatively lighting up with identical transitions rates as measured via two-color FRET. These anti-correlated fluctuations of FRET to Cy5 and to Cy5.5 were fully synchronized within the 20-ms time resolution used, demonstrating that the Cy5 arm is moving toward the Cy3 arm at the same time as the Cy5.5 arm is moving away from the Cy3 arm. While this was not a surprising result, such a conclusion cannot be drawn from regular two-color FRET studies. In the second experiment, we moved the dye positions so that strong FRET to Cy5 occurs only if a parallel conformation is populated (Figure 11.6B). The data show anti-correlated signal fluctuations of Cy3 and Cy5.5 but no change in Cy5, proving that the parallel conformations are not even transiently populated within the 6-ms time resolution used [132].

11.4.3
Spontaneous Branch Migration Observed with a Single Step Resolution

In a more recent study [15], we modified the central base sequence of the non-migratable HJ, junction 7, to obtain a mono-migratable junction with two possible branch points (Figure 11.7A). In one branch point (**U**) all four arms are 11 bp in length. In the other branch point (**M**), the vertical arms are 12 bp long and the horizontal arms are 10 bp long. In the presence of Mg^{2+}, each branch point would have two stacking conformations; hence four different states are expected. This was indeed confirmed from smFRET traces (Figure 11.7B). The molecule shows slow two-state fluctuations interrupted by a brief period with much more rapid two-state fluctuations (zoom-in shown in Figure 11.7C). In comparison, we have never observed more than two FRET states from non-migratable junctions. In the rapid fluctuation phase E values for the higher FRET state are lower than those of the slow fluctuation phase (Figure 11.7D). Since simple geometric considerations suggest that the E in the high FRET states should be higher for the branch point **U**, we tentatively assigned the slowly fluctuating mode to **U** and the rapidly fluctuating mode to **M**. This assignment was confirmed by a hydroxyl radical cleavage experiment that located the dominant branch point. This result shows that we have the FRET resolution to detect even a single base-pair branch migration and that multiple conformer changes occur in each branch point before branch migration. The lifetime of each conformer decreased at lower Mg^{2+} concentrations with no change in relative bias. Similar behavior was observed from four other sequences and no parallel states were observed when dyes were moved to two opposing arms. Thus these mono-migratable HJs behave very much like non-migratable HJs when they are residing in one branch point, implying that the migratability is only a small perturbation in the dynamic structural properties of the HJ.

The average number of conformer transitions before a branch migration step remained constant as we varied Mg^{2+}. Consequently the lifetime of each branch point is directly proportional to the average lifetime of stacked conformers within that branch point, indicating that the stability of stacked conformations, determined by the DNA sequences, governs the branch migration kinetics. Similar observations

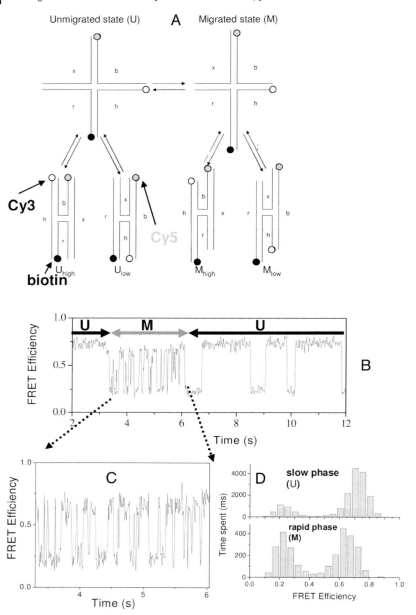

Figure 11.7 Mon-migratable HJ. (A) Each of the two brabch points (**U** and **M**) will fold into two conformations. For this particular sequence one branch point (**U**) is 20 times more populated than the other (**M**) due to sequence-dependent stability. (B) Single-molecule FRET trace shows two phases, a slowly fluctuating phase with a bias to the high FRET state (**U**) and a rapidly fluctuating phase without a bias (**M**, detail is shown in C). (C) 50 mM Mg²⁺ was used to slow down the reactions here but measurements at lower concentrations showed that only the absolute rates increased without any change in relative populations of branch points and conformations. (D) FRET histograms within slow and rapid phases show that two branch points have distinct FRET values for the high FRET conformations. Adapted from [15].

were made from five different DNA sequences implying that the result is not specific to a particular sequence. The bias between the two adjacent branch points can be as high as 40-fold, a surprising result considering the number of base pairs and their nature (GC versus AT etc.) do not change after branch migration. This suggests that local branch migration may be much faster than previously thought, and the ensemble estimate of the branch migration rate must be affected greatly by the presence of ultra-stable branch points. This also raises the possibility that spontaneous branch migration of a transiently protein-free HJ *in vivo* can be substantial so that these stable branch points are preferentially accessed by junction-resolving enzymes or other recombination machinery.

11.5
Outlook

I have summarized some of the key findings from our single-molecule FRET studies of helicase and Holliday junction. We were very lucky in that previous biophysical and biochemical studies, for example by our collaborators Tim Lohman and David Lilley, guided the design and interpretation of our single-molecule experiments to place our results on firm ground. Despite the extensive knowledge base previously available, our single-molecule data provided some remarkable surprises, for example that a helicase protein can repetitively shuttle on the same DNA segment. One immediate extension would be to study the interaction between the helicases and Holliday junctions. In fact, many recombination proteins that catalyze branch migration of HJ are classified as helicases and it will be very interesting to investigate how the unique properties of the HJ revealed by the single-molecule studies can influence the activities of enzyme that recognize and act on the HJ.

References

1 Weiss, S. (1999) *Science*, **283**, 1676–1683.

2 Weiss, S. (2000) *Nat. Struct. Biol.*, **7**, 724–729.

3 Forster, T. (ed.) (1965) *Modern Quantum Chemistry* (Ed. O. Sinanoglu), Academic, New York.

4 Stryer, L. and Haugland, R.P. (1967) *Proc. Natl. Acad. Sci. USA*, **58**, 719–726.

5 Selvin, P.R. (2000) *Nat. Struct. Biol.*, **7**, 730–734.

6 Deniz, A.A., Laurence, T.A., Dahan, M., Chemla, D.S., Schultz, P.G. and Weiss, S. (2001) *Annu. Rev. Phys. Chem.*, **52**, 233–253.

7 Ha, T. (2001) *Methods*, **25**, 78.

8 Ha, T. (2001) *Curr. Opin. Struct. Biol.*, **11**, 287–292.

9 Ha, T. (2004) *Biochemistry*, **43**, 4055–4063.

10 Ha, T., Enderle, T., Ogletree, D.F., Chemla, D.S., Selvin, P.R. and Weiss, S. (1996) *Proc. Natl Acad. Sci. USA*, **93**, 6264–6268.

11 Deniz, A.A., Dahan, M., Grunwell, J.R., Ha, T.J., Faulhaber, A.E., Chemla, D.S., Weiss, S. and Schultz, P.G. (1999) *Proc. Natl Acad. Sci. USA*, **96**, 3670–3675.

12 Lee, J.Y., Okumus, B., Kim, D.S. and Ha, T. (2005) *Proc. Natl Acad. Sci. USA*, **102**, 18938–18943.

13 Murphy, M.C., Rasnik, I., Cheng, W., Lohman, T.M. and Ha, T. (2004) *Biophys. J.*, **86**, 2530–2537.

14 McKinney, S.A., Declais, A.-C., Lilley, D.M.J. and Ha, T. (2003) *Nat. Struct. Biol.*, **10**, 93–97.

15 McKinney, S.A., Freeman, A.D., Lilley, D.M. and Ha, T. (2005) *Proc. Natl Acad. Sci. USA*, **102**, 5715–5720.

16 Buranachai, C., McKinney, S.A. and Ha, T. (2006) *Nano Lett.* **6**, 496–500.

17 Clamme, J.P. and Deniz, A.A. (2005) *Chemphyschem*, **6**, 74–77.

18 Laurence, T.A., Kong, X., Jager, M. and Weiss, S. (2005) *Proc. Natl Acad. Sci. USA*, **102**, 17348–17353.

19 Ha, T., Zhuang, X.W., Kim, H.D., Orr, J.W., Williamson, J.R. and Chu, S. (1999) *Proc. Natl Acad. Sci. USA*, **96**, 9077–9082.

20 Zhuang, X.W., Bartley, L.E., Babcock, H.P., Russell, R., Ha, T.J., Herschlag, D. and Chu, S. (2000) *Science*, **288**, 2048–2051.

21 Zhuang, X.W., Kim, H., Pereira, M.J.B., Babcock, H.P., Walter, N.G. and Chu, S. (2002) *Science*, **296**, 1473–1476.

22 Bokinsky, G., Rueda, D., Misra, V.K., Rhodes, M.M., Gordus, A., Babcock, H.P., Walter, N.G. and Zhuang, X. (2003) *Proc. Natl Acad. Sci. USA*, **100**, 9302–9307.

23 Rueda, D., Bokinsky, G., Rhodes, M.M., Rust, M.J., Zhuang, X. and Walter, N.G. (2004) *Proc. Natl Acad. Sci. USA*, **101**, 10066–10071.

24 Xie, Z., Srividya, N., Sosnick, T.R., Pan, T. and Scherer, N.F. (2004) *Proc. Natl Acad. Sci. USA*, **101**, 534–539.

25 Tan, E., Wilson, T.J., Nahas, M.K., Clegg, R.M., Lilley, D.M.J. and Ha, T. (2003) *Proc. Natl Acad. Sci. USA*, **100**, 9308.

26 Nahas, M.K., Wilson, T.J., Hohng, S., Lilley, D.M.J. and Ha, T. (2004) *Nat. Struct. Mol. Biol.*, **11**, 1107–1113.

27 Hohng, S., Wilson, T.J., Tan, E., Clegg, R.M., Lilley, D.M. and Ha, T. (2004) *J. Mol. Biol.*, **336**, 69–79.

28 Hodak, J.H., Fiore, J.L., Nesbitt, D.J., Downey, C.D. and Pardi, A. (2005) *Proc. Natl Acad. Sci. USA*, **102**, 10505–10510.

29 Brasselet, S., Peterman, E.J.G., Miyawaki, A. and Moerner, W.E. (2000) *J. Phys. Chem. B*, **104**, 3676–3682.

30 Pljevaljcic, G., Millar, D.P. and Deniz, A.A. (2004) *Biophys. J.*, **87**, 457–467.

31 Kim, H.D., Nienhaus, G.U., Ha, T., Orr, J.W., Williamson, J.R. and Chu, S. (2002) *Proc. Natl Acad. Sci. USA*, **99**, 4284–4289.

32 Deniz, A.A., Laurence, T.A., Beligere, G.S., Dahan, M., Martin, A.B., Chemla, D.S., Dawson, P.E., Schultz, P.G. and Weiss, S. (2000) *Proc. Natl Acad. Sci. USA*, **97**, 5179–5184.

33 Lipman, E.A., Schuler, B., Bakajin, O. and Eaton, W.A. (2003) *Science*, **301**, 1233–1235.

34 Schuler, B., Lipman, E.A. and Eaton, W.A. (2002) *Nature*, **419**, 743–747.

35 Diez, M., Zimmermann, B., Borsch, M., Konig, M., Schweinberger, E., Steigmiller, S., Reuter, R., Felekyan, S., Kudryavtsev, V., Seidel, C.A. and Graber, P. (2004) *Nat. Struct. Mol. Biol.*, **11**, 135–141.

36 Zimmermann, B., Diez, M., Zarrabi, N., Graber, P. and Borsch, M. (2005) *EMBO J.* **24**, 2053–2063.

37 Rasnik, I., Myong, S., Cheng, W., Lohman, T.M. and Ha, T. (2004) *J. Mol. Biol.*, **336**, 395–498.

38 Myong, S., Rasnik, I., Joo, C., Lohman, T.M. and Ha, T. (2005) *Nature*, **437**, 1321–1325.

39 Joo, C., McKinney, S.A., Nakamura, M., Rasnik, I., Myong, S. and Ha, T. (2006) *Cell*, **126**, 515–527.

40 Kapanidis, A.N., Margeat, E., Laurence, T.A., Doose, S., Ho, S.O., Mukhopadhyay, J., Kortkhonjia, E., Mekler, V., Ebright, R.H. and Weiss, S. (2005) *Mol. Cell*, **20**, 347–356.

41 Blanchard, S.C., Gonzalez, R.L., Kim, H.D., Chu, S. and Puglisi, J.D. (2004) *Nat. Struct. Mol. Biol.*, **11**, 1008–1014.

42 Blanchard, S.C., Kim, H.D., Gonzalez, R.L., Jr., Puglisi, J.D. and Chu, S. (2004) *Proc. Natl Acad. Sci. USA*, **101**, 12893–12898.

43 Ha, T.J., Ting, A.Y., Liang, J., Caldwell, W.B., Deniz, A.A., Chemla, D.S., Schultz,

P.G. and Weiss, S. (1999) *Proc. Natl Acad. Sci. USA*, **96**, 893–898.

44 Cognet, L., Harms, G.S., Blab, G.A., Lommerse, P.H.M. and Schmidt, T. (2000) *Appl. Phys. Lett.*, **77**, 4052–4054.

45 Rasnik, I., McKinney, S.A. and Ha, T. (2005) *Acc. Chem. Res.*, **38**, 542–548.

46 Sofia, S.J., Premnath, V. and Merrill, E.W. (1998) *Macromolecules*, **31**, 5059.

47 Prime, K.L. and Whitesides, G.M. (1991) *Science*, **252**, 1164.

48 Ha, T., Rasnik, I., Cheng, W., Babcock, H.P., Gauss, G., Lohman, T.M. and Chu, S. (2002) *Nature*, **419**, 638–641.

49 Kuzmenkina, E.V., Heyes, C.D. and Nienhaus, G.U. (2005) *Proc. Natl Acad. Sci. USA*, **102**, 15471–15476.

50 Kuzmenkina, E.V., Heyes, C.D. and Nienhaus, G.U. (2006) *J. Mol. Biol.*, **357**, 313–324.

51 Margeat, E., Kapanidis, A.N., Tinnefeld, P., Wang, Y., Mukhopadhyay, J., Ebright, R.H. and Weiss, S. (2006) *Biophys. J.*, **90**, 1419–1431.

52 Blainey, P.C., van Oijen, A.M., Banerjee, A., Verdine, G.L. and Xie, X.S. (2006) *Proc. Natl Acad. Sci. USA*, **103**, 5752–5757.

53 van Oijen, A.M., Blainey, P.C., Crampton, D.J., Richardson, C.C., Ellenberger, T. and Xie, X.S. (2003) *Science*, **301**, 1235–1238.

54 Lee, J.B., Hite, R.K., Hamdan, S.M., Xie, X.S., Richardson, C.C. and van Oijen, A.M. (2006) *Nature*, **439**, 621–624.

55 Cosa, G., Zeng, Y., Liu, H.W., Landes, C.F., Makarov, D.E., Musier-Forsyth, K. and Barbara, P.F. (2006) *J. Phys. Chem. B Condens. Matter Mater. Surf. Interfaces, Biophys.*, **110**, 2419–2426.

56 Liu, H.W., Cosa, G., Landes, C.F., Zeng, Y., Kovaleski, B.J., Mullen, D.G., Barany, G., Musier-Forsyth, K. and Barbara, P.F. (2005) *Biophys. J.*, **89**, 3470–3479.

57 Roy, R., Kozlov, A.G., Lohman, T.M. and Ha, T. (in preparation).

58 Okumus, B., Wilson, T.J., Lilley, D.M.J. and Ha, T. (2004) *Biophys. J.*, **87**, 2798–2806.

59 Dohoney, K.M. and Gelles, J. (2001) *Nature*, **409**, 370–374.

60 Bianco, P.R., Brewer, L.R., Corzett, M., Balhhorn, R., Yeh, Y., Kowalczykowski, S.C. and Baskin, R.J. (2001) *Nature*, **409**, 374–378.

61 Abdel-Monem, M., Durwald, H. and Hoffmann-Berling, H. (1976) *Eur. J. Biochem.*, **65**, 441–449.

62 Abdel-Monem, M. and Hoffmann-Berling, H. (1976) *Eur. J. Biochem.*, **65**, 431–440.

63 Lohman, T.M. and Bjornson, K.P. (1996) *Annu. Rev. Biochem.*, **65**, 169–214.

64 Hall, M.C. and Matson, S.W. (1999) *Mol. Microbiol.*, **34**, 867–877.

65 Patel, S.S. and Picha, K.M. (2000) *Annu. Rev. Biochem.*, **69**, 651–697.

66 von Hippel, P.H. and Delagouette, E. (2001) *Cell*, **104**, 177–190.

67 Eoff, R.L. and Raney, K.D. (2005) *Biochem. Soc. Trans.*, **33**, 1474–1478.

68 Cordin, O., Banroques, J., Tanner, N.K. and Linder, P. (2006) *Gene*, **367**, 17–37.

69 Sung, P., Bailly, V., Weber, C., Thompson, L.H., Prakash, L. and Prakash, S. (1993) *Nature*, **365**, 852–855.

70 Ellis, N.A., Groden, J., Ye, T.Z., Straughen, J., Lennon, D.J., Ciocci, S., Proytcheva, M. and German, J. (1995) *Cell*, **83**, 655–666.

71 Gray, M.D., Shen, J.C., Kamath-Loeb, A.S., Blank, A., Sopher, B.L., Martin, G.M., Oshima, J. and Loeb, L.A. (1997) *Nat. Genet.*, **17**, 100–103.

72 Kitao, S., Shimamoto, A., Goto, M., Miller, R.W., Smithson, W.A., Lindor, N.M. and Furuichi, Y. (1999) *Nat. Genet.*, **22**, 82–84.

73 Hickson, I.D. (2003) *Nat. Rev. Cancer*, **3**, 169–178.

74 von Hippel, P.H. (2004) *Nat. Struct. Mol. Biol.*, **11**, 494–496.

75 Korolev, S., Hsieh, J., Gauss, G.H., Lohman, T.M. and Waksman, G. (1997) *Cell*, **90**, 635–647.

76 Velankar, S.S., Soultanas, P., Dillingham, M.S., Subramanya, H.S. and Wigley, D.B. (1999) *Cell*, **97**, 75–84.

77 Cheng, W., Hsieh, J., Brendza, K.M. and Lohman, T.M. (2001) *J. Mol. Biol.*, **310**, 327–350.

78 Cheng, W., Brendza, K.M., Gauss, G.H., Korolev, S., Waksman, G. and Lohman, T.M. (2002) *Proc. Natl Acad. Sci. USA*, **99**, 16006–16011.

79 Brendza, K.M., Cheng, W., Fischer, C.J., Chesnik, M.A., Niedziela-Majka, A. and Lohman, T.M. (2005) *Proc. Natl Acad. Sci. USA*, **102**, 10076–10081.

80 Kim, J.L., Morgenstern, K.A., Griffith, J.P., Dwyer, M.D., Thomson, J.A., Murcko, M.A., Lin, C. and Caron, P.R. (1998) *Structure*, **6**, 89–100.

81 Bernstein, D.A., Zittel, M.C. and Keck, J.L. (2003) *EMBO J.*, **22**, 4910–4921.

82 Singleton, M.R., Dillingham, M.S., Gaudier, M., Kowalczykowski, S.C. and Wigley, D.B. (2004) *Nature*, **432**, 187–193.

83 Singleton, M.R., Sawaya, M.R., Ellenberger, T. and Wigley, D.B. (2000) *Cell*, **101**, 589–600.

84 Li, D., Zhao, R., Lilyestrom, W., Gai, D., Zhang, R., DeCaprio, J.A., Fanning, E., Jochimiak, A., Szakonyi, G. and Chen, X.S. (2003) *Nature*, **423**, 512–518.

85 Rasnik, I., Myong, S. and Ha, T. (2006) *Nucleic Acids Res.*

86 Perkins, T.T., Li, H.W., Dalal, R.V., Gelles, J. and Block, S.M. (2004) *Biophys. J.*, **86**, 1640–1648.

87 Dumont, S., Cheng, W., Serebrov, V., Beran, R.K., Tinoco, I., Jr., Pyle, A.M. and Bustamante, C. (2006) *Nature*, **439**, 105–108.

88 Dessinges, M.N., Lionnet, T., Xi, X.G., Bensimon, D. and Croquette, V. (2004) *Proc. Natl Acad. Sci. USA*, **101**, 6439–6444.

89 Dennis, C., Fedorov, A., Kas, E., Salome, L. and Grigoriev, M. (2004) *EMBO J.*, **23**, 2413–2422.

90 Dawid, A., Croquette, V., Grigoriev, M. and Heslot, F. (2004) *Proc. Natl Acad. Sci. USA*, **101**, 11611–11616.

91 Amit, R., Gileadi, O. and Stavans, J. (2004) *Proc. Natl Acad. Sci. USA*, **101**, 11605–11610.

92 Bustamante, C., Chemla, Y.R., Forde, N.R. and Izhaky, D. (2004) *Annu. Rev. Biochem.*, **73**, 705–748.

93 Scott, J.F., Eisenberg, S., Bertsch, L.L. and Kornberg, A. (1977) *Proc. Natl Acad. Sci. USA*, **74**, 193–197.

94 Sandler, S.J. (2000) *Genetics*, **155**, 487–497.

95 Marians, K.J. (2004) *Phil. Trans. R. Soc. Lond. B Biol. Sci.*, **359**, 71–77.

96 Holliday, R. (1964) *Genet. Res.*, **5**, 282–304.

97 Kowalczykowski, S.C. (2000) *Trends Biochem. Sci.*, **25**, 156–165.

98 Lilley, D.M.J. (2000) *Quart. Rev. Biophys.*, **33**, 109–159.

99 Lilley, D.M.J. and White, M.F. (2001) *Nat. Rev. Mol. Cell Biol.*, **2**, 433–443.

100 Murchie, A.I., Clegg, R.M., von Kitzing, E., Duckett, D.R., Diekmann, S. and Lilley, D.M. (1989) *Nature*, **341**, 763–766.

101 Clegg, R.M., Murchie, A.I.H. and Lilley, D.M.J. (1994) *Biophys. J.*, **66**, 99–109.

102 Clegg, R.M., Murchie, A.I., Zechel, A., Carlberg, C., Diekmann, S. and Lilley, D.M. (1992) *Biochemistry*, **31**, 4846–4856.

103 Cooper, J.P. and Hagerman, P.J. (1989) *Proc. Natl Acad. Sci. USA*, **86**, 7336–7340.

104 Churchill, M.E., Tullius, T.D., Kallenbach, N.R. and Seeman, N.C. (1988) *Proc. Natl Acad. Sci. USA*, **85**, 4653–4656.

105 Duckett, D.R., Murchie, A.I., Diekmann, S., von Kitzing, E., Kemper, B. and Lilley, D.M. (1988) *Cell*, **55**, 79–89.

106 Chen, J.H., Churchill, M.E., Tullius, T.D., Kallenbach, N.R. and Seeman, N.C. (1988) *Biochemistry*, **27**, 6032–6038.

107 Nowakowski, J., Shim, P.J., Stout, C.D. and Joyce, G.F. (2000) *J. Mol. Biol.*, **300**, 93–102.

108 Nowakowski, J., Shim, P.J., Prasad, G.S., Stout, C.D. and Joyce, G.F. (1999) *Nat. Struct. Biol.*, **6**, 151–156.

109 Eichman, B.F., Vargason, J.M., Mooers, B.H.M. and Ho, P.S. (2000) *Proc. Natl Acad. Sci. USA*, **97**, 3971–3976.

110 Ortiz-Lombardia, M., Gonzalez, A., Eritja, R., Aymami, J., Azorin, F. and Coll, M. (1999) *Nat. Struct. Biol.*, **6**, 913–917.

111 Eichman, B.F., Mooers, B.H.M., Alberti, M., Hearst, J.E. and Ho, P.S. (2001) *J. Mol. Biol.*, **308**, 15–26.

112 Shlyakhtenko, L.S., Potaman, V.N., Sinden, R.R. and Lyubchenko, Y.L. (1998) *J. Mol. Biol.*, **280**, 61–72.

113 Lushnikov, A.Y., Bogdanov, A. and Lyubchenko, Y.L. (2003) *J. Biol. Chem.*, **278**, 43130–43134.

114 Sha, R.J., Liu, F.R. and Seeman, N.C. (2002) *Biochemistry*, **41**, 5950–5955.

115 Nollmann, M., Stark, W.M. and Byron, O. (2004) *Biophys. J.*, **86**, 3060–3069.

116 Pikkemaat, J.A., Overmars, F.J.J., Dreeftromp, C.M., Vandenelst, H., Vanboom, J.H. and Altona, C. (1996) *J. Mol. Biol.*, **262**, 349–357.

117 Overmars, F.J.J. and Altona, C. (1997) *J. Mol. Biol.*, **273**, 519–524.

118 Overmars, F.J.J., Lanzotti, V., Galeone, A., Pepe, A., Mayol, L., Pikkemaat, J.A. and Altona, C. (1997) *Eur. J. Biochem.*, **249**, 576–583.

119 Carlstrom, G. and Chazin, W.J. (1996) *Biochemistry*, **35**, 3534–3544.

120 Chen, S.M. and Chazin, W.J. (1994) *Biochemistry*, **33**, 11453–11459.

121 Miick, S.M., Fee, R.S., Millar, D.P. and Chazin, W.J. (1997) *Proc. Natl Acad. Sci. USA*, **94**, 9080–9084.

122 Seeman, N.C. (1982) *J. Theor. Biol.*, **99**, 237–247.

123 Kallenbach, N.R., Ma, R.I. and Seeman, N.C. (1983) *Nature*, **305**, 829–831.

124 Hsu, P.L. and Landy, A. (1984) *Nature*, **311**.

125 von Kitzing, E., Lilley, D.M.J. and Diekmann, S. (1990) *Nucleic Acids Res.* **18**, 2671–2683.

126 Torbet, J. (1992) *Methods Enzymol.*, **211**, 518–532.

127 Mao, C.D., Sun, W.Q. and Seeman, N.C. (1999) *J. Am. Chem. Soc.*, **121**, 5437–5443.

128 Murchie, A.I., Portugal, J. and Lilley, D.M. (1991) *EMBO J.*, **10**, 731–738.

129 Li, X.J., Wang, H. and Seeman, N.C. (1997) *Biochemistry*, **36**, 4240–4247.

130 Grainger, R.J., Murchie, A.I.H. and Lilley, D.M.J. (1998) *Biochemistry*, **37**, 23–32.

131 Joo, C., McKinney, S.A., Lilley, D.M.J. and Ha, T. (2004) *J. Mol. Biol.*, **341**, 739–751.

132 Hohng, S., Joo, C. and Ha, T. (2004) *Biophys. J.*, **87**, 1328–1337.

12

High-speed Atomic Force Microscopy for Nano-visualization of Biomolecular Processes

Toshio Ando, Takayuki Uchihashi, Noriyuki Kodera, Daisuke Yamamoto,
Masaaki Taniguchi, Atsushi Miyagi, and Hayato Yamashita

12.1
Introduction

Single-molecule observations by fluorescence microscopy provide us with translational or rotational information about individual fluorescent spots emitted from fluorophores attached to molecules. Such information is valuable for dissecting the dynamic behavior of the labeled biomolecules at work. However, it must be inferred from the observed fluorescent spots how the labeled molecules are in fact behaving. In order to understand biomolecular behavior, we should directly observe single molecules at nanometer spatial and millisecond temporal resolution.

The atomic force microscope (AFM) [1] made it possible to observe the nanometer-scale world in liquids. Although it can visualize the structure of unstained biomolecules under physiological solution conditions, it takes minutes to get an image, far too slow to observe dynamic biomolecular processes. This slow imaging rate is due to the fact that AFM employs mechanical scanning to detect the sample height at each pixel. It is quite difficult to quickly move a mechanical device of macroscopic size with sub-nanometer accuracy without producing unwanted vibrations. Various efforts carried out in the past decade have improved the imaging rate of AFM (e.g. [2–8]). Current high-speed AFM can capture images on video at ~30 frames/s with a scan range of ~250 nm and scan lines of ~100, without significantly disturbing weak biomolecular interactions [9]. Recent studies have demonstrated that this new microscope can reveal biomolecular processes such as myosin V walking along actin tracks (N. Kodera *et al.*, unpublished data) and association/dissociation dynamics of chaperonin GroEL-GroES that occurs in a negatively cooperative manner (D. Yamamoto *et al.*, unpublished data). These studies clearly indicate that high-speed AFM has great potential to reveal how and what structural changes in individual molecules occur when exercising their physiological functions. However, high-speed AFM technology is still immature. The imaging rate needs to

Single Molecule Dynamics in Life Science. Edited by T. Yanagida and Y. Ishii
Copyright © 2009 WILEY-VCH Verlag GmbH & Co. KGaA, Weinheim
ISBN: 978-3-527-31288-7

be increased further and the tip-sample interaction force has to be reduced in order to expand the scope of biomolecular processes to which high-speed AFM can be applied.

Because high-speed AFM is rather new, its technical issues are not commonly known while only a few successful imaging experiments have been performed. Therefore, in this chapter, we focus mainly on the fundamental issues of high-speed AFM and the various techniques involved, which will, we hope, facilitate the entry of young researchers to this new field. In addition, some descriptions on substratum preparations required for imaging biomolecular processes are given. Examples of image data are also presented (video images can be seen at http://www.s.kanazawa-u.ac.jp/phys/biophys/roadmap.htm).

12.2
AFM Set-up and Operation

A general AFM set-up is given in Figure 12.1. AFM forms an image by touching the sample surface with a sharp tip attached to the free end of a soft cantilever beam while the sample stage is being scanned horizontally in two dimensions. This contact causes the cantilever to deflect. The deflection is often detected by an optical lever method where a collimated laser beam is focused onto the cantilever and reflected back into closely spaced photodiodes whose photocurrents are fed into a differential amplifier. The output of the differential amplifier is proportional to the cantilever deflection. During raster scan of the sample stage, the stage is moved in the z-direction to maintain a constant deflection of the cantilever (hence, a constant tip-sample

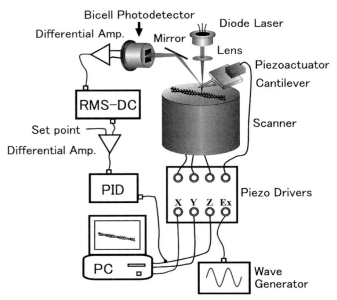

Figure 12.1 Schematic of tapping mode AFM set-up.

interaction force), which is performed by feedback operation. The resulting three-dimensional movement of the sample stage traces the sample surface. A computer constructs the topography image, usually from electric signals that are used to drive the sample stage scanner in the z-direction. In the operation mode described above, the cantilever tip is always in contact with the sample, exerting a relatively large, undesirable lateral force on the sample due to a large cantilever spring constant in the lateral direction. To avoid this problem, tapping-mode AFM was invented [10] in which the cantilever is oscillated in the z-direction at (or near) its resonant frequency. The oscillation amplitude is reduced by a repulsive interaction between the tip and sample. The amplitude signal is generated usually by an rms-to-dc converter and is maintained at a constant level (set point) by feedback operation.

12.3
Imaging Rate and Feedback Bandwidth

The frame acquisition time (T) is expressed by

$$T = 2NL/V_s \tag{12.1}$$

where L is the scan size, N the number of scan lines, and V_s the scan speed of the sample stage in the x-direction. The maximum scan speed is limited by feedback bandwidth. Suppose that the sample has a sinusoidal shape with a periodicity λ. Then, the sample stage is moved in the z-direction at a feedback frequency $f(= V_s/\lambda)$ to maintain the cantilever's oscillation amplitude. Due to the "chasing after" nature of the feedback operation, there is always a time delay in the feedback loop. Feedback bandwidth (f_B) is usually defined by the feedback frequency at which the sample topography is traced with a 45° phase delay. For a feedback bandwidth f_B, the shortest possible frame acquisition time, T_m, is approximated by

$$T_m = 2NL/(\lambda f_B) \tag{12.2}$$

Of course, this is just a rough estimate as the very fragile samples are disrupted by extra tip-sample interactions resulting from the 45° phase delay.

The 45° phase delay corresponds to a closed-loop time delay $\Delta\tau = 1/(8f)$. In the feedback loop, there are various devices with finite time delays (see Figure 12.1). In tapping mode AFM, the delay mainly consists of the time (τ_a) required to measure the cantilever's oscillation amplitude, the response time (τ_c) of the cantilever, the response time (τ_s) of the z-scanner, the integral time (τ_I) of error signals in the feedback controller, and the parachuting time (τ_p). Here, "parachuting" means that the cantilever tip detaches completely from the sample surface at a steep down-hill region of the sample and cannot quickly land on the surface again. The minimum τ_a is given by $1/(2f_c)$, where f_c is the cantilever's fundamental resonant frequency. Cantilevers and the z-scanner are second-order resonant systems. Therefore τ_c and τ_s are expressed by $Q_c/(\pi f_c)$ and $Q_s/(\pi f_s)$, respectively, where f_s is the resonant frequency of the z-scanner, and Q_c and Q_s are respectively the quality factors of the cantilever and the z-scanner. τ_I and τ_p are functions of various parameters, of which approximate analytical expressions will be given later. From these quantities, and from a fact that the closed-loop phase delay is approximately twice the open-loop phase delay feedback bandwidth, f_B, is approximately expressed by

$$f_B = \frac{\alpha f_c}{8} \Big/ \left[1 + \frac{2Q_c}{\pi} + \frac{2f_c Q_s}{\pi f_s} + 2f_c(\tau_I + \tau_p) \right] \qquad (12.3)$$

where α is a factor by which the derivative operation contained in feedback control compensates for a phase delay in the feedback loop. According to our experience, $\alpha = \sim2.8$. From Equation 12.2 and the relationship $\Delta\tau = \alpha/(8f)$ (here, the compensation effect is taken into consideration), we can estimate the feedback bandwidth and the minimum time delay for a given imaging condition. For video-rate imaging (30 frames/s) with $L = 200$ nm, $N = 100$, and $\lambda = 10$ nm, we require $f_B = 121$ kHz and $\Delta\tau = 2.89\,\mu$s.

12.4
Feedback Operation and Parachuting

To maintain the amplitude of an oscillating cantilever at a constant level while the sample stage is being raster-scanned in the xy-directions, the detected amplitude is compared with the set point amplitude. Their difference (error signal) is fed to a proportional-integral-derivative (PID) feedback controller. The PID output is fed to a voltage amplifier to drive the z-piezcactuator. This is repeated until the error signal is minimized. To reduce the tapping force exerted by the oscillating tip onto the sample, the set point amplitude should be set close to the cantilever's free oscillation amplitude. However, under this condition, the tip tends to detach completely from the sample surface, especially at a steep down-hill region of the sample. Once detached, the error signal is constant (i.e. saturated at a small level) irrespective of how far the tip is separated from the sample surface. The gain parameters of the PID controller can be increased to reduce the parachuting time. However, such large gains in turn produce an overshoot in an up-hill region of the sample, which promotes parachuting around the top region of the sample and leads to instability in the feedback operation.

As seen above, parachuting is problematic, especially for high-speed bioAFM in which the tapping force has to be minimized. During parachuting, the information of sample topography is completely lost. The parachuting time is a function of various parameters such as the maximum sample-height, h_0, the peak-to-peak free oscillation amplitude, $2A_0$, of the cantilever, the amplitude set point, r, the cantilever's resonant frequency and the phase delay, φ, in the feedback operation.

Let us find a condition under which parachuting occurs. When the phase of the feedback operation is delayed by φ, a cantilever tip senses the "residual sample topography" $\Delta S(t)$ as a function of time. $\Delta S(t)$ is expressed as

$$\Delta S(t) = \frac{h_0}{2} [\sin(2\pi ft) - \sin(2\pi ft - \varphi)] = h_0 \sin\frac{\varphi}{2} \cos\left(2\pi ft - \frac{\varphi}{2}\right) \qquad (12.4)$$

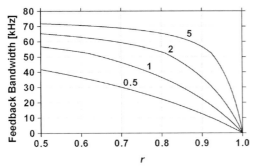

Figure 12.2 Dependence of the amplitude set point on the feedback bandwidth of tapping mode AFM. The number attached to each line is the ratio of the cantilever's peak-to-peak oscillation amplitude to the sample height. This dependence is theoretically derived.

Parachuting occurs when the maximum height of the residual topography, $h_0 \sin (\varphi/2)$, is larger than the difference between the cantilever's peak-to-peak free oscillation amplitude and the set point amplitude, i.e. $h_0 \sin(\varphi/2) > 2A_0(1 - r)$. An analytical expression for the parachuting time has been derived [6]:

$$\tau_p \approx [(\tan\beta)/\beta - 1]/f_c \qquad (12.5)$$

where β is $\cos^{-1}[2A_0(1 - r)/3h_0\sin(\varphi/2)]$, and the feedback gain is set to a level at which an error signal corresponding to the separation distance, $2A_0(1 - r)$, diminishes roughly in a single period of the cantilever oscillation. For example, for $h = 4A_0$, $r = 0.9$, $\varphi = \pi/4$, and $f_c = 1\,\text{MHz}$, τ_p is estimated to be $1.21\,\mu\text{s}$, which significantly reduces the feedback bandwidth.

The main component of PID control is integral. It is difficult to estimate theoretically the integral time constant (τ_I) with which the optimum feedback control is attained. We obtained experimentally $\tau_I = 4h_0 \sin (\pi/8)/A_0 f_c$. The feedback bandwidth as a function of the set point, r, and $2A_0/h_0$ is shown in Figure 12.2.

12.5
Key Devices for High-Speed AFM

Equations 12.3 and 12.4 give us a quantitative guideline for developing a high-speed tapping mode AFM. In order to achieve high-speed AFM, we need the following devices and techniques: small cantilevers with a high-resonant frequency and a small spring constant, a fast amplitude detector, a sample stage scanner with high resonant frequencies, active damping techniques for reducing the quality factor of the scanner, and a technique to minimize the parachuting time. In addition, we have to optimize the optical lever method for detecting deflection of a small cantilever.

12.5.1
Small Cantilevers and Related Devices

The small cantilevers that have most recently been developed by Olympus are manufactured from silicon nitride and have dimensions of 6 μm long, 2 μm wide, and 90 nm thick. The resonant frequency is 3.5 MHz in air and 1.2 MHz in water, the spring constant is about 0.2 N/m, and the quality factor is 2 ∼ 3 in water. The tip has a beak-like shape. The apex radius, ∼17 nm, is not small enough to produce a high-resolution image. We usually use electron-beam-deposition (EBD) to form a sharp tip extending from the original tip. Small pieces of naphthalene crystals are placed into a small container, the top of which is perforated with holes of ∼0.1 mm in diameter. The container is placed in the scanning electron microscope chamber and cantilevers are placed just above the container's holes. The spot mode electron beam irradiates the cantilever tip to produce a needle on the tip with a growth rate of ∼50 nm/s. The EBD tip is further sharpened by plasma etching in argon gas, which reduces the apex radius to ∼4 nm. Sometimes it is heated further at ∼600° in a vacuum to strengthen the EBD tip. We use the optical lever system developed in 2001 [2] for detecting deflection of a small cantilever. The previous photodiode was replaced with a four-segmented Si PIN photodiode (3 pF, 40 MHz). The photodiode amplifier has a bandwidth of 20 MHz. We also use the fast amplitude detector developed in 2001 [2]. The peak and bottom voltages of the input sinusoidal signals are held by a sample/hold circuit for every half-oscillation cycle. Their difference is output as an amplitude signal. The sample/hold timing signals are usually derived from the input signals themselves. However, external signals that are synchronized with the cantilever excitation signals can be used, which is sometimes useful for maximizing the detection sensitivity of the tip-sample interaction. Since the amplitude signals are produced at every half-cycle of oscillation, the amplitude read time is 0.42 μs. The cantilever's response time is about 0.66 μs in water.

12.5.2
Scanner

The scanner is the device that is most difficult to optimize for high-speed scanning. High-speed scans of mechanical devices with macroscopic dimensions tend to produce unwanted vibrations. Three techniques are required to minimize unwanted vibrations: (a) a technique to suppress the impulsive forces that are produced by quick displacement of the actuators, (b) a technique to increase the resonant frequencies, and (c) an active damping technique to reduce the quality factor .

The first issue was solved by a counterbalancing technique [2, 3]. For example, for a z-scanner that moves at much higher frequencies than the x- and y-scanners, two identical piezoactuators are placed at a supporting base in the opposite direction and displaced simultaneously with the same length. An alternative method is to support a piezoactuator at both ends with flexures. This method was recently and effectively applied to the x- and z-scanners (unpublished data).

The resonant frequency of a piezoactuator is determined almost solely by its maximum displacement (in other words, by its length). However, it can be effectively extended by an inverse compensation method as described later. The structural resonant frequency is enhanced by the use of a compact structure and a material that has a large Young's modulus to density ratio. However, a compact structure tends to produce interferences between the three-scan axes. A ball-guide stage [2] is one choice for avoiding such interferences. An alternative method is to use flexures (blade springs) that are flexible enough to be displaced but rigid enough in the directions perpendicular to the displacement axis [11, 12]. It should be noted that the scanner mechanics, except for piezoacutators, has to be produced by monolithic processing in order to minimize the number of resonant elements.

Active damping of the x- and y-scanners is easy as their scan speed is not high and their scan waves are known beforehand. Therefore, feedforward control for active damping can be implemented in a digital mode. The Fourier transform of the x-scanner displacement in an isosceles triangle, $X(t)$, with a periodicity of T_x is given by

$$F_x(\omega) = 2\pi X_0 \left[\frac{1}{2}\delta(\omega) - \frac{2}{\pi^2} \sum_{k=-\infty}^{+\infty} \frac{1}{k^2} \delta(\omega - k\omega_0) \right] \quad (k : \text{odd}), \tag{12.6}$$

where X_0 is the maximum displacement and $\omega_0 = 2\pi/T_x$. Suppose that the transfer function, $G_x(i\omega)$, of the x-scanner is experimentally measured, then, the inverse Fourier transform of $F_x(\omega)/G_x(i\omega)$ gives the driving signal, $X'(t)$, to move the x-scanner exactly in $X(t)$. However, in practice, we need only the first ~15 terms of $F_x(\omega)$. Thus, the driving signal is expressed as

$$X'(t) = \frac{X_0}{2G_x(0)} - \frac{4X_0}{\pi^2} \sum_{k=1}^{29} \frac{1}{k^2} \frac{1}{|G_x(ik\omega_0)|} \cos[k\omega_0 t - \Phi(k\omega_0)] \quad (k : \text{odd}),$$

$$\tag{12.7}$$

where $\Phi(\omega)$ is the phase of the transfer function $G_x(i\omega)$. This feedforward damping method works very well.

The above feedforward control method used to dampen the x-scanner cannot be applied to the z-scanner since its scan waves are unpredictable. The active Q-control is well known as an active damping technique and has been often used to control the quality factor of cantilevers [13–16]. When this control is applied to the z-scanner, its displacements have to be detected. However, it is difficult to do so. Kodera *et al.* [5] developed a new method in which instead of detecting the displacements, output signals from an electric circuit characterized with the same transfer function as the z-scanner were used to dampen the z-scanner. With this technique, we achieved a bandwidth of 150 kHz and a quality factor of 0.5, which resulted in a response time of 1.1 µs. This method worked well for a z-scanner with a simple transfer function but not for one with multiple resonant peaks. For active damping, an alternative method can be used. The z-scanner is driven through a circuit with a transfer function $1/G(s)$, where G (s) is the transfer function of the z-scanner. However, for a complicated G(s), it is very difficult to design an electric circuit characterized by $1/G(s)$. Instead, we invented a circuit that can automatically produce a transfer function that is approximately the same as the inverse transfer function of a given transfer function [9, 17]. This approximation becomes better with the use of operational amplifiers having higher bandwidths. This method works not only for reducing the quality factor but also for extending the apparent resonant frequency, so long as the driver for the piezoactuators has enough gain at high frequencies. Using this method and holding a piezoactuator (resonant frequency, 370 kHz) at its corners, we achieved a bandwidth of 500 kHz for a z-scanner which can be maximally displaced by 1 μm (unpublished data).

12.5.3
Dynamic PID Control

Various efforts have been made to increase the AFM scan speed. However, not much attention has been directed towards reducing the tip-sample interaction force. This reduction is quite important for biological AFM imaging. The most ideal scheme is the use of non-contact AFM (nc-AFM) but, to date, high-speed nc-AFM has not been exploited. It is unknown if the high-speed and non-contact conditions can reconcile with each other. We discuss this matter later. There are several methods to reduce the force in tapping mode: (a) using softer cantilevers, (b) enhancing the quality factor of small cantilevers, (c) using a shallower amplitude set point (i.e. r is close to 1) . However, none of these methods appear compatible with high-speed scanning. Softer cantilevers can be obtained only by sacrificing the resonant frequency. The most advanced small cantilevers developed seem to have reached their limit in balancing high resonant frequency with a small spring constant. Although the tapping force decreases with increasing Q in the cantilever so too does its response speed.

The last possibility, a shallower amplitude set point, promotes "parachuting" during which the error signal is saturated at $2A_0(1 - r)$, and therefore, the parachuting time is prolonged with increasing r, resulting in decrease in the feedback bandwidth. This difficult issue was resolved by the invention of a new PID controller known as a "dynamic PID controller" whose gains were automatically changed depending on the cantilever's oscillation amplitude [6]. Briefly, a threshold level is set between the peak-to-peak free oscillation amplitude, $2A_0$, and the set point amplitude, $2A_0r$. When the cantilever oscillation amplitude exceeds this threshold level, the feedback gain is increased, which either shortens the parachuting time or avoids it entirely. The dynamic PID controller can avoid parachuting in fact even when r is increased to ~0.9. Therefore, the feedback bandwidth becomes almost independent of r so long as r is set at less than ~0.9 (Figure 12.3). Thus, high-speed scanning and gentle handling of the sample do not necessarily conflict.

12.6
Bioimaging

Various attempts to capture biological processes have been made in order to test high-speed AFM. Imaging studies have also been undertaken to establish

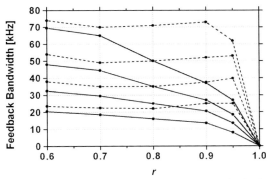

Figure 12.3 Effect of dynamic PID control on the feedback bandwidth of tapping mode AFM. These data were experimentally obtained. Solid lines: a conventional PID controller was used; broken lines: a dynamic PID controller was used. The solid-line curves and the dotted-line curves are aligned from top to bottom according to the ratios $2A_0/h_0 = 5, 2, 1,$ and 0.5. A z-scanner with a bandwidth of 150 kHz was used.

sample and substratum preparation methods necessary to image biomolecular processes. These efforts have steadily improved the quality of captured images, the imaging rate, and the magnitude of the tip-sample interaction force. Imaging experiments performed most recently captured dynamic protein–protein inter-actions on video with an imaging performance capable of revealing molecular mechanisms. We describe below some imaging experiments in chronological order.

In 2001, we reported the first image data successively captured at 80 ms/frame. The sample was myosin V weakly attached to a mica surface in a solution containing ATP [2]. Although the images were noisy, the swinging-lever-like movement was visible. Since the ATPase rate of myosin V in the absence of ATP is low, this movement was not observed repeatedly, which made interpretation of the data inconclusive. Soon after, we attempted to observe the gliding movement of actin filaments on a myosin V-coated surface. However, actin filaments did not appear in the scanning region. The reason was that actin filaments were easily detached from anchored myosin V by the scanning cantilever tip. In 2002, we developed a method to combine UV-flash photolysis of caged-compounds with high-speed AFM. A strong UV-flash bent a cantilever significantly, resulting in a strong impact between the tip and the substratum and hence damaging the tip. Therefore, many flashes of attenuated UV were applied while the sample stage was being scanned in "y-return" after its slight withdrawal from the cantilever in the z-direction. We applied this method to observe conformational changes in myosin V that occurred synchro-nously with UV irradiation. Because of the synchronicity, the observed changes can be interpreted as those really induced by ATP binding. In fact, immediately after UV irradiation, the head portion bent around the head–neck junction and returned to its original straight form (Figure 12.4a). This technique was also applied to the

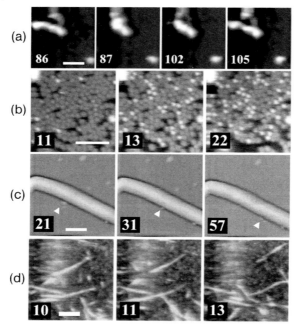

Figure 12.4 Various biological processes captured by high-speed AFM. The number attached to each image indicates the frame number (F#). (a) Myosin V head bending upon UV-flash photolysis of caged ATP. F#86: before UV flash; F#87–105: after UV flash. Scale bar, 30 nm; imaging rate, 80 ms/frame. (b) GroEL binding to GroEL upon UV-flash photolysis of caged ATP. F#11: before UV flash; F#13: after the first UV flash; F#22: after the second UV flash. Scale bar, 100 nm; imaging rate, 1 s/frame. (c) Movement of kinesin–gelsolin along a microtubule. Arrow heads indicate kinesin–gelsolin. Scale bar, 100 nm; imaging rate, 0.64 s/frame. (d) Actin filaments gliding over a surface that is densely coated with myosin V. Scale bar, 200 nm; imaging rate, 1 s/frame.

observation of GroES binding to GroEL. In this observation, GroEL was densely coated onto a mica surface in an end-up orientation with GroES floating in an ATP-contained solution. We were able to observe GroES binding immediately after UV flashes (Figure 12.4b).

In 2003, we developed the first version of the dynamic PID controller to reduce the tapping force. Using this controller, we imaged at 0.5 s/frame unidirectional movement of a chimera kinesin along a microtubule in an ATP-containing solution [18]. For this kinesin, the C-terminal tail ends were replaced with gelosolin to avoid strong attachment of its intrinsic tail ends to the mica surface. The observed kinesin moved unidirectionally along a microtubule while attaching weakly to the mica surface (Figure 12.4c). Kinesin moving along a microtubule without touching the mica surface was not observed. This is because tip-sample interaction was still too strong even when applying the first version of our dynamic PID controller. The tip removed kinesin attached only to the microtubules. In the absence of the dynamic PID controller, microtubules were destroyed. Similar results were seen with actin and myosin. With the dynamic PID control, actin filaments gliding on a mica surface densely coated with myosin V were captured on video (Figure 12.4d) [4, 12], whereas a low myosin V density resulted in minimal observation of the gliding movement.

With an improved dynamic PID controller together with a newly-developed optical deflection sensor with low noise, the set point could be set at >0.9 and thus actin filaments gliding on a surface sparsely coated with myosin V were successfully imaged [4, 12]. By chance, a short actin filament entered the observed region. Its entire length was within the region allowing all myosin V molecules interacting with this filament to be identified. The interacting myosin V heads were oriented in one direction, similar to the well-known arrow-head structure in muscles. From this oriented structure, the polarity of the actin filament was identified. The filament moved towards the minus (pointed) end, which was the natural direction (Figure 12.5). However, conformational changes in the interacting myosin V heads were not evident during the unidirectional movement of the filament.

Recently, a prototype high-speed AFM, an improved version of our first design [2, 3], has become commercially available (Nano Live Vision™ manufactured by Olympus and distributed by RIBM). Its users recently filmed the ATP-

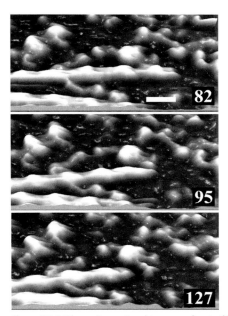

Figure 12.5 Three-dimensional images of actin filament sliding movement captured by high-speed AFM. The number on each image indicates the frame number. Scale bar 30 nm, imaging rate, 180 ms/frame.

dependent binding of GroES to GroEL immobilized onto a mica surface in an end-up orientation [19]. Additionally, they observed the formation and dissociation of a streptavidin-biotinylated DNA complex [20] and the one-dimensional diffusion of a restriction enzyme along a DNA strand followed by DNA cleavage [21].

The dynamic processes mentioned above were already known or expected from a series of biochemical and biophysical studies, meaning these filmed images did not offer extremely new insights into the respective molecular mechanisms. Nevertheless, because this technique is new and unproven, known or expected biological processes need to be confirmed by high-speed AFM imaging before applying the technique to broader studies. In addition, techniques to prepare samples and substrates for their attachment need to be developed. These are often different from their static imaging counterparts. Upon proving its utility for known molecular processes, high-speed AFM will become a reliable tool for unexplored biological processes.

Using high-speed AFM, we have been seeking to image single myosin V molecules walking along actin filaments. Single myosin V molecules move processively along actin tracks [22]. The hand-over-hand walking of myosin V is already established [23–26] but its detailed behavior is still unknown. For AFM studies, the myosin V tail was removed by digestion [27] because it tended to attach to the mica surface. However, in a low ionic solution, the truncated myosin V (HMM) still tended to attach to the mica surface. Therefore, we elevated the ionic strength, although this lowered the myosin V affinity for actin. Due to the weak affinity, the oscillating cantilever tip with the usual free amplitude ($\sim$5 nm) disturbed the actin–myosin V interaction. Therefore, we reduced the free amplitude to $\sim$1 nm, sacrificing the feedback bandwidth. Under these conditions, we successfully captured the walking movement of myosin V on video at 0.1 s/frame and observed that during the process the leading and trailing heads altered their positions with a walking stride of $\sim$72 nm (N. Kodera *et al.*, unpublished data). However, some level of myosin V affinity for the mica surface still remained in the high-ionic solution. Its affinity acted as a mechanical load against myosin V walking, and therefore, reduced the probability of observing myosin V moving processively.

We have also been seeking to image GroEL–GroES interaction dynamics in an ATP-containing solution where biotinylated GroEL [28] is immobilized on streptavidin 2D-crystal sheets in a side-on orientation. Due to this orientation, both the GroEL rings were accessible to GroES floating in the solution. Because floating GroES did not interfere with imaging, a high concentration of GroES could be used, in contrast to the situation with single-molecule fluorescence microscopy. The negative cooperativity between the two GroEL rings (e.g. [29]) was confirmed. GroEL alternated its rings between the GroES-associated and -dissociated states. However, interestingly, releasing one GroES-associated complex and forming another did not necessarily occur simultaneously. Two controversial intermediates (e.g. [30]), bare GroEL and GroEL–(GroES)$_2$, were detected just prior to the switching (D. Yamamoto *et al.*, unpublished data).

12.7
Other Type of Imaging

12.7.1
Phase-contrast Imaging

Tapping mode AFM (also known as amplitude-modulation AFM (AM-AFM)) has the capability of imaging compositional variations in heterogeneous surfaces in addition to surface topography [31–33]. The phase difference between the excitation signal and the cantilever oscillation is affected by several surface properties. With energy-conservative tip-sample interaction, the resonant frequency shifts by approximately $-0.5kf_c/k_c$. The frequency shift results in a phase shift as the excitation frequency is fixed. For a given frequency shift, the phase shift increases with Q. With conventional cantilevers, the frequency shift is generally around 50 Hz. Therefore, phase-contrast imaging had been possible only with a large Q (hence only at a low imaging-rate). Because the ratio f_c/k_c with the most advanced small cantilevers is $\sim$1000 times larger than that with conventional cantilevers, we can expect a large shift of $\sim$50 kHz. Therefore, even with a small Q, a relatively large phase shift occurs. Thus, we do not need to detect the phase shift using very sensitive yet slow phase detectors such as lock-in amplifiers. In order to explore the possibility of high-speed phase-contrast imaging, a fast phase detector was developed by Uchihashi *et al.* [34] based on a previous design [35]. This can detect phase shifts within a single oscillation cycle and importantly at any time within a cycle. Images of styrene-butadiene-styrene block copolymer films having different contrasts were obtained in liquids depending on the detection timing within the oscillation cycle. Remarkably, this phase imaging was carried out at $\sim$80 ms/frame (Figure 12.6) [34]. This new technique for

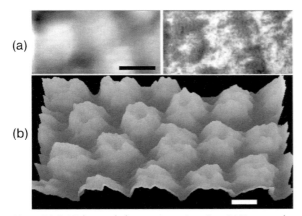

(a)

(b)

Figure 12.6 High-speed phase-contrast imaging. (a) Topography image (left) and phase contrast image (right) of a commercial test sample captured simultaneously at an imaging rate of 78 ms/frame. The nature of the sample is unknown. Scale bar, 100 nm. (b) Phase contrast image of GroEL attached to a mica surface in an end-up orientation. Scale bar, 10 nm; imaging rate, 0.5 s/frame.

phase-contrast imaging will allow us to study dynamic changes in the physicochemical nature of protein molecules in action.

12.7.2
Recognition Imaging

By attaching to the cantilever tip a molecule that binds specifically to complementary molecules contained in a multi-component sample, phase-contrast imaging can reveal the arrangement of specific molecules in the sample. However, to date the studies for this type of recognition imaging is minimal. Hinterdorfer and his colleagues recently developed a recognition imaging technique using a different approach [36, 37]. A cantilever tip that is functionalized with a probe molecule through a flexible short-linker is oscillated at a frequency below resonance. Its oscillation is split into lower and upper components. The lower-component oscillation signals reflect the repulsive tip-sample interaction and provide topography images. The upper-component signals reflect the association events between the probe molecule and its counterpart and provide recognition images. Thus, topographic and recognition images can be obtained simultaneously, which has been demonstrated with systems such as antibody–antigen and biotin–avidin. This method is very useful to identify locations of specific molecules in the topography image of a multi-component sample.

12.8
Substratum

The choice of the substratum on which a sample is placed is very important for AFM observation of biomolecular processes. Preparations of substrata with very low roughness have been explored extensively for observing still images by AFM. However, these preparations have been devised to attach samples firmly onto them and are therefore inapplicable for studying biomolecular processes. For our interests, various properties are required for a substratum including (a) appropriate binding affinity for the sample, which ensures the retention of its physiological function, (b) selective attachment of a specific component in a multi-component sample, (c) attachment of molecules in a desired orientation. Mica (natural muscovite or synthetic fluorophlogopite) has been frequently used as a source substratum because its surface is flat at the atomic level over a large area . It has net negative charges and is therefore quite hydrophilic. Bare mica surfaces adsorb various proteins by electrostatic interaction. Except for some cases (such as GroEL attachment in an end-up orientation), its binding orientation is not unique, meaning that selective attachment of a specific species cannot be expected. Furthermore, its use sometimes compromises the sample.

If possible, it is best to have a surface to which the sample never attaches. Thus, the surface can be modified so that only a specific species of molecule attaches through a linker in a desired orientation. A membrane surface with zwitterionic polar head groups such as phosphatidyl choline (PC) and phosphatidyl ethanolamine (PE) is known to resist protein adsorption (e.g. [38, 39]). Streptavidin is also useful for specific attachment of biotinylated protein because it also resists non-specific protein adsorption. Mica surface-supported planar lipid bilayers [40] can be easily formed from liposomes because their strong hydrophilic interaction disrupts the liposomes [41]. Various lipids with functional groups that are attached to polar groups (e.g. biotin attached to PE; Ni-NTA attached to PS) are available and can be used to accomplish the specific attachment of proteins labeled with biotin or his-tags to planar lipid bilayers. DOPC lipids are useful for preparing streptavidin 2D crystals when used together with biotinylated lipids [42, 43]. These lipids contain an unsaturated hydrocarbon on each of the two alkyl chains, which causes bending of the chains, and therefore weakens the interaction between neighboring DOPCs. This weak interaction lowers its phase-transition temperature to $\sim-20°C$, which affords a large fluidity to the planar bilayer at room temperature and thus facilitates 2D-crystal formation of streptavidin (Figure 12.7a). The densely packed streptavidin does not diffuse easily. If less diffusion is necessary, the packed streptavidin can be crosslinked with glutaraldehyde, which does not affect its ability to bind to biotin. We used this surface for the selective attachment of biotinylated GroEL in a side-on orientation. DPPC contains no unsaturated hydrocarbons in the alkyl chains, therefore its phase-transition temperature is high (~41 °C) making DPPC appropriate for preparing planar bilayers with a low fluidity. For example, when planar bilayers are formed with DPPC at a high temperature (~60 °C) together with a certain fraction of DPPE–biotin, streptavidin that is sparsely attached to the surface barely diffuses at room temperature (Figure 12.7b). When DOPE–biotin is used together with DPPC, the sparsely attached streptavidin diffuses at a moderate speed (Figure 12.7c).

12.9
Future Prospects

High-speed AFM imaging studies are a nascent field. Therefore, the number of researchers and imaging data that have been successfully captured on video are quite limited. However, it seems very certain that this new microscopy will be used widely in the biological sciences. High-speed AFM will become commercially available probably in 3–5 years. Presently, though, the force exerted between the oscillating tip and sample is barely able to preserve weak protein–protein associations. Moreover, the imaging rate is not high enough to study fast biological processes. For high-speed AFM to become truly useful for a wide variety of biological systems, reduction in the force and increase in the imaging rate are essential.

How much better can the imaging rate and tip-sample interaction force be? The cantilever is the factor limiting both requirements. Generally, to improve the resonant frequency the rigidity must be compromised and vice versa. High-speed nc-AFM will allow the use of more rigid cantilevers with higher resonant frequencies without concerns about disrupting delicate samples. It also has the potential to image

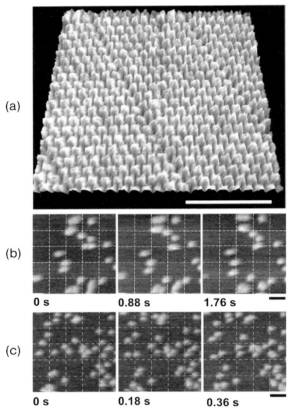

Figure 12.7 Streptavidin on mica-supported planar lipid bilayers. (a) Streptavidin 2D crystals formed on bilayers of DOPC + DOPE–biotin. Scale bar, 30 nm. (b) Streptavidin on bilayers of DPPC + DPPE–biotin. Scale bar, 40 nm; imaging rate, 0.18 s/frame. (c) Streptavidin on bilayers of DPPC + DOPE–biotin. Scale bar, 40 nm; imaging rate, 0.18 s/frame.

dynamic nanostructures on living cells, currently an impossible task due to the extremely soft nature of living cell membranes. To date, nc-AFM has been realized only in a vacuum where cantilevers can have a large Q. However, the cantilever's response speed decreases with increasing Q. Therefore, it is incompatible with high-speed imaging. We require a non-contact condition that is compatible with cantilevers having a small Q. Although still unproven, one possibility is the use of ultrasonic interference between the cantilever tip and the sample by exciting them ultrasonically at different frequencies f_1 and f_2 ($f_1, f_2 \gg f_c$). The difference, $|f_1 - f_2|$, is set close to the cantilever's resonant frequency. This configuration (Figure 12.8) is the same as that employed for scanning near-field ultrasound holography (SNFUH) that has recently been developed for high-resolution sub-surface imaging [44].

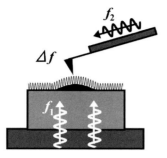

Figure 12.8 Schematic of proposed high-speed ncAFM set-up. Ultrasonic waves are launched from the bottom of the sample as well as from the cantilever. The frequencies, f_1 and f_2, are much higher than the fundamental resonant frequency, f_c, of the cantilever, and their difference, $\Delta f \equiv |f_1 - f_2|$, is similar to f_c. The two ultrasonic waves interfere with each other and produce acoustic waves with a frequency of Δf, which excites the cantilever. The wave front of ultrasonic waves with frequency f_1, is formed at the sample surface and sensed by the cantilever which is not in contact with the surface but is close to it.

When the sample in a solution consists only of protein molecules attached to a uniform substratum, the wave-front of the ultrasound propagated through the substratum may trace the sample topography. This wave front can probably be detected by the cantilever tip, which is close to but not in contact with the sample surface.

A second possibility for high-speed, non-contact imaging may derive from ion-conductance scanning probe microscopy (ICSPM), which has already satisfied the non-contact condition [45]. Due to the progress in fabrication techniques to produce very sharp glass capillaries with a small pore at the end, the spatial resolution has reached a few nm [46]. Immobile protein molecules of ∼14 nm on living cell membranes have been successfully imaged [47]. However, in order to materialize high-speed ICSPM, we need to find a method to increase the bandwidth of ion-conductance detection because ionic currents through a small pore are very low.

Is high-speed recognition imaging even possible? The effective concentration of a probe molecule attached to a cantilever tip depends on the tether length. With a tether length of 2 nm, the concentration becomes ∼50 mM, which is high enough for the association reaction to take place in 20 μs for a system with a typical association rate constant of $1 \times 10^6 \, \mathrm{M}^{-1} \, \mathrm{s}^{-1}$. This suggests that recognition imaging at a moderate rate (∼0.1 s/frame) is possible, so long as the cantilever oscillation amplitude is small.

At present, we have no technology that allows us to study the structural dynamics of intracellular organelles at high spatial and temporal resolution. Recently, a far-field fluorescence imaging technique (STED microscopy) with diffraction-unlimited resolution has been developed based on stimulated emission depletion of fluorophores [48, 49]. Its high spatial resolution has been demonstrated by resolving the arrangements of densely packed 40-nm beads and supramolecular aggregates in a cell membrane. However, it is unlikely that this new microscopy will attain a high temporal resolution without sacrificing the spatial resolution because of the limited number of photons collected in a short time bin. The recently demonstrated intracellular imaging by SNFUH [44] may be a possibility for 3D imaging of intracellular structures at high spatial and temporal resolution. It seems easy to combine SNFUH with the high-speed scanning techniques developed so far. It is still unclear whether this new imaging mode has the necessary resolution in the z-direction. However, the images obtained with ultrasonic waves launched at different angles should contain information along the z-axis. Therefore, there may be a method to reconstitute a 3D image from multiple images obtained using different launching angles.

References

1 Binnig, G. Quate, C.F. and Gerber, Ch. (1986) Atomic force microscope. *Phys. Rev. Lett.*, **56**, 930–933.

2 Ando, T., Kodera, N., Takai, E., Maruyama, D., Saito, K. and Toda, A. (2001) A high-speed atomic force microscope for studying biological macromolecules. *Proc. Natl Acad. Sci. USA*, **98**, 12468–12472.

3 Ando, T., Kodera, N., Takai, E., Maruyama, D., Saito, K. and Toda, A. (2002) A high-speed atomic force microscope for studying biological macromolecules in action. *Jpn. J. Appl. Phys.*, **41**, 4851–4856.

4 Ando, T., Uchihashi, T., Kodera, N., Miyagi, A., Nakakita, R., Yamashita, H. and Sakashita, M. (2006) High-speed atomic force microscopy for studying the dynamic behavior of protein molecules at work. *Jpn. J. Appl. Phys.*, **45B**, 1897–1903.

5 Kodera, N., Yamashita, H. and Ando, T. (2005) Active damping of the scanner for high-speed Atomic Force Microscopy. *Rev. Sci. Instrum.*, **76**, 053708 (5 pp).

6 Kodera, N., Sakashita, M. and Ando, T. (2006) A dynamic PID controller for high-speed atomic force microscopy. *Rev. Sci. Instrum.*, **77**, 083704 (7 pp).

7 Viani, M.B., Schäffer, T.E., Paloczi, G.T., Pietrasanta, L.I., Smith, B.L., Thompson, J.B., Richter, M., Rief, M., Gaub, H.E., Plaxco, K.W., Cleland, A.N., Hansma, H.G. and Hansma, P.K. (1999) Fast imaging and fast force spectroscopy of single biopolymers with a new atomic force microscope designed for small cantilevers. *Rev. Sci. Instrum.*, **70**, 4300–4303.

8 Vianni, M.B., Pietrasanta, L.I., Thompson, J.B., Chand, A., Gebeshuber, I.C., Kindt, J.H., Richter, M., Hansma, H.G. and Hansma, P.K. (2000) Probing protein–protein interactions in real time. *Nature Struct. Biol.*, **7**, 644–647.

9 Yamashita, H., Uchihashi, T., Kodera, N., Miyagi, A., Yamamoto, D. and Ando, T. (2007) Tip-sample distance control using photo-thermal actuation of a small cantilever for high-speed atomic force microscopy. *Rev. Sci. Instrum.*, **78**, 083702 (5 pp).

10 Zhong, Q., Inniss, D., Kjoller, K. and Elings, V.B. (1993) Fractured polymer/scilica fiber surfaces studied by tapping mode atomic force microscopy. *Sur. Sci. Lett.*, **290**, L688–L692.

11 Kindt, J.H., Fantner, G.E., Cutroni, J.A. and Hansma, P.K. (2004) Rigid design of fast scanning probe microscopes using finite element analysis. *Ultramicroscopy*, **100**, 259–265.

12 Ando, T., Uchihashi, T., Kodera, N., Miyagi, A., Nakakita, R., Yamashita, H. and Matada, K. (2005) High-speed AFM for studying the dynamic behavior of protein molecules at work. *e-J. Surf. Sci. Nanotech.*, **3**, 384–392.

13 Anczykowski, B., Cleveland, J.P., Kruger, D., Elings, V. and Fuchs, H. (1998) Analysis of the interaction in dynamic mode SFM by means of experimental data and computer simulation. *Appl. Phys. A*, **66**, S885–S889.

14 Tamayo, J., Humphris, A.D.L. and Miles, M.J. (2000) Piconewton regime dynamic

force microscopy in liquid. *Appl. Phys. Lett.*, **77**, 582–584.

15 Tamayo, J., Humphris, A.D.L., Owen, R.J. and Miles, M.J. (2002) High-Q dynamic force microscopy in liquid and its application to living cells. *Biophys. J.*, **817**, 526–537.

16 Sulchek, T., Hsieh, R., Adams, J.D., Yaralioglu, G.G., Minne, S.C., Quate, C.F., Cleveland, J.P., Atalar, A. and Adderton, D.M. (2000) High-speed tapping mode imaging with active Q control for atomic force microscopy. *Appl. Phys. Lett.*, **76**, 1473–1475.

17 Morita, S., Yamada, H. and Ando, T. 2007. Japan AFM roadmap 2006. *Nanotechnol.*, **18**, 08401 (10 pp).

18 Ando, T., Kodera, N., Naito, Y., Kinoshita, T., Furuta, K. and Toyoshima, Y.Y. (2003) A high-speed atomic force microscope for studying biological macromolecules in action. *Chem. Phys. Chem.*, **4**, 1196–1202.

19 Yokokawa, M., Wada, C., Ando, T., Sakai, N., Yagi, A., Yoshimura, S.H. and Takeyasu, K. (2006) Fast-scanning atomic force microscopy reveals the ATP/ADP-dependent conformational changes of GroEL. *EMBO J.*, **25**, 4567–4576.

20 Kobayashi, M., Sumitomo, K. and Torimitsu, K. (2007) Real-time imaging of DNA-streptavidin complex formation in solution using a high-speed atomic force microscope. *Ultramicroscopy*, **107**, 184–190.

21 Yokokawa, M., Yoshimura, S.H., Naito, Y., Ando, T., Yagi, A., Sakai, N. and Takeyasu, K. (2006) Fast-scanning atomic force microscopy reveals the molecular mechanism of DNA cleavage by ApaI endonuclease. *IEE Proc Nanobiotechnol.*, **153**, 60–66.

22 Sakamoto, T., Amitani, I., Yokota, E. and Ando, T. (2000) Direct observation of processive movement by individual myosin V molecules. *Biochem. Biophys. Res. Commun.*, **272**, 586–590.

23 Yildiz, A., Forkey, J.N., McKinney, S.A., Ha, T., Goldman, Y.E. and Selvin, P.R. (2003) Myosin V walks hand-over-hand: single fluorophore imaging with 1.5 nm localization. *Science*, **300**, 2061–2065.

24 Forkey, J.N., Quinlan, M.E., Shaw, M.A., Corrie, J.E.T. and Goldman, Y.E. (2003) Three-dimensional structural dynamics of myosin V by single-molecule fluorescence polarization. *Nature*, **422**, 399–404.

25 Warshaw, D.M., Kennedy, G.G., Work, S.S., Krementsova, E.B., Beck, S. and Trybus, K.M. (2005) Differential labeling of myosin V heads with quantum dots allows direct visualization of hand-over-hand processivity. *Biophys. J.*, **88**, L30–L32.

26 Syed, S., Snyder, G.E., Franzini-Armstrong, C., Selvin, P.R. and Goldman, Y.E. (2006) Adaptability of myosin V studied by simultaneous detection of position and orientation. *EMBO J.*, **25**, 1795–1803.

27 Koide, H., Kinoshita, T., Tanaka, Y., Tanaka, S., Nagura, N., Meyer zu Hörste, G. and Ando, T. (2006) Identification of the specific IQ motif of myosin V from which calmodulin dissociates in the presence of Ca^{2+}. *Biochemistry*, **45**, 11598–11604.

28 Taguchi, H., Ueno, T., Tadakuma, H., Yoshida, M. and Funatsu, T. (2001) Single-molecule observation of protein–protein interactions in the chaperonin system. *Nature Biotech.*, **19**, 861–865.

29 Sigler, P.B., Xu, Z., Rye, H.S., Burston, S.G., Fenton, W.A. and Horwich, A.L. (1998) Structure and function in GroEL-mediated protein folding. *Annu. Rev. Biochem.*, **67**, 581–608.

30 Sparrer, H. and Buchner, J. (1997) How GroES regulates binding of nonnative protein to GroEL. *J. Biol. Chem.*, **272**, 14080–14086.

31 Magonov, S.N., Elings, V. and Whangbo, M.H. (1997) Phase imaging and stiffness in tapping-mode atomic force microscopy. *Sur. Sci.*, **375**, L385–L391.

32 Bar, G., Thomann, Y., Brandsch, R., Cantow, H.J. and Whangbo, M.G. (1997) Factors affecting the height and phase images in tapping mode atomic force

microscopy. Study of phase-separated polymer blends of poly(ethene-co-styrene) and poly(2,6-dimethyl-1,4-phenylene oxide). *Langmuir*, **13**, 3807–3812.

33 Tamayo, J. and Garciá, R. (1996) Deformation, contact time, and phase contrast in tapping mode scanning force microscopy. *Langmuir*, **12**, 4430–4435.

34 Uchihashi, T., Ando, T. and Yamashita, H. (2006) Fast phase imaging in liquids using a rapid scan atomic force microscope. *Appl. Phys. Lett.*, **89**, 213112 (3 pp).

35 Stark, M. and Guckenberger, R. (1999) Fast low-cost phase detection setup for tapping-mode atomic force microscopy. *Rev. Sci. Instrum.*, **70**, 3614–3619.

36 Stroh, C., Wang, H., Bash, R., Ashcroft, B., Nelson, J., Gruber, H., Lohr, D., Lindsay, S.M. and Hinterdorfer, P. (2004) Single-molecule recognition imaging microscopy. *Proc. Natl. Acad. Sci. USA*, **101**, 12503–12507.

37 Hinterdorfer, P. and Dufrêne, Y.F. (2006) Detection and localization of single molecular recognition events using atomic force microscopy. *Nature Methods*, **3** 347–355.

38 Zhang, S.F., Rolfe, P., Wright, G., Lian, W., Milling, A.J., Tanaka, S. and Ishihara, K. (1998) Physical and biological properties of compound membranes incorporating a copolymer with a phosphorylcholine head group. *Biomat.*, **19**, 691–700.

39 Vadgama, P. (2005) Surface biocompatibility. *Annu. Rep. Prog. Chem., Sect. C: Phys. Chem.*, **101**, 14–52.

40 Sackmann, E. (1996) Supported membranes: Scientific and practical applications. *Science*, **271**, 43–48.

41 Reviakine, I. and Brisson, A. (2000) Formation of supported phospholipid bilayers from unilamellar vesicles investigated by atomic force microscopy. *Langmuir*, **16**, 1806–1815.

42 Scheuring, S., Müller, D.J., Ringler, P., Heymann, J.B. and Engel, A. (1999) Imaging streptavidin 2D crystals on biotinylated lipid monolayers at high resolution with the atomic force microscope. *J. Microsc.*, **193**, 28–35.

43 Reviakine, I. and Brisson, A. (2001) Streptavidin 2D crystals on supported phospholipid bilayers: Toward constructing anchored phospholipid bilayers. *Langmuir*, **17**, 8293–8299.

44 Shekhawat, G. and Dravid, V.P. (2005) Nanoscale imaging of buries structures via scanning near-field ultrasound holography. *Science*, **310**, 89–92.

45 Hansma, P.K., Drake, B., Marti, O., Gould, S.A.C. and Prater, C.B. (1989) The scanning ion-conductance microscope. *Science*, **243**, 641–643.

46 Ying, L., Bruckbauer, A., Zhou, D., Gorelik, J., Schevchuk, A., Lab, M., Korchev, Y. and Klenerman, D. (2005) The scanned nanopipette: a new tool for high resolution bioimaging and controlled deposition of biomolecules. *Phys. Chem. Chem. Phys.*, **7**, 2859–2866.

47 Schevchuk, A.I., Frolenkov, G.I., Sanchez, D., James, P.S., Freedman, N., Lab, M.J., Jones, R., Klenerman, D. and Korchev, Y.F. (2006) Imaging proteins in membranes of living cells by high-resolution scanning ion conductance microscopy. *Angew. Chem. Int. Ed.*, **45**, 2212–2216.

48 Hell, S.W., Dyba, M. and Jakobs, S. (2004) Concepts for nanoscale resolution in fluorescence microscopy. *Curr. Opin. Neurobiol.*, **14**, 599–609.

49 Donnert, G., Keller, J., Medda, R., Andrei, M.A., Rizzoli, S.O., Lührmann, R., Jahn, R., Eggeling, C. and Hell, S.W. (2006) Macromolecular-scale resolution in biological fluorescence microscopy. *Proc. Natl. Acad. Sci. USA*, **103**, 11440–11445.

13
Force-clamp Spectroscopy of Single Proteins

Lorna Dougan, Jasna Brujic, and Julio M. Fernandez

13.1
Introduction

On a fundamental level, in order to investigate the properties of any physical system, from crystalline materials to glasses and even single molecules, the system needs to be perturbed and its response subsequently observed over time. The resulting relaxation phenomena carry the most important information about the intrinsic properties of the system. In our experiments, we use mechanical force to drive the system and observe the end-to-end length of a protein in response to the perturbation. In response to an applied stretching force, a wide variety of proteins are known to unfold and refold [1]. Since force is present in many biomechanical processes, it is a natural variable for probing the protein's physical properties and even further, its energy landscape [2]. For example, the biological activity of proteins is closely connected to their mechanical properties, particularly in proteins with a mechanical function, such as the immunoglobulin and fibronectin-like modules found in muscle fibers [3]. It is therefore probable that forced unfolding of proteins occurs *in vivo*, for example in the extension of titin [4, 5] and during the action of chaperones [6–8]. While it is clear that living organisms and their components are precisely adapted to mechanical environments by evolution, the mechanisms of supporting stress on the microscopic scale are less well understood due to the complexity of the systems.

By contrast, in bulk biochemistry experiments, folded proteins are typically perturbed out of their native states by a variety of denaturants, including chemicals such as urea and/or large temperature changes. Unlike mechanical forces, these denaturants do not occur naturally in biological settings. Bulk experiments measure the average rates of the unfolding or refolding reactions from an ensemble of molecules. In these conventional experiments the pathways are averaged out in the ensemble and modeled by simple two-state reactions, such that the details of the underlying energy landscape are smeared out. An important consideration when investigating protein unfolding and refolding is the reaction coordinate along which the measurements are taken and interpreted. Molecules undergoing thermal or chemical bulk denaturation explore a wide range of unfolded, molten globule structures without a well-defined reaction

Single Molecule Dynamics in Life Science. Edited by T. Yanagida and Y. Ishii
Copyright © 2009 WILEY-VCH Verlag GmbH & Co. KGaA, Weinheim
ISBN: 978-3-527-31288-7

coordinate, but are nevertheless close in size (radius of gyration) to the native states of the protein. The native contacts therefore remain relatively close to one another, which may explain the apparent two-state cooperativity in the folding reactions in bulk biochemistry experiments. Since both thermal and chemical perturbations drive the proteins to very different initial unfolded states, it is difficult to examine the diversity of folding reactions. Interestingly, even in the bulk some experiments have broadly characterized the heterogeneity of the protein landscape, revealing that different pathways become dominant depending on the folding conditions [9–18]. What is still lacking is a consideration of the microscopic unfolding and refolding pathways under the same external conditions to probe the dynamic diversity of the states explored by the same protein. Single-molecule AFM techniques are used to apply a denaturing force along a well-defined reaction coordinate (end-to-end length) driving proteins to a fully extended unfolded state. This level of control allows us to examine statistically the folding pathways of the protein in question.

It should be noted that there is already a body of single-molecule literature showing evidence of dynamic disorder on multiple timescales that is explored by proteins and enzymes in solution in their native states [19–23]. These results are obtained using fluorescence and infrared spectroscopy to investigate the conformational changes of the individual molecules over time, or the activity in enzymes. The observed complex kinetics are signatures of multiple energy minima in the protein landscape, corresponding to slightly different conformational sub-states of the molecule. The details of the energy fluctuations in the landscape can therefore be determined by following each individual reaction by single-molecule experimentation and performing a statistical analysis of the obtained distributions [24–29].

In this chapter we first introduce the technique of force spectroscopy AFM and the single-molecule fingerprint achieved using polyprotein engineering. We then present the recently developed constant force mode of the AFM (force-clamp spectroscopy) as a tool to explore the force-dependent protein unfolding kinetics in Section 13.3. A novel statistical analysis of such force-clamp data reveals deviations from two-state kinetics and is discussed in terms of a disordered free energy landscape in Section 13.4. Using this insightful technique in Section 13.5 we next present protein refolding trajectories that reveal the complexity in the folding pathways explored under a stretching force. We relate these results to simplistic physical models of protein folding and uncover the need for the development of more sophisticated theoretical frameworks for the understanding of this important process. Finally, we expand the capabilities of the force-clamp technique to observe individual chemical reactions in a single protein in Section 13.6, revealing both the effect of force on chemical kinetics as well as the structure of the reaction transition state.

13.2
Single-protein AFM Techniques

The advent of single-protein Atomic Force Microscopy (AFM) techniques combined with protein engineering techniques has made it possible, for the first time, to examine the mechanical properties of both native and engineered tandem modular

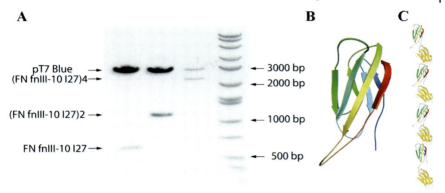

Figure 13.1 Construction of [10]FNIII fibronectin polyprotein. (A) Agarose gel stained with ethidium bromide showing cDNAs of the digested monomer construct ([10]FNIII-I27) in the first lane, the dimer in the second lane ([10]FNIII-I27)$_2$ and the tetramer in the third lane ([10]FNIII-I27)$_4$. The top band always shows the pT7 Blue vector. The last lane shows the size standard. (B) Diagram of the β-sandwich structure of the [10]FNIII fibronectin module. (C) Schematic representation of the final polyprotein ([10]FNIII-I27)$_4$.

proteins. The first force spectroscopy AFM experiments were performed using single proteins [30], however, these experiments lacked any fingerprint for an unambiguous unfolding event. This problem was solved by ligating multiple copies of a single protein module and expressing the resultant gene in bacteria (Figure 13.1). The engineering of proteins made from tandem repeats of an identical module, polyproteins, has then permitted a module-by-module investigation of the mechanical properties of native proteins. Polyproteins provide a consistent fingerprint, which then allows us to identify the molecule of interest from other background interactions [31]. Furthermore, the construction of engineered polyproteins makes it possible to carry out extensive mutagenesis experiments [32–34]. More recently, a simpler approach has been demonstrated where pairs of cysteine residues are introduced by mutagenesis at various locations throughout the protein structure, thereafter, polyproteins are simply obtained through the spontaneous oxidation of the cysteine residues between protein monomers [35, 36]. There are many modular proteins that perform their function in tandem, such as the immunoglobulin modules in the muscle protein titin [37] or multiple ubiquitin modules in protein degradation [38]. While the use of polyproteins over the past 8 years has permitted the rapid development of single-protein AFM techniques, their use is not without controversy [39–42] and ideally, the single-protein AFM techniques will evolve to a point where the use of protein monomers can be reliably recorded.

13.2.1
Force-extension Spectroscopy

In a force extension experiment a single polyprotein is stretched between the tip of a cantilever and a flat substrate (gold) that is mounted on a piezoelectric positioner

(Figure 13.2A). As the distance between the tip and substrate increases with constant velocity, extension of the molecule generates a restoring force that is measured from the deflection of a pre-calibrated cantilever. This system allows spatial manipulation of less than a nanometer and can measure forces of only a few picoNewtons up to hundreds of picoNewtons.

The resulting force-extension curve of a polyprotein has the characteristic appearance of a saw-tooth pattern (Figure 13.2B). This pattern results from the sequential extension and unfolding of the protein modules, which serves as a fingerprint for the individual single molecules. The peak force reached before an unfolding event measures the mechanical stability of the protein module, while the spacing between peaks is a measure of the increased contour length of the protein as it unfolds. The saw-tooth pattern is described by the worm-like chain (WLC) model of polymer elasticity, although it is unlikely that this simple polymer picture applies to proteins. This model expresses the relationship between the force and extension of a protein using two fitting parameters: its persistence length (protein stiffness) and its contour length (maximum end-to-end length).

Force-extension experiments have been used extensively in probing the mechanical behavior of many different proteins and have begun to challenge some of the simplified thermodynamic descriptions of proteins obtained from bulk experiments that are prevalent in the literature [1, 33, 34, 43–53]. However, force-extension experiments lack the ability to accurately measure force-dependent parameters since the force varies dynamically throughout the experiment. Computational studies have served as an important guide to the experiments, and have made significant contributions to our understanding of protein folding [49, 54–62]. Nevertheless, there was a need to develop an experimental tool which could probe force-dependent parameters. The development of a new force spectroscopy technique in which the force can be kept constant was crucial to gain real insight into the folding process in proteins. Therefore, in this chapter, we have focused on the force-clamp technique and highlighted some of the key breakthroughs it has allowed in our understanding of the physics and chemistry of protein folding.

13.2.2
Force-clamp Spectroscopy

In force-clamp spectroscopy a single protein molecule is held at a constant stretching force, such that the unfolding and refolding processes can be observed as a function of time [28, 50]. The cantilever is kept at a constant deflection (force) for a few seconds with a feedback response time of 4–6 ms. Stretching a polyprotein at a high constant force results in a well-defined series of step increases in length, marking the unfolding and extension of individual modules in the chain (Figure 13.2C). The size of the observed steps is directly correlated to the number of amino acids released by each unfolding event, which corresponds to 20 nm in the case of ubiquitin at a constant force of 100 pN (Figure 13.3A). The observed staircase therefore serves as a fingerprint of the single molecule. The cantilever picks up molecules at random points on the surface, such that the number of modules in the chain, **N**, exposed to

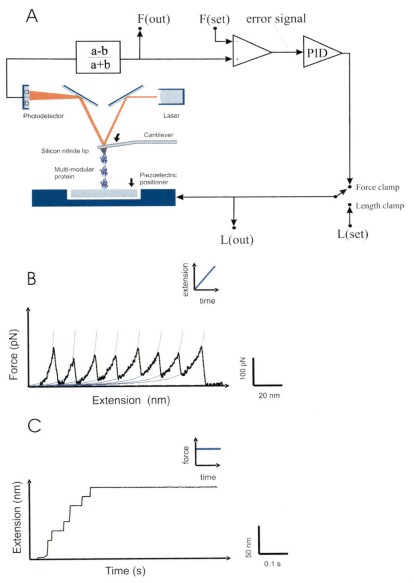

Figure 13.2 Two modes of single-molecule force spectroscopy. (A) A simplified diagram of the AFM, showing the laser beam reflecting on the cantilever, to a photodiode detector. The photodiode signal is calibrated in picoNewtons of force. A polyprotein is pulled at constant velocity by means of a piezoelectric actuator the increasing pulling force triggers the unfolding of a module. Continued pulling repeats the cycle resulting in a force-extension curve (B) with a characteristic "sawtooth pattern". (C) When pulling is done under feedback, the piezoelectric actuator abruptly adjusts the extension of the polyprotein to keep the pulling force at a constant value (force-clamp). Unfolding now results in a staircase-like elongation of the protein as a function of time.

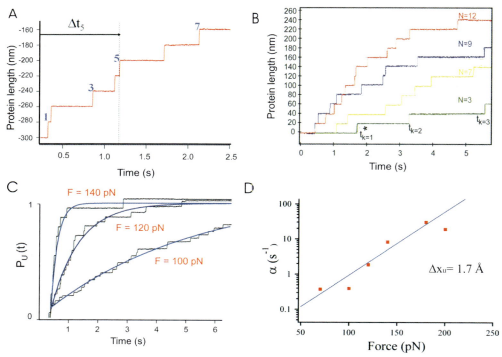

Figure 13.3 Unfolding of a ubiquitin polyprotein at a constant stretching force. (A) A typical length versus time recording (red trace) obtained by stretching a ubiquitin polyprotein at a constant force of 110 pN. The polyprotein elongates in steps of 20 nm, marking the unfolding of individual ubiquitins in the chain. The dwell time (Δt_k) of each unfolding event of order **k** (numbered 1–7 in this example) can be accurately measured. (B) Unfolding trajectories over time for protein chains of different lengths, **N**, at constant force. In a stochastic process, the dwell times on average depend on the total number of modules in the chain and on the order **k** in the sequence. (C) Averaging and normalizing several trajectories such as those in B, at each of the three forces shown, gives the overall probability of unfolding as a function of time (black traces), independent of protein length. The blue lines correspond to single exponential fits with unfolding time constants of $\alpha_u = 7.7\,s^{-1}$ at 140 pN, $1.9\,s^{-1}$ at 120 pN and $0.4\,s^{-1}$ at 100 pN, respectively. (D) The unfolding rate depends exponentially on the stretching force. A semi-logarithmic plot of the unfolding rate, α_u, as a function of the stretching force F is shown. A fit to the data using an Arrhenius term [63], where $\alpha_u(F) = \alpha_0 \exp(F\Delta x/k_B T)$, gives values of the unfolding rate at zero force, $\alpha_0 = 0.015\,s^{-1}$ and $\Delta x = 1.7\,\text{Å}$.

force can be anything up to the engineered protein length ($\mathbf{N} = 12$), resulting in a number of unfolding trajectories (Figure 13.3B). An ensemble of such trajectories allows for accurate investigation of the force and time dependency of protein unfolding, independent of the length of the chain.

Averaging a few unfolding trajectories provides a measure of the unfolding probability as a function of time for each stretching force (Figure 13.3C) which can be approximated by a single exponential fit analogous to bulk measurements. Moreover, the linear relationship observed in the semi-logarithmic plot of the unfolding rate, α, as a function of the pulling force, F (Figure 13.3D) directly demonstrates that the unfolding rate of polyubiquitin is exponentially dependent on the pulling force. An Arrhenius term can be fit by the data to quantify the time to rupture a bond under a mechanical stretching force [63]. This approximation models a two-state process with a single unfolding energy barrier in the reaction coordinate of the end-to-end length. This model then yields a measure of the distance to the transition state along the force induced reaction coordinate, $\Delta x = 0.17$ nm, beyond which the mechanical stability of the protein is lost. The distance is comparable to the length of a hydrogen bond in water, which is in excellent agreement with the result of computer simulations that identify the crucial hydrogen bonds that stabilize the protein's transition state under force [62]. It should be noted that the Arrhenius term [63], while currently used widely, assumes a force-independent distance to the transition state. Other models have recently been proposed which offer a more detailed analysis of the force-induced transitions [24, 64].

In rare instances in the unfolding trajectory we are able to capture mechanically stable intermediate structures that add up to the expected step size of 20 nm, suggesting some diversity in the unfolding energy barriers. Interestingly, collecting a much larger data set also begins to show important deviations from the fits to the two-state behavior, as the less-traveled unfolding pathways become statistically significant. We investigate the stochastic dynamics of unfolding and the success of the two-state model in the next section.

13.3
Order Statistics in Unfolding

Using the force-clamp technique we are able to test the diversity in the pathways and the correlation between the protein modules in each chain by investigating the kinetics of unfolding in further detail.

We first develop an analysis method for analyzing the unfolding dwell times from the force-clamp technique. Since the number of modules in each chain varies up to the engineered protein length ($N = 12$) (Figure 13.3B) and the dwell times depend on the order of the event **k** we need to apply order statistics to the data to investigate correlations between the modules and the heterogeneity in the pathways.

Even without correlations (Markov process) and given a single reaction pathway, statistics deems that the dwell times, **t**, to the unfolding events depend on the order number of the event, **k**, and the chain length, **N**. Indeed, the probability of observing **k** unfolding events out of **N** folded protein modules in time **t** should follow a binomial distribution,

$$P(t, N, k) = \alpha \frac{N!}{(k-1)!(N-k)!} (1 - e^{-\alpha t})^{(k-1)} (e^{-\alpha t})^{(N-k+1)} \tag{13.1}$$

where **α** is the unfolding rate constant for the ensemble. With an extensive pool of data we showed that this distribution could not account for all the unfolding events of ubiquitin. In order to investigate the source of the outliers in the obtained distribution, we next devised a maximum likelihood method at the single-molecule level.

If the unfolding process is Markovian, implying that the unfolding events are independent of one another, yet involve different unfolding pathways (i.e. non-homogeneous), the diversity in the pathways can be estimated using the maximum-likelihood method (MLM) [65] to allocate a rate constant, α, to each single molecule chain (Figure 13.4). The probability of a single module unfolding in time **t** is assumed to be $p = 1 - e^{(-\alpha t)}$ as supported by the observation that each unfolding event takes place in a single step on the timescale of the experiment. Therefore, for a given α, from the experimental trajectory of a sequence of **k** events unfolding at times $\mathbf{t_k}$ (Figure 13.4) and the appropriate binomial counting shown in Equation 13.1, the unfolding probability is given by the product of probabilities,

$$P(t_1, \ldots, t_{k_{max}} | \alpha, N) = \alpha^{k^{max}} \frac{N!}{(N-k_{max})!} e^{-\alpha(N-k_{max}-1)t_{det}} \prod_{k=1}^{k_{max}} e^{-\alpha t_k}. \tag{13.2}$$

The probability map for each unfolding trajectory is calculated over a wide range of α, from 0.01 to $100\,\mathrm{s}^{-1}$, as well as over the possible range of **N** (Figure 13.4) for the particular trace shown. Since molecules detach from the cantilever at random times, **N** must lie between the number of observed steps and the engineered protein length. The maximum likelihood method uses the unfolding probability, $\mathbf{P(t_1 \ldots t_{kmax})}$ in

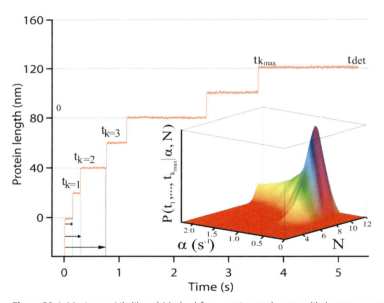

Figure 13.4 Maximum Likelihood Method for analyzing protein unfolding. A typical polyprotein unfolding trajectory that exhibits six consecutive unfolding events under a constant force of 110 pN is shown here. From the measured dwell times, $\mathbf{t_k}$, and the detachment time, $\mathbf{t_{det}}$, we estimate the most likely rate constant, α, and length of the chain, N, using the equation for the probability of unfolding as a function of time. The inset shows the shape of the probability function and the peak isolated for $N = 10$ and $\alpha = 0.5\,\mathrm{s}^{-1}$ for this trajectory.

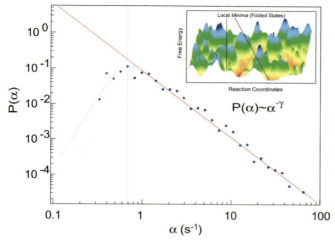

$P(\alpha) \sim \alpha^{-\gamma}$

Figure 13.5 Power law distribution of unfolding rates: a histogram of the most likely unfolding rates for each polyprotein chain in the ensemble plotted on a log-log plot. Beyond the peak, the rates are distributed in a power law with a decay coefficient $\gamma = 1.8$. The rates slower than the peak carry a larger error as the molecules often detach from the cantilever at long times. The inset shows the tentative energy landscape with multiple energy minima in the native state ensemble, which could give rise to such broad kinetics according to the Trapp model for glasses [74].

Equation 13.2 to predict the most likely α, N as the maximum in the obtained probability map. The maxima of the probability maps for all trajectories give rise to a distribution of $P_{max}(\alpha)$ for the whole set of data (Figure 13.5).

We obtained a very broad distribution of unfolding rates that follows a power law over more than two decades [27]. Such a broad distribution could not be explained by the errors in the MLM or by the noise in the constant force. Rather, this result is indicative of an unanticipated degree of complexity in the unfolding pathways of ubiquitin, supporting the view that the energy landscape is much more complex than was previously thought. It is the rare events that dominate the shape of the distribution, which implies that only a large statistical ensemble of unfolding events could reveal this complex physical picture.

13.4
Disordered Free Energy Landscape

The system's complex kinetics, as observed in the power law decay of the rate constants, is a consequence of the underlying roughness of the free energy landscape [66]. Therefore, a plausible explanation for the observed heterogeneity is to consider the diversity in the conformations within the native state ensemble [67, 68] under a constant force. Thermal motion of the folded protein explores local fluctuations in the secondary structure on an energy scale of a few $k_B T$ [29, 69], where k_B is the Boltzmann constant, which is crucial to the optimal functioning of the protein [19]. Understanding these fluctuations is therefore of fundamental biological importance [70], yet the details of the underlying distribution of conformations is experimentally challenging.

Both X-ray crystallography and NMR structures bias the results towards the average structure of the minimum energy conformations [71], such that the less-visited metastable states are overlooked. The diversity in conformational energies of the native state is further neglected by solution studies of protein folding/unfolding, which use the two-state model to distinguish between the average pathways under different experimental conditions using phi-value analysis [72].

On the other hand, the scale-free nature of the measured rate distribution, P(α), characterized by the power law coefficient, can be interpreted by a rough energy landscape in which the energy states, **E**, are distributed with an exponential distribution **P(E)**, where the energy scale $\bar{E} = 6.7\,k_B T$. This result is based on the assumption that the

$$P(E) = \frac{\tau_0^{-.15}}{k_b T} e^{-E/\bar{E}} \qquad (13.3)$$

protein is hopping over a multitude of unfolding energy barriers via an Arrhenius process [73]. This broad energy distribution reveals a wealth of conformational states available to the protein which is not observed in the reaction coordinate of the end-to-end length. Relatively small perturbations (below 1 nm) in the length of the protein seem to have significantly different free energies within this model. This interpretation, analogous to many other complex glassy systems [74], has previously been encountered in the literature examining folded-protein dynamics at the single-molecule level, including enzyme kinetics and activity [19–23, 75]. Structurally, more detailed NMR relaxation experiments reveal deviations from the two-state model for individual residues in a protein [76], and a separate work detects more spatially diverse structures in the same protein ubiquitin than was previously anticipated [18]. Using force-clamp AFM we have provided the first detailed signature of such complex behavior encountered in the unfolding process [27]. In the next section we reveal the exciting new breakthroughs in protein folding achieved through force-clamp spectroscopy.

13.5
Protein Folding

A major goal of biomolecular science has been to understand the protein folding problem. In the last few decades, significant progress has been made, both from the fundamental biochemical perspective [72, 77–79] and from the practical aspect of predicting protein structure from sequence data [80], and even designing artificial proteins [81]. From the physics perspective, there has also been a growing appreciation of the need to use appropriate statistical mechanical tools to characterize the free energy landscape of a protein [25, 26, 82, 83].

Many theories have emerged as a result, largely supported by numerical simulations of model protein systems [84–91]. However, what has been lacking is an experimental tool which can probe protein folding at the single-molecule level and has the ability to test the physical models proposed. For the first time, unique insight into the folding trajectory of a single protein can be gained using force-clamp spectroscopy, marking the beginning of rigorous tests of the microscopic physical mechanisms involved in protein folding. In the next section we will introduce the force quench experiment, followed by a discussion of one particular physical model for protein folding.

13.5.1
The Force Quench Experiment

The "force quench" experiment monitors the end-to-end length of a single poly-protein during reversible unfolding–folding cycles [28]. In the example shown (Figure 13.6A) the polyprotein is composed of nine repeats of the small protein ubiquitin, held at a constant force of 110 pN for a fixed time of 4 s. This applied force results in the probabilistic unfolding of individual ubiquitin modules, as described in Section 13.2.2. After 4 s we reduce or "quench" the force from 110 to 15 pN. The

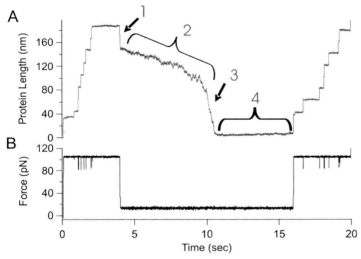

Figure 13.6 Force quench experiment reveals the folding trajectory of a single polyprotein. (A) The folding pathway of ubiquitin is directly measured by force-clamp spectroscopy. The end-to-end length of a protein is shown as a function of time. The folding trajectory can be divided into four distinct regions. The first stage (1) is due to the elastic recoil of the unfolded polyprotein as the force is reduced. The second stage (2) in the folding trajectory begins at the end of the rapid elastic recoil and is marked by a noticeable increase in length fluctuations. This stage relaxes with a slow rate of collapse. The beginning of the third stage (3) is observed by an abrupt change in the slope of the trajectory, where the molecule contracts to a conformation of an end-to-end length comparable to the native state. Stages (2) and (3) vary greatly between trajectories and cannot always be distinguished. The final stage (4) is where the protein acquires its mechanical stability through the formation of the native contacts [28]. (B) The corresponding applied force is also shown as a function of time.

spontaneous contraction of the protein down to the folded end-to-end length corresponds to the folding trajectory of the mechanically unfolded polyprotein. To confirm that the protein has folded, we again raise the force to 110 pN and the polyprotein again unfolds to its initial unfolded length in steps of 20 nm (Figure 13.6A).

The folding trajectories we observe can be divided into four distinct regions, marked by abrupt changes in the slope of the collapse (Figure 13.6A). Furthermore, we have determined that the collapse time of a protein is highly dependent on the quenching force [28]. Figure 13.7A illustrates the variation in trajectories for the polyprotein ubiquitin when quenched from a high force (trajectory 1), where the protein fails to fold on the timescale of an experiment, to a low force (trajectory 5)

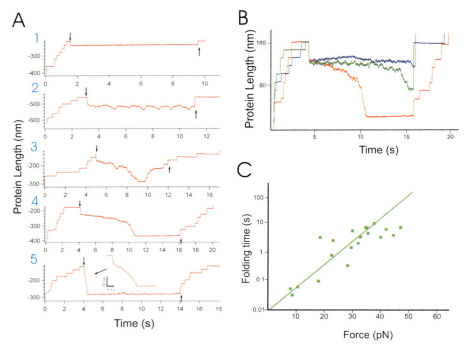

Figure 13.7 Complexity of the folding energy landscape. (A) The folding pathway of ubiquitin at a number of different quench forces is directly measured by force-clamp spectroscopy. The end-to-end length of a protein is shown as a function of time. As the force is quenched to lower forces (trajectories 1–5 where trajectory 1 corresponds to a high quench force of 50 pN while trajectory 5 corresponds to a lower quench force of 23 pN), the time taken to refold becomes smaller on average. At a high quench force (1–2) the protein may fail to fold on the timescale of the experiment.

(B) The folding pathway is probabilistic. The three folding trajectories shown begin from the same initial conditions and are quenched to the same force, however very different trajectories are clearly observed. The blue trace illustrates a protein which failed to fold on the timescale of the experiment, the green trace shows a protein which attempted to fold while the red trace shows a successfully folded protein. (C) Average folding times are exponentially dependent on the quench force with a $\alpha_0 = 100 \text{ s}^{-1}$ and $\Delta x_f = 8.2 \text{ Å}$, using an Arrhenius term [63].

were folding is successful. These force-clamp experiments have provided evidence that the probability of observing a folding event as well as its duration is strongly force dependent. Additionally, it should be noted that even under the same experimental conditions different folding trajectories are observed. For example, in Figure 13.7B, three distinct folding trajectories are shown for the polyprotein ubiquitin which have been unfolded and then quenched to the same force. In the majority of cases the protein fails to fold on the timescale of the experiments (blue trace), while in some cases the protein attempts to fold (green trace). Only in a small percentage of cases is the protein observed to fold (red trace). Therefore the application of force greatly reduces the folding probability of the protein on the timescale of the experiment.

To observe the force dependency of protein folding we grouped the experimental data from many force quench experiments at the longest range of contour lengths, for which the effect of force is most evident (150 to 200 nm). We observed that the duration of the folding collapse is exponentially dependent on the quench force (Figure 13.7C), as observed in the unfolding process (Figure 13.3C). An Arrhenius term can again be fit by the data to model a two-state process with a single folding energy barrier in the reaction coordinate of the end-to-end length [63]. The distance to the folded transition state, $\Delta x_f = 8.2\,\text{Å}$, is a distance comparable to the length-scales expected for long range hydrophobic forces [92] which are widely believed to play a crucial role in protein folding [93]. It should be noted that his length is considerably longer than that in the unfolding process ($\Delta x_u = 1.7\,\text{Å}$).

Another important difference between the unfolding and the refolding process is that the modules unfold stochastically, independently of one another, while the contraction in the protein length during folding appears cooperative between the modules. This asymmetry has been the topic of much debate in the literature, suggesting protein aggregation [40] or entropic masking [39]. Aggregation was ruled out by the second unfolding of the same modules in response to the second pull [41]. Entropic masking of stepwise folding was rejected by the fact that the folding trajectories scale with the number of modules in the chain **N** [42]. In other words, a single module refolding trajectory has the same end-to-end length profile as a chain of modules multiplied by **N**. We can therefore consider the different stages in folding as being present for any number of modules in the chain and compare them with theories for even single monomer proteins.

13.5.2
Developing a Model for Protein Folding

Many theoretical models have been developed which attempt to explain the pathways of protein folding [54, 84–91]. The generally accepted model for folding of proteins in the bulk is a two-state picture, in which intermediates are only rarely encountered under specific folding conditions. These data predict the height of the energy barrier to folding and the structures associated with the energy minima in this simple landscape deduced from point mutation experiments. They also predict the relative position of the transition state and since the reaction coordinate is not known this relative position is dimensionless. It should be noted that the trajectory of a protein that folds after chemical denaturation involves changes in the end-to-end length of only a couple of nanometers, very close to the radius of gyration of the native state. For example, in the presence of 5 M urea, small angle X-ray scattering has shown that a protein chain's radius of gyration expands from ~16 to 35 Å, an expansion of only 2.1 nm [94]. A further study revealed the high persistence of native-like topology in a denatured protein in 8 M urea [95]. Therefore, force quench experiments monitor folding trajectories over much longer length-scales, exploring new regions of the folding landscape.

Since the mechanical and thermal/chemical studies of protein folding involve very different endpoints they are not directly comparable. At present a number of physical models exist which attempt to explain the folding timescales observed in thermal denaturation experiments. For example, multipathway folding mechanisms, discovered using minimal protein models in conjunction with scaling arguments, have been used to obtain timescales for protein folding [91]. These scaling laws have been derived for proteins in bulk solution which have undergone a temperature perturbation. While we would expect that similar phenomena may be at work in a mechanically perturbed system, we anticipate that the timescales predicted using these scaling laws will be incorrect. There is therefore a need to develop physical models which consider force as a perturbation in the unfolding and subsequent refolding of proteins.

Nonetheless, it is interesting to compare the relative timescales of the distinct stages observed in the force quench folding trajectory (Figure 13.6) with that of the scaling laws discovered using minimal protein models [91]. The predicted timescales describe the stages of folding as a function of the number of amino acids, formulated in analogy with the principles of polymer physics together with an imposed directed search for native contacts. According to this model there are two distinct mechanisms by which proteins reach the native state [91]. The first route is a direct pathway which involves a specific collapse followed by a direct pathway or nucleation to a native state conformation. The alternative route or the "indirect route" involves three key stages, each with a particular timescale. These scaling laws can now be compared with experimental protein folding trajectories for the first time.

The first stage in the force-clamp folding trajectory of the polyprotein ubiquitin (Figure 13.6) is observed to have a timescale of ~10 ms. According to entropic recoil [91] this stage is the non-specific collapse of the protein to a compact conformation on a timescale τ_{NSC} given by

$$\tau_{NSC} \propto \frac{\eta a}{\gamma} N^2 \qquad (13.4)$$

where η is the viscosity of the solution, a is the persistence length of the protein, γ is the surface tension and N is the number of amino acids in the protein. For the protein ubiquitin in aqueous solution this predicts $\tau_{NSC} = 0.032\,\mu s$, which is too fast to be detected by the 10-ms time resolution of the force-clamp technique. It is clear that the non-specific collapse of the protein is the fastest stage in the folding trajectory (Figure 13.6), in agreement with the scaling laws predictions. According to this minimal model [91] the second stage in the folding mechanism involves the kinetic ordering of the foldable chain, describing the protein's search among the space of unfolded structures, such that the sequence directs the folding reaction towards structures that are native-like. The search among the compact structures leading to one of the minimum energy structures is thought to proceed by a diffusive process in the rugged energy landscape. The timescale for diffusion, $\tau_{KO,}$ and the subsequent kinetic ordering is given by

$$\tau_{KO} \approx \tau_D N^{\zeta} \tag{13.5}$$

where ζ is a dynamical folding exponent and τ_D is a time constant. Numerical analysis has found that a good estimate for ζ is 3 for heteropolymers while the exponent is much higher, up to $\zeta = 6$ for model proteins with native interactions [96]. The time constant τ_D is thought to correspond roughly to the timescale for local dihedral angle transitions, which is obtained from simulations to be $\sim 10^{-8}$ s [91]. The search for a structure close to the native state corresponds to combined stages 2 and 3 (Figure 13.6), during which the protein attains the end-to-end length of the folded protein under a low stretching force. The timescale for folding of ubiquitin for this stage of the folding trajectory is then predicted to be 0.4 ms from Equation 13.5. Clearly this timescale is vastly different from the timescales observed in the force-clamp trajectory (Figure 13.6) in which a diffusional search for a native-like conformation can take of the order of seconds and is highly dependent on the magnitude of the applied force (Figure 13.7A).

The final stage in this minimal model is the activated transition from one of the many native-like minimum energy structures to a native-state conformation [91]. The timescale for the barrier activated transition, τ_{AT}, is given by

$$\tau_{AT} \approx \tau_{corr} e^{0.6\sqrt{N}} \tag{13.6}$$

where τ_{corr} is defined as the correlation time for harmonic fluctuations in the measured reaction coordinate [97]. This equation is based on a model in which the protein diffuses through a rough energy landscape with multiple energy minima. These transitions, involving the formation and rupture of native and non-native contacts to establish the folded transition state, are presumed to take place during stage 4, below the length resolution of our force-clamp technique. We therefore turn to force-extension experiments to directly investigate the N dependence on the folding times at zero force in the same protein molecule since the scaling laws for the multipathway mechanism have been derived in the absence of force [91].

We have utilized protein engineering to control the contour length and therefore the number of amino acids unraveled in a single protein [98]. Force extension curves were obtained for a polyprotein for repeated unfolded and refolding cycles. In each cycle the protein was first unfolded, resulting in a sawtooth pattern of unfolding peaks. The force was then dropped to zero and the protein was given a period of time to refold. Subsequently, the protein was again mechanically unfolded. The fraction of refolded proteins was obtained for a particular delay period, for a wide range of waiting times. From this experimental data the average refolding time at zero force

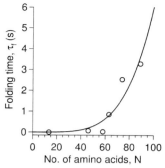

Figure 13.8 Folding time dependence on chain length. A plot of the folding time, τ_F, versus the number of amino acids, N, is shown here. The solid blue line shows the fit to a barrier-activated process, $\tau_F = A\exp(0.6N^\gamma)$, where $A = 0.002$ s and $\gamma = 0.57$ [98].

could then be found, as well as its dependence on the number of amino acids. These experiments revealed that the exponential scaling law for a barrier-activated process, where $\tau_f = \tau_{corr} \exp(0.6 \, N^{0.57}/k_BT)$, fits the data very well (Figure 13.8) with $\tau_{corr} = 2$ ms [98]. It is interesting to note the similarity between the form of the experimentally-derived scaling law for the activated transition and that from the mutipathway mechanism model (Equation 13.6). These force-extension refolding experiments, performed at zero force, are therefore in good agreement with current physical models.

The trend observed in the force extension experiments supports the view of protein folding in a complex energy landscape, in which the conformational degrees of freedom are affected by the size of the protein chain. This result has been confirmed by numerous numerical simulations on model proteins of varying lengths. However, previous ensemble folding experiments from bulk studies have argued that the folding times are solely determined by the native state topology, exhibiting little or no correlation with chain length [99]. These experiments were performed on a wide variety of proteins, where the distinct amino acid sequence and topology of the native structure may have played a dominant role in the folding timescales.

These results emphasize the importance of single-molecule experiments in improving our understanding of the physical picture underlying the folding process. Physical models provide clear theoretical predictions for the folding mechanisms of proteins. There is an urgent requirement for the development of physical models which incorporate force as a perturbation along a well-defined reaction coordinate. It is also evident that the folding process is not simply driven by entropy but rather is a result of a subtle interplay between the enthalpic contributions of residues in the protein chain and the entropy of the surrounding solvent environment [92, 100–103]. By studying how the folding trajectories respond to a variety of physical–chemical conditions and protein engineering it will be possible to uncover the physical phenomena underlying each stage in the protein folding trajectories.

13.6
Force as a Probe of Protein Chemistry

In the previous sections we have shown how force-clamp spectroscopy can be used to probe the transition states of protein folding and unfolding reactions. An Arrhenius term [63] offers a simple relationship between reaction kinetics and the physical distance along a force-induced reaction coordinate to reach the transition state, Δx. The length-scale of this distance to the transition state was found to be between $\Delta x_f \sim$ 8 Å [28] and 60 Å [104]) for protein folding, in rough agreement with the expected role of long range hydrophobic forces [92]. For unfolding this distance was found to be much shorter, in the range of $\Delta x_u \sim 1.7$ to 2.5 Å [50]. This distance is comparable to the size of a water molecule, suggesting an important role for the solvent in the hydrogen bond rupture necessary for unfolding [49]. However, it should be noted that protein unfolding and refolding are complex processes, potentially involving hundreds to thousands of atoms. Thus, it is difficult to precisely determine how an individual interaction contributes to this transition state structure. In this section we show that force-clamp spectroscopy can also be used to probe a simple system, composed of only a few atoms, to carefully monitor the transition state structure of a chemical reaction. We examine the reduction of individual disulfide bonds in a protein molecule, identifying not only a transition state structure on a sub-Ångstrom scale, but also for the first time finding how mechanical force can influence chemical kinetics [105].

A recent review by Beyer and Clausen-Schaumann covers the role of mechanical forces in catalyzing chemical reactions [106]. The authors noted that a widespread difficulty in previous studies was that the reaction of interest could never be consistently oriented with respect to the applied mechanical force and thus, the influence of mechanical forces on these chemical reactions could not be studied quantitatively [106]. Recently, we have shown that force-clamp spectroscopy can overcome these barriers to directly measure the effect of a mechanical force on the kinetics of a chemical reaction [105]. In these experiments, a disulfide bond was engineered into a well-defined position within the structure of the I27 immunoglob- ulin module of human cardiac titin (Figure 13.9A). Disulfide bonds are covalent linkages formed between thiol groups of cysteine residues. These bonds are common in many extracellular proteins and are important both for mechanical and thermo- dynamic stability. The reduction of these bonds by other thiol-containing compounds via an uncomplicated S_N2-type mechanism [107–109] is common both *in vivo* and *in vitro*; a commonly used agent is the dithiol reducing agent dithiothreitol (DTT). Using force-clamp spectroscopy to extend single polyproteins of the modified I27 we first observe partial unfolding of individual modules up to the disulfide bond ($\Delta L1$ in Figure 13.9). The disulfide bond in I27 presents a covalent barrier to the complete unfolding of the protein module. In the presence of DTT, this disulfide bond can be reduced, allowing for extension of the residues trapped behind the disulfide bond (ΔL_2 in Figure 13.9). Thus single disulfide reduction events are easily identified by observation of steps of length ΔL_2 in the unfolding trajectories. To measure the rate of reduction at a particular force and a particular concentration of reducing agent we obtained an ensemble of single-molecule unfolding trajectories of the type seen in

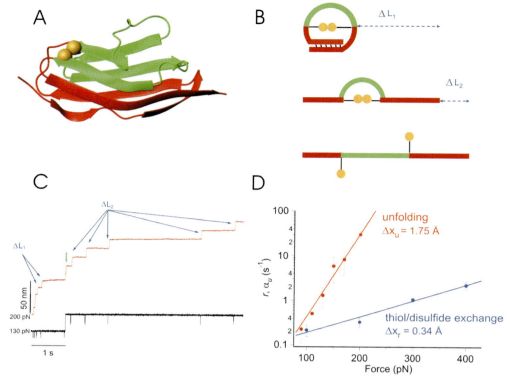

Figure 13.9 Probing protein chemistry (A) A disulfide-bonded protein. An engineered disulfide bond is introduced between two residues in the protein (yellow). This results in 46 unsequestered residues (red) and 43 sequestered residues (green) (B) Applying a mechanical force first triggers the unfolding and extension ΔL_1 of the protein up to the position of the disulfide bond. If DTT is present in the solution, disulfide bond reduction can occur, allowing for the extension of the trapped residues ΔL_2. (C) Force-clamp experiment showing a double-pulse protocol which separates the unsequestered unfolding ΔL_1 ($\sim$10.5 nm) from disulfide reduction events ΔL_2 ($\sim$14.2 nm) in the presence of DTT (D) Force sensitivity and bond lengthening in the thiol/disulfide exchange chemical reaction: a semi-logarithmic plot of the rate of thiol/disulfide exchange, r (blue circles) and of the unfolding rate, τ_U (red circles) as a function of the pulling force. The solid blue line is a fit which yields a distance to the transition state for disulfide bond reduction of $\Delta x_r = 0.34$ Å while the solid red line yields a value of $\Delta x_U = 1.75$ Å for the unfolding distance to the transition state of I27 [105].

Figure 13.9C. We found that over a range of 100 to 400 pN of applied force the rate of disulfide bond reduction was accelerated 10-fold, demonstrating that mechanical force can indeed catalyze this chemical reaction.

The observed force dependence of the rate of disulfide bond reduction by DTT was found to be much less sensitive than the rate of I27 unfolding (Figure 13.9D). The weaker force dependence indicates a much shorter value of Δx for disulfide bond reduction ($\Delta x_r = 0.34$ Å in Figure 13.9D) [105]. Indeed, it was remarkable that the measured distance to the transition state of this S_N2-type chemical reaction was in close agreement with disulfide bond lengthening at the transition state of thiol–disulfide exchange as found by DFT calculations [110]. This result indicates that the force dependence of the observed reaction kinetics is governed by the detected sub-Ångstrom length changes between the two sulfur atoms at the reaction transition state. Such a phenomenon may apply to many other chemical reactions. Thus it seems that force-clamp spectroscopy will be a powerful tool both to determine the effect of force on chemical kinetics in general as well to directly probe the structure of chemical transition states.

13.7
Conclusions

Force-clamp spectroscopy has been highly successful in using mechanical force to probe the physics and chemistry of proteins. In this chapter we have attempted to demonstrate the robustness, reliability and ease of use of this technique to study recombinant proteins. Force-clamp spectroscopy can be used to perturb the energy landscape of a protein at a constant force along a well-defined reaction coordinate, revealing fine details of the transition states of folding, unfolding and even of protein chemical reactions. These transition states range from nanometers down to the sub-Ångström scale, revealing the underlying physics involved in these processes. By applying a constant force a number of force-dependent parameters can be obtained with confidence. While the force-clamp spectroscopy technique has developed rapidly over the last few years, the instrumentation is still limited to a bandwidth of a few hundred hertz. The most significant development that we can expect to see in the next few years is a force-clamp spectrometer with a megahertz bandwidth. Such a development will permit a far more detailed examination of the energy landscape of a protein, following their conformational dynamics with single bond resolution. Now that the single-molecule field is far from its infancy, the challenge is to develop theoretical models of protein folding to incorporate force, as well as to consider the interactions driving the microscopic mechanism of protein folding. The force-clamp technique therefore unites fields as diverse as protein biochemistry, statistical mechanics far from equilibrium and the protein folding community.

Acknowledgments

We would like to express our gratitude to S. Garcia-Maynes and A. P. Wiita for helpful discussions and critical reading of the chapter.

References

1 Oberhauser, A. and Carrión-Vázquez, M. (2008) Mechanical Biochemistry of Proteins One Molecule at a Time. *J. Biol. Chem.*, **283**, 6617–6621.

2 Ritort, F. (2006) Single-molecule experiments in biological physics: methods and applications. *Journal of Physics – Condensed Matter*, **18**, R531–R583.

3 Linke, W.A., Kulke, M., Li, H.B., Fujita-Becker, S., Neagoe, C., Manstein, D.J., Gautel, M. and Fernandez, J.M. (2002) PEVK domain of titin: An entropic spring with actin-binding properties. *Journal of Structural Biology*, **137**, 194–205.

4 Labeit, S. and Kolmerer, B. (1995) Titins – giant proteins in charge of muscle ultrastructure and elasticity. *Science*, **270**, 293–296.

5 Linke, W.A., Ivemeyer, M., Olivieri, N., Kolmerer, B., Ruegg, J.C. and Labeit, S. (1996) Towards a molecular under-standing of the elasticity of titin. *Journal of Molecular Biology*, **261**, 62–71.

6 Neupert, W. and Brunner, M. (2002) The protein import motor of mitochondria. *Nature Reviews Molecular Cell Biology*, **3**, 555–565.

7 Shtilerman, M., Lorimer, G.H. and Englander, S.W. (1999) Chaperonin function: Folding by forced unfolding. *Science*, **284**, 822–825.

8 Valpuesta, J.M., Martin-Benito, J., Gomez-Puertas, P., Carrascosa, J.L. and Willison, K.R. (2002) Structure and function of a protein folding machine: the eukaryotic cytosolic chaperonin CCT. *FEBS Letters*, **529**, 11–16.

9 Khorasanizadeh, S., Peters, I.D., Butt, T.R. and Roder, H. (1993) Folding and stability of a tryptophan-containing mutant of ubiquitin. *Biochemistry*, **32**, 7054–7063.

10 Krantz, B.A. and Sosnick, T.R. (2000) Distinguishing between two-state and three-state models for ubiquitin folding. *Biochemistry*, **39**, 11696–11701.

11 Krantz, B.A. and Sosnick, T.R. (2001) Engineered metal binding sites map the heterogeneous folding landscape of a coiled coil. *Nature Structural Biology*, **8**, 1042–1047.

12 Larios, E., Li, J.S., Schulten, K., Kihara, H. and Gruebele, M. (2004) Multiple probes reveal a native-like intermediate during low-temperature refolding of ubiquitin. *Journal of Molecular Biology*, **340**, 115–125.

13 Leeson, D.T., Gai, F., Rodriguez, H.M., Gregoret, L.M. and Dyer, R.B. (2000) Protein folding and unfolding on a complex energy landscape. *Proceedings of the National Academy of Sciences of the United States of America*, **97**, 2527–2532.

14 Lipman, E.A., Schuler, B., Bakajin, O. and Eaton, W.A. (2003) Single-molecule measurement of protein folding kinetics. *Science*, **301**, 1233–1235.

15 Sosnick, T.R., Dothager, R.S. and Krantz, B.A. (2004) Differences in the folding transition state of ubiquitin indicated by phi and psi analyses. *Proceedings of the National Academy of Sciences of the United States of America*, **101**, 17377–17382.

16 Wright, C.F., Lindorff-Larsen, K., Randles, L.G. and Clarke, J. (2003) Parallel protein-unfolding pathways revealed and mapped. *Nature Structural Biology*, **10**, 658–662.

17 Kneller, G.R. (2005) Quasielastic neutron scattering and relaxation processes in proteins: analytical and simulation-based models. *Physical Chemistry Chemical Physics*, **7**, 2641–2655.

18 Lindorff-Larsen, K., Best, R.B., DePristo, M.A., Dobson, C.M. and Vendruscolo, M. (2005) Simultaneous determination of protein structure and dynamics. *Nature*, **433**, 128–132.

19 Itoh, K. and Sasai, M. (2004) Dynamical transition and proteinquake in photoactive yellow protein. *Proceedings of the National Academy of Sciences of the United States of America*, **101**, 14736–14741.

20 Min, W., Luo, G.B., Cherayil, B.J., Kou, S.C. and Xie, X.S. (2005) Observation of a power-law memory kernel for fluctua-tions within a single protein molecule. *Physical Review Letters*, **94**, 194–204.

21 van Oijen, A.M., Blainey, P.C., Crampton, D.J., Richardson, C.C., Ellenberger, T. and Xie, X.S. (2003) Single-molecule kinetics of lambda exonuclease reveal base dependence and dynamic disorder. *Science*, **301**, 1235–1238.

22 Xue, Q.F. and Yeung, E.S. (1995) Differences in the chemical-reactivity of individual molecules of an enzyme. *Nature*, **373**, 681–683.

23 Yang, H., Luo, G.B., Karnchanaphanurach, P., Louie, T.M., Rech, I., Cova, S., Xun, L.Y. and Xie, X.S. (2003) Protein conformational dynamics probed by single-molecule electron transfer. *Science*, **302**, 262–266.

24 Dudko, O.K., Hummer, G. and Szabo, A. (2006) Intrinsic rates and activation free energies from single-molecule pulling experiments. *Physical Review Letters*, **96**, 108101–108104.

25 Hummer, G. and Szabo, A. (2003) Kinetics from nonequilibrium single-molecule pulling experiments. *Biophysical Journal*, **85**, 5–15.

26 Jarzynski, C. (1997) Equilibrium free-energy differences from nonequilibrium measurements: A master-equation approach. *Physical Review E*, **56**, 5018–5035.

27 Brujic, J., Hermans, R.I., Walther, K.A. and Fernandez, J.M. (2006) Single-molecule force spectroscopy reveals signatures of glassy dynamics in the energy landscape of ubiquitin. *Nature Physics*, **2**, 282–286.

28 Fernandez, J.M. and Li, H. (2004) Force-clamp spectroscopy monitors the folding trajectory of a single protein. *Science*, **303**, 1674–1678.

29 Nevo, R., Brumfeld, V., Kapon, R., Hinterdorfer, P. and Reich, Z. (2005) Direct measurement of protein energy landscape roughness. *EMBO Reports*, **6**, 482–486.

30 Mitsui, K., Hara, M. and Ikai, A. (1996) Mechanical unfolding of alpha(2.-macroglobulin molecules with atomic force microscope. *FEBS Letters*, **385**, 29–33.

31 Fisher, T.E., Carrion-Vazquez, M., Oberhauser, A.F., Li, H.B., Marszalek, P.E. and Fernandez, J.M. (2000) Single molecule force spectroscopy of modular proteins in the nervous system. *Neuron*, **27**, 435–446.

32 Li, H.B., Carrion-Vazquez, M., Oberhauser, A.F., Marszalek, P.E. and Fernandez, J.M. (2000) Point mutations alter the mechanical stability of immunoglobulin modules. *Nature Structural Biology*, **7**, 1117–1120.

33 Li, L.W., Huang, H.H.L., Badilla, C.L. and Fernandez, J.M. (2005) Mechanical unfolding intermediates observed by single-molecule force spectroscopy in a fibronectin type III module. *Journal of Molecular Biology*, **345**, 817–826.

34 Williams, P.M., Fowler, S.B., Best, R.B., Toca-Herrera, J.L., Scott, K.A., Steward, A. and Clarke, J. (2003) Hidden complexity in the mechanical properties of titin. *Nature*, **422**, 446–449.

35 Dietz, H. and Rief, M. (2006) Protein structure by mechanical triangulation. *Proceedings of the National Academy of Sciences of the United States of America*, **103**, 1244–1247.

36 Dietz, H., Bertz, M., Schlierf, M., Berkemeier, F., Bornschlögl, T., Junker, J.P. and Rief, M. (2006) Cysteine engineering of polyproteins for single-molecule force spectroscopy. *Nature Protocols*, **1**, 80–84.

37 Li, H.B., Linke, W.A., Oberhauser, A.F., Carrion-Vazquez, M., Kerkviliet, J.G., Lu, H., Marszalek, P.E. and Fernandez, J.M. (2002) Reverse engineering of the giant muscle protein titin. *Nature*, **418**, 998–1002.

38 Ciechanover, A. and Schwartz, A.L. (1998) The ubiquitin–proteasome pathway: The complexity and myriad functions of proteins death. *Proceedings of the National Academy of Sciences of the United States of America*, **95**, 2727–2730.

39 Best, R.B. and Hummer, G. (2005) Comment on "Force-clamp spectroscopy monitors the folding trajectory of a single protein". *Science*, **308**, 498.

40 Sosnick, T.R. (2004) Comment on "Force-clamp spectroscopy monitors the folding trajectory of a single protein". *Science*, **306**, 411.

41 Fernandez, J.M., Li, H.B. and Brujic, J. (2004) Response to comment on "Force-clamp spectroscopy monitors the folding trajectory of a single protein". *Science*, **306**, 411.

42 Brujic, J. and Fernandez, J.W. (2005) Response to comment on "Force-clamp spectroscopy monitors the folding trajectory of a single protein". *Science*, **308**, 498.

43 Dougan, L. and Fernandez, J.M. (2007) Tandem repeating modular proteins avoid aggregation in single molecule force spectroscopy experiments. *J. Phys. Chem. A*, **111**, 12402–12408.

44 Brockwell, D.J., Paci, E., Zinober, R.C., Beddard, G.S., Olmsted, P.D., Smith, D.A., Perham, R.N. and Radford, S.E. (2003) Pulling geometry defines the mechanical resistance of a beta-sheet protein. *Nature Structural Biology*, **10**, 731–737.

45 Binz, H.K., Stumpp, M.T., Forrer, P., Amstutz, P. and Pluckthun, A. (2003) Designing repeat proteins: Well-expressed, soluble and stable proteins from combinatorial libraries of consensus ankyrin repeat proteins. *Journal of Molecular Biology*, **332**, 489–503.

46 Tang, K.S., Fersht, A.R. and Itzhaki, L.S. (2003) Sequential unfolding of ankyrin repeats in tumor suppressor p16. *Structure*, **11**, 67–73.

47 Zweifel, M.E. and Barrick, D. (2001) Studies of the ankyrin repeats of the *Drosophila melanogaster* Notch receptor. 1. Solution conformational and hydrodynamic properties. *Biochemistry* **40**, 14344–14356.

48 Zweifel, M.E. and Barrick, D. (2001) Studies of the ankyrin repeats of the Drosophila melanogaster Notch receptor. 2. Solution stability and cooperativity of unfolding. *Biochemistry*, **40**, 14357–14367.

49 Brockwell, D.J., Beddard, G.S., Paci, E., West, D.K., Olmsted, P.D., Smith, D.A. and Radford, S.E. (2005) Mechanically unfolding the small, topologically simple protein L. *Biophysical Journal*, **89**, 506–519.

50 Schlierf, M., Li, H. and Fernandez, J.M. (2004) The unfolding kinetics of ubiquitin captured with single-molecule force-clamp techniques. *Proceedings of the National Academy of Sciences of the United States of America*, **101**, 7299–7304.

51 Schwaiger, I., Kardinal, A., Schleicher, M., Noegel, A.A. and Rief, M. (2004) A mechanical unfolding intermediate in an actin-crosslinking protein. *Nature Structural & Molecular Biology*, **11**, 81–85.

52 Fowler, S.B., Best, R.B., Herrera, J.L.T., Rutherford, T.J., Steward, A., Paci, E., Karplus, M. and Clarke, J. (2002) Mechanical unfolding of a titin Ig domain: Structure of unfolding intermediate revealed by combining AFM, molecular dynamics simulations, NMR and protein engineering. *Journal of Molecular Biology*, **322**, 841–849.

53 Bullard, B., Garcia, T., Benes, V., Leake, M.C., Linke, W.A. and Oberhauser, A.F. (2006) The molecular elasticity of the insect flight muscle proteins projectin and kettin. *Proceedings of the National Academy of Sciences of the United States of America*, **103**, 4451–4456.

54 Mirny, L. and Shakhnovich, E. (2001) Protein folding theory: From lattice to all-atom models. *Annual Review of Biophysics and Biomolecular Structure*, **30**, 361–396.

55 Snow, C.D., Sorin, E.J., Rhee, Y.M. and Pande, V.S. (2005) How well can simulation predict protein folding kinetics, and thermodynamics? *Annual Review of Biophysics and Biomolecular Structure*, **34**, 43–69.

56 Berendsen, H.J.C. (1998) Protein folding – A glimpse of the holy grail? *Science*, **282**, 642–643.

57 Lange, O.F. and Grubmuller, H. (2006) Collective Langevin dynamics of conformational motions in proteins. *Journal of Chemical Physics*, **124**, 214903–217918.

58 Autenrieth, F., Tajkhorshid, E., Schulten, K. and Luthey-Schulten, Z. (2004) Role of water in transient cytochrome c(2) docking. *Journal of Physical Chemistry B*, **108**, 20376–20387.

59 Lu, H. and Schulten, K. (1999) Steered molecular dynamics simulation of conformational changes of immuno-globulin domain I27 interpret atomic force microscopy observations. *Chemical Physics*, **247**, 141–153.

60 Lu, H. and Schulten, K. (1999) Steered molecular dynamics simulations of force-induced protein domain unfolding. *Proteins – Structure Function and Genetics*, **35**, 453–463.

61 Isralewitz, B., Gao, M. and Schulten, K. (2001) Steered molecular dynamics and mechanical functions of proteins. *Current Opinion in Structural Biology*, **11**, 224–230.

62 Lu, H. and Schulten, K. (2000) The key event in force-induced unfolding of titin's immunoglobulin domains. *Biophysical Journal*, **79**, 51–65.

63 Bell, G.I. (1978) Models for specific adhesion of cells to cells. *Science*, **200**, 618–627.

64 Schlierf, M. and Rief, M. (2006) Single-molecule unfolding force distributions reveal a funnel-shaped energy landscape. *Biophysical Journal*, **90**, L33–L35.

65 Colquhoun, D. and Sakmann, B. (1981) Fluctuations in the microsecond time range of the current through single acetylcholine-receptor ion channels. *Nature*, **294**, 464–466.

66 Ansari, A., Berendzen, J., Bowne, S.F., Frauenfelder, H., Iben, I.E.T., Sauke, T.B., Shyamsunder, E. and Young, R.D. (1985) Protein states and protein quakes. *Proceedings of the National Academy of Sciences of the United States of America*, **82**, 5000–5004.

67 Baysal, C. and Atilgan, A.R. (2005) Relaxation kinetics and the glassiness of native proteins: Coupling of timescales. *Biophysical Journal*, **88**, 1570–1576.

68 Roy, M., Chavez, L.L., Finke, J.M., Heidary, D.K., Onuchic, J.N. and Jennings, P.A. (2005) The native energy landscape for interleukin-1 beta. Modulation of the population ensemble through native-state topology. *Journal of Molecular Biology*, **348**, 335–347.

69 Hyeon, C.B. and Thirumalai, D. (2003) Can energy landscape roughness of proteins and RNA, be measured by using mechanical unfolding experiments? *Proceedings of the National Academy of Sciences of the United States of America*, **100**, 10249–10253.

70 Karplus, M., Gao, Y.Q., Ma, J.P., van der Vaart, A. and Yang, W. (2005) Protein structural transitions and their functional role. *Philosophical Transactions of the Royal Society of, London Series a – Mathematical Physical and Engineering Sciences*, **363**, 331–355.

71 Best, R.B., Clarke, J. and Karplus, M. (2005) What contributions to protein side-chain dynamics are probed by NMR experiments? A molecular dynamics simulation analysis. *Journal of Molecular Biology*, **349**, 185–203.

72 Fersht, A.R. and Sato, S. (2004) Phi-value analysis and the nature of protein-folding transition states. *Proceedings of the National Academy of Sciences of the United States of America*, **101**, 7976–7981.

73 Frauenfelder, H., Sligar, S.G. and Wolynes, P.G. (1991) The energy landscapes and motions of proteins. *Science*, **254**, 1598–1603.

74 Monthus, C. and Bouchaud, J.P. (1996) Models of traps and glass pheno-menology. *Journal of Physics a – Mathematical and General*, **29**, 3847–3869.

75 Luo, G.B., Andricioaei, I., Xie, X.S. and Karplus, M. (2006) Dynamic distance disorder in proteins is caused by trapping. *Journal of Physical Chemistry B*, **110**, 9363–9367.

76 Sadqi, M., Fushman, D. and Munoz, V. (2006) Atom-by-atom analysis of global downhill protein folding. *Nature*, **442**, 317–321.

77 Matouschek, A., Kellis, J.T., Serrano, L. and Fersht, A.R. (1989) Mapping the transition-state and pathway of protein folding by protein engineering. *Nature*, **340**, 122–126.

78 Religa, T.L., Markson, J.S., Mayor, U., Freund, S.M.V. and Fersht, A.R. (2005) Solution structure of a protein denatured state and folding intermediate. *Nature*, **437**, 1053–1056.

79 Sato, S., Religa, T.L. and Fersht, A.R. (2006) Phi-analysis of the folding of the B domain of Protein A using multiple optical probes. *Journal of Molecular Biology*, **360**, 850–864.

80 Socolich, M., Lockless, S.W., Russ, W.P., Lee, H., Gardner, K.H. and Ranganathan, R. (2005) Evolutionary information for specifying a protein fold. *Nature*, **437**, 512–518.

81 Russ, W.P., Lowery, D.M., Mishra, P., Yaffe, M.B. and Ranganathan, R. (2005) Natural-like function in artificial WW domains. *Nature*, **437**, 579–583.

82 Collin, D., Ritort, F., Jarzynski, C., Smith, S.B., Tinoco, I. and Bustamante, C. (2005) Verification of the Crooks fluctuation theorem and recovery of RNA folding free energies. *Nature*, **437**, 231–234.

83 Dudko, O.K., Filippov, A.E., Klafter, J. and Urbakh, M. (2003) Beyond the conventional description of dynamic force spectroscopy of adhesion bonds. *Proceedings of the National Academy of Sciences of the United States of America*, **100**, 11378–11381.

84 Bryngelson, J.D. and Wolynes, P.G. (1987) Spin-glasses and the statistical-mechanics of protein folding. *Proceedings of the National Academy of Sciences of the United States of America*, **84**, 7524–7528.

85 Cho, S.S., Levy, Y. and Wolynes, P.G. (2006) P versus Q: Structural reaction coordinates capture protein folding on smooth landscapes. *Proceedings of the National Academy of Sciences of the United States of America*, **103**, 586–591.

86 Onuchic, J.N., LutheySchulten, Z. and Wolynes, P.G. (1997) Theory of protein folding: The energy landscape perspective. *Annual Review of Physical Chemistry*, **48**, 545–600.

87 Wolynes, P.G., Onuchic, J.N. and Thirumalai, D. (1995) Navigating the Folding Routes. *Science*, **267**, 1619–1620.

88 Bicout, D.J. and Szabo, A. (2000) Entropic barriers, transition states, funnels, and exponential protein folding kinetics: A simple model. *Protein Science*, **9**, 452–465.

89 Hagen, S.J., Hofrichter, J., Szabo, A. and Eaton, W.A. (1996) Diffusion-limited contact formation in unfolded cytochrome c: Estimating the maximum rate of protein folding. *Proceedings of the National Academy of Sciences of the United States of America*, **93**, 11615–11617.

90 Guo, Z.Y. and Thirumalai, D. (1995) Kinetics of protein-folding – nucleation mechanism, time scales, and pathways. *Biopolymers*, **36**, 83–102.

91 Thirumalai, D. (1995) From minimal models to real proteins: time scales for protein folding kinetics. *Journal of Physics I France*, **5**, 1457–1467.

92 Israelachvili, J.N., Pashley, R.M., Perez, E. and Tandon, R.K. (1981) Forces between hydrophobic surfaces in aqueous-electrolyte and surfactant solutions containing common airborne impurities. *Colloids and Surfaces*, **2**, 287–291.

93 Tanford, C. (1978) Hydrophobic effect and organization of living matter. *Science*, **200**, 1012–1018.

94 Alexandrescu, A.T., Abeygunawardana, C. and Shortle, D. (1994) Structure and dynamics of a denatured 131-residue fragment of Staphylococcal nuclease – a heteronuclear NMR study. *Biochemistry*, **33**, 1063–1072.

95 Shortle, D. and Ackerman, M.S. (2001) Persistence of native-like topology in a denatured protein in 8 M urea. *Science*, **293**, 487–489.

96 Gutin, A.M., Abkevich, V.V. and Shakhnovich, E.I. (1996) Chain length scaling of protein folding time. *Physical Review Letters*, **77**, 5433–5436.

97 Socci, N.D. and Onuchic, J.N. (1995) Kinetic and thermodynamic analysis of protein-like heteropolymers – Monte-Carlo histogram technique. *Journal of Chemical Physics*, **103**, 4732–4744.

98 Koti, A.S.R, Brujic, J., Huang, H.H., Wiita, A.P., Lu, H., Walther, K.A., Carrion-Vazquez, M., Li, H. and Fernandez, J.M. (2006) Contour length and refolding rate of a small protein controlled by engineered disulfide bonds. *Biophysical Journal*, (in press).

99 Plaxco, K.W. and Gross, M. (2001) Unfolded, yes, but random? Never! *Nature Structural Biology*, **8**, 659–660.

100 Forbes, J.G., Jin, A.J., Ma, K., Gutierrez-Cruz, G., Tsai, W.L. and Wang, K.A. (2005) Titin PEVK segment: charge-driven elasticity of the open and flexible polyampholyte. *Journal of Muscle Research and Cell Motility*, **26**, 291–301.

101 Hoang, T.X., Trovato, A., Seno, F., Banavar, J.R. and Maritan, A. (2004) Geometry and symmetry presculpt the free-energy landscape of proteins. *Proceedings of the National Academy of Sciences of the United States of America*, **101**, 7960–7964.

102 Israelachvili, J. and Wennerstrom, H. (1996) Role of hydration and water structure in biological and colloidal interactions. *Nature*, **379**, 219–225.

103 Pappu, R.V., Srinivasan, R. and Rose, G.D. (2000) The Flory isolated-pair hypothesis is not valid for polypeptide chains: implications for protein folding. *Proceedings of the National Academy of Sciences of the United States of America*, **97**, 12565–12570.

104 Cecconi, C., Shank, E.A., Bustamante, C. and Marqusee, S. (2005) Direct observation of the three-state folding of a single protein molecule. *Science*, **309**, 2057–2060.

105 Wiita, A.P., Ainavarapu, S.R.K., Huang, H.H. and Fernandez, J.M. (2006) Force-dependent chemical kinetics of disulfide bond reduction observed with single-molecule techniques. *Proceedings of the National Academy of Sciences of the United States of America*, **103**, 7222–7227.

106 Beyer, M.K. and Clausen-Schaumann, H. (2005) Mechanochemistry: The mechanical activation of covalent bonds. *Chemical Reviews*, **105**, 2921–2948.

107 Keire, D.A., Strauss, E., Guo, W., Noszal, B. and Rabenstein, D.L. (1992) Kinetics and equilibria of thiol disulfide interchange reactions of selected biological thiols and related molecules with oxidized glutathione. *Journal of Organic Chemistry*, **57**, 123–127.

108 Pappas, J.A. (1977) Theoretical studies of reactions of sulfur-sulfur bond. 1. General heterolytic mechanisms. *Journal of the American Chemical Society*, **99**, 2926–2930.

109 Rosenfield, R.E., Parthasarathy, R. and Dunitz, J.D. (1977) Directional preferences of nonbonded atomic contacts with divalent sulphur. 1. Electrophiles and nucleophiles. *Journal of the American Chemical Society*, **99**, 4860–4862.

110 Fernandes, P.A. and Ramos, M.J. (2004) Theoretical insights into the mechanism for thiol/disulfide exchange. *Chemistry-A European Journal*, **10**, 257–266.

Index

Single Molecule Dynamics in Life Science. Edited by T. Yanagida and Y. Ishii
Copyright © 2009 WILEY-VCH Verlag GmbH & Co. KGaA, Weinheim
ISBN: 978-3-527-31288-7